socio ↑

perception    action
cognition        P   r   c
    emotional

thesis: an educational model which functions to **trigger** those prerequisites **over** E's & O's.

# Performance Psychology

Intra - Inter - grup - socio

pescud rel col.

per — activ
cog. — emotiv

bio - psycho - socio

V
V
Synthesis

brut

mico
rso
macro

# Performance Psychology
Perception, Action, Cognition, and Emotion

*Edited by*

Markus Raab
Babett Lobinger
Sven Hoffmann
Alexandra Pizzera
Sylvain Laborde
German Sport University Cologne
Institute of Psychology
Cologne, Germany

AMSTERDAM • BOSTON • HEIDELBERG • LONDON
NEW YORK • OXFORD • PARIS • SAN DIEGO
SAN FRANCISCO • SINGAPORE • SYDNEY • TOKYO

Academic Press is an imprint of Elsevier

Academic Press is an imprint of Elsevier
125 London Wall, London EC2Y 5AS, UK
525 B Street, Suite 1800, San Diego, CA 92101-4495, USA
225 Wyman Street, Waltham, MA 02451, USA
The Boulevard, Langford Lane, Kidlington, Oxford OX5 1GB, UK

Copyright © 2016 Elsevier Inc. All rights reserved.

No part of this publication may be reproduced or transmitted in any form or by any means, electronic or mechanical, including photocopying, recording, or any information storage and retrieval system, without permission in writing from the publisher. Details on how to seek permission, further information about the Publisher's permissions policies and our arrangements with organizations such as the Copyright Clearance Center and the Copyright Licensing Agency, can be found at our website: www.elsevier.com/permissions.

This book and the individual contributions contained in it are protected under copyright by the Publisher (other than as may be noted herein).

**Notices**
Knowledge and best practice in this field are constantly changing. As new research and experience broaden our understanding, changes in research methods, professional practices, or medical treatment may become necessary.

Practitioners and researchers may always rely on their own experience and knowledge in evaluating and using any information, methods, compounds, or experiments described herein. In using such information or methods they should be mindful of their own safety and the safety of others, including parties for whom they have a professional responsibility.

To the fullest extent of the law, neither the Publisher nor the authors, contributors, or editors, assume any liability for any injury and/or damage to persons or property as a matter of products liability, negligence or otherwise, or from any use or operation of any methods, products, instructions, or ideas contained in the material herein.

**Library of Congress Cataloging-in-Publication Data**
A catalog record for this book is available from the Library of Congress

**British Library Cataloguing-in-Publication Data**
A catalogue record for this book is available from the British Library

ISBN: 978-0-12-803377-7

For information on all Academic Press publications visit our website at http://store.elsevier.com

*Publisher:* Nikki Levy
*Acquisition Editor:* Nikki Levy
*Editorial Project Manager:* Barbara Makinster
*Production Project Manager:* Caroline Johnson
*Designer:* Mark Rogers

Typeset by TNQ Books and Journals
www.tnq.co.in

# Contents

Contributors xiii
Preface xv

## Section A
## What is Performance Psychology?

### 1. The Building Blocks of Performance: An Overview
*Markus Raab*

| | |
|---|---|
| Perception | 4 |
| Action | 7 |
| Cognition | 7 |
| Emotion | 8 |
| References | 9 |

### 2. Theoretical Framework of Performance Psychology: An Action Theory Perspective
*Jürgen R. Nitsch and Dieter Hackfort*

| | |
|---|---|
| **Definition and Scope of Performance Psychology** | 12 |
| Performance and Psychology | 12 |
| Structure of Performance Orientation | 13 |
| Characteristics of Peak Performance | 14 |
| The Action Paradigm—A Meta-Theoretical Perspective on Performance | 15 |
| Intention—The Organizing Principle of Action | 17 |
| The Action Space and Its Situational Configuration | 18 |
| The Functional Architecture of Actions | 20 |
| Functionality of Emotional Processes with Special Reference to Performance | 23 |
| Multifacetedness and Functional Complexity | 23 |
| Functional Disturbances | 24 |
| Options in Emotional Processing | 25 |
| **Conclusion** | 26 |
| **References** | 26 |

3. **Measurement Considerations in Performance Psychology**
   *Gershon Tenenbaum and Edson Filho*

   | | |
   |---|---|
   | Measurement Considerations in Performance Psychology | 31 |
   | Types of Measures | 32 |
   | Quantitative Measures | 32 |
   | Qualitative Measures | 34 |
   | Measurement and Theory Development | 36 |
   | Alternative Models | 36 |
   | The Principle of Parsimony | 36 |
   | Two-Parameter Model for Capturing the Cognitive–Affective–Behavioral Linkage in Performance Psychology | 37 |
   | New Trends in Performance Measurement | 40 |
   | Summary | 42 |
   | References | 42 |

4. **Applications within Performance Psychology**
   *Terry Clark and Aaron Williamon*

   | | |
   |---|---|
   | Functions of Applied Performance Psychology | 46 |
   | The Road to Excellence | 47 |
   | Developing an Ecologically Appropriate Performance Psychology Program in Music | 50 |
   | Considerations for Implementing Performance Psychology | 53 |
   | Phase 1—Orientation | 53 |
   | Phase 2—Activity Analysis | 54 |
   | Phase 3—Individual/Team Assessment | 54 |
   | Phase 4—Conceptualization | 55 |
   | Phase 5—Psychological Skills Training | 56 |
   | Phase 6—Implementation | 56 |
   | Phase 7—Evaluation | 57 |
   | Conclusion | 58 |
   | References | 58 |

# Section B
# Performance Phenomena of Cognitive–Action Interaction

5. **Bridging the Gap between Action and Cognition: An Overview**
   *Babett H. Lobinger*

   | | |
   |---|---|
   | Cognition and Action | 68 |
   | The Yips in Golf | 70 |
   | Neurological Origin: Focal Dystonia | 71 |
   | Psychological Origin: Choking | 72 |

| | Motor Origin: Dynamic Stereotype | 72 |
| | Diagnosing and Treating the Yips | 73 |
| | The Education of Soccer Coaches | 75 |
| | Talent Identification and Assessment in High-Performance Sports | 78 |
| | Conclusion | 82 |
| | References | 82 |

## 6. Improving Performance by Means of Action–Cognition Coupling in Athletes and Coaches

*Gloria B. Solomon*

| The Profession of Coaching | 88 |
| Purpose | 89 |
| Development of Coaching Expertise | 89 |
| History of Leadership in Sport | 89 |
| Behavioral Theories | 90 |
| Situational Theories | 90 |
| Multidimensional Model of Leadership | 91 |
| Member Characteristics and Coach Behavior | 92 |
| Coach Behavior and Satisfaction | 93 |
| Coach Behavior and Performance | 94 |
| Summary of Leadership in Sport | 94 |
| **Athlete Perceptions of Successful Coaching** | 94 |
| **Expectancy Effects in Competitive Sport** | 96 |
| Step 1—Coach Develops Expectations for Athlete Performance | 96 |
| Step 2—Expectations Influence Coaching Behaviors | 97 |
| Step 3—Perceptions of Coach Behavior Affects Athletes | 97 |
| Step 4—Athlete Performance Conforms to Coach Expectations | 98 |
| **Summary and Conclusion** | 98 |
| References | 98 |

## 7. Music Performance: Expectations, Failures, and Prevention

*Eckart Altenmüller and Christos I. Ioannou*

| **Communalities and Differences between Sports and Music Performance** | 104 |
| **Performance Failures in Musicians** | 105 |
| Music Performance Anxiety | 105 |
| Choking under Pressure | 106 |
| Dynamic Stereotype | 107 |
| Musician's Dystonia | 108 |
| **Improving Performance in Musicians** | 110 |
| **A Heuristic Model of Performance Failures in Musicians** | 112 |
| **Conclusion: Some Implications for Prevention** | 114 |
| Acknowledgment | 115 |
| References | 115 |

8. Motor Imagery and Mental Training in Older Adults

*Michael Kalicinski, Monika Thomas and Babett H. Lobinger*

| | |
|---|---|
| Motor Imagery as a Prerequisite for Mental Training | 122 |
| Motor Imagery and Mental Training in Older Adults | 123 |
| Mental Balance Training for Postural Control | 124 |
| Method | 125 |
| Intervention Procedures | 126 |
| Results and Discussion | 128 |
| Conclusion and Future Research | 130 |
| References | 130 |

# Section C
# Dysfunctional Learning, Errors, and Other Performance Phenomena of Perception–Cognition Interactions

9. Bridging the Gap between Perception and Cognition: An Overview

*Sven Hoffmann*

| | |
|---|---|
| Neurophysiological Implementation of Cognitive Mechanisms Guiding Perception and Action | 136 |
| Prefrontal Cortex and Attentional Selection | 137 |
| "The Winner Takes It All" | 140 |
| The Role of Dopamine | 142 |
| Cognitive Models: Drift Diffusion | 142 |
| Conclusion and Outlook | 146 |
| References | 147 |

10. Performance and Error Monitoring: Causes and Consequences

*Sven Hoffmann*

| | |
|---|---|
| Theories of Error Processing | 152 |
| Methodological Issues When Measuring Response-Related EEG Activity | 156 |
| Predicting Errors | 160 |
| Predicting Behavioral Adaptation | 160 |
| Individual Differences and Error Monitoring | 162 |
| Conclusion | 164 |
| References | 165 |

11. Committing Errors as a Consequence of an Adverse Focus of Attention

*Daniel Schneider*

| | |
|---|---|
| "Serious" Errors in the Course of Perception | 169 |
| Selective Attention as Biased Competition | 170 |

|       | Selective Attention and Change Blindness | 173 |
|---|---|---|
|       | Perceptual Errors under Conditions of Inattention | 176 |
|       | Electrophysiological Evidence for the Role of Selective Attention in Change Detection | 179 |
|       | Selective Attention and Visual Awareness | 181 |
|       | Conclusion | 184 |
|       | References | 184 |

### 12. Lifestyle and Interventions for Improving Cognitive Performance in Older Adults

*Patrick D. Gajewski and Michael Falkenstein*

| Cognitive Leisure Activity | 191 |
|---|---|
| Cognitive Training: Intervention Studies | 191 |
| Physical Training: Intervention Studies | 195 |
| Conclusion | 198 |
| References | 199 |

## Section D
## Self-Other Perceptions and Other Performance Phenomena of Perception-Action Interactions

### 13. Bridging the Gap between Perception and Action: An Overview

*Alexandra Pizzera*

| How Does Perception Affect Action? | 208 |
|---|---|
| How Does Action Affect Perception? | 213 |
| Online and Offline Effects of the Perception–Action Link | 215 |
| Conclusion and Future Research | 217 |
| References | 218 |

### 14. Capturing Motion for Enhancing Performance: An Embodied Cognition Perspective on Sports and the Performing Arts

*Vassilis Sevdalis and Clemens Wöllner*

| Overview of Theoretical Approaches | 224 |
|---|---|
| Overview of Empirical Research | 225 |
| Research in the Performing Arts | 226 |
| Research in Sports | 229 |
| Conclusion | 230 |
| References | 232 |

### 15. Auditory Action Perception

*Christian Kennel and Alexandra Pizzera*

| Auditory Perception | 235 |
|---|---|
| Auditory Action Perception | 236 |

| | |
|---|---|
| Bidirectional Action Perception Coupling | 236 |
| Internal Model and Reafferences | 238 |
| Empirical Evidence | 239 |
| Auditory Action Perception with Natural Movement Sounds | 241 |
| Auditory Action Perception with Artificial Movement Sounds (Sonification) | 243 |
| Outlook in Applied Fields | 246 |
| Conclusion and Future Research | 247 |
| References | 248 |

### 16. Visual Perception in Expert Action

*Rita de Oliveira*

| | |
|---|---|
| State of Research | 254 |
| Temporal Aspects of Visual Perception and Action | 254 |
| Spatial Aspects of Visual Perception and Action | 258 |
| Applied Science: The Visual Control of Basketball Shooting | 261 |
| Reinterpretation of Previous Studies | 263 |
| Methodological Considerations | 264 |
| Future Research | 266 |
| Acknowledgment | 267 |
| References | 267 |

## Section E
## Performance under Pressure of Individuals or Teams and Other Performance Phenomena of Emotion–Cognition Interactions

### 17. Bridging the Gap between Emotion and Cognition: An Overview

*Sylvain Laborde*

| | |
|---|---|
| Emotions and Other Affective Phenomena | 276 |
| Cognition: A Necessary Distinction between Executive and Non-Executive Functions | 277 |
| Performance-Oriented Theories Bridging the Gap between Emotion and Cognition | 278 |
| General Emotion–Performance Theories | 278 |
| Performance-Oriented Emotion–Cognition Theories | 280 |
| A Critical View of the Theories Reviewed | 283 |
| Content of Section E | 286 |
| Conclusion | 286 |
| References | 287 |

18. **Performing under Pressure: Influence of Personality-Trait-Like Individual Differences**

   *Emma Mosley and Sylvain Laborde*

   | | |
   |---|---|
   | PTLIDs and the Influence on Performance under Pressure | 292 |
   | Competitive Trait Anxiety | 293 |
   | Trait Emotional Intelligence | 294 |
   | Hardiness | 295 |
   | Mental Toughness | 295 |
   | Optimism and Pessimism | 296 |
   | Perfectionism | 298 |
   | Reinvestment | 299 |
   | Resilience | 300 |
   | Sensation Seeking (Risk Taking) | 301 |
   | PTLID Summary | 302 |
   | Future Research Directions within PTLID Research | 302 |
   | Integrating and Combining PTLIDs | 302 |
   | PTLIDs: An Interactionist Approach | 304 |
   | The Blanket Approach: Guidelines for Use and Concluding Remarks | 306 |
   | References | 307 |

19. **The Influence of Hormonal Stress on Performance**

   *Franziska Lautenbach and Sylvain Laborde*

   | | |
   |---|---|
   | Instructions for Getting Your Dream Job Based on Cortisol Research | 315 |
   | Cortisol—What Is It and What Does It Do? | 316 |
   | State and Trait Influences on Cortisol | 317 |
   | Cortisol and Performance | 318 |
   | Cortisol and Sports Performance | 319 |
   | Cortisol and Cognitive Performance | 320 |
   | Cortisol and Motor Performance | 322 |
   | Cortisol–Performance Framework | 323 |
   | Summary and Outlook | 325 |
   | References | 325 |

20. **Performing under Pressure: High-Level Cognition in High-Pressure Environments**

   *K. Werner*

   | | |
   |---|---|
   | Problem Solving | 330 |
   | Creativity | 330 |
   | Decision-Making | 331 |
   | Comparing Problem Solving and Creativity | 332 |
   | Comparing Decision-Making and Problem Solving/Creativity | 333 |

| | |
|---|---|
| **Emotional Influence on Higher Cognition** | 333 |
| Problem Solving | 334 |
| Creativity | 335 |
| Decision-Making | 336 |
| **Conclusion and Directions for Future Research** | 336 |
| Future Research on Pressure and High-Level Cognition | 337 |
| Future Research on Affective States and High-Level Cognition | 337 |
| **References** | 339 |
| | |
| Index | 341 |

# Contributors

**Eckart Altenmüller** Institute of Music Physiology and Musicians' Medicine, Hanover University of Music, Drama and Media, Hanover, Germany

**Terry Clark** Centre for Performance Science, Royal College of Music, London, UK

**Rita de Oliveira** School of Applied Sciences, London South Bank University, London, UK

**Michael Falkenstein** Leibniz Research Centre for Working Environment and Human Factors (IfADo), Dortmund, Germany; Institute for Working, Learning and Aging (ALA), Bochum, Germany

**Edson Filho** School of Psychology, University of Central Lancashire, Lancashire, UK

**Patrick D. Gajewski** Leibniz Research Centre for Working Environment and Human Factors (IfADo), Dortmund, Germany

**Dieter Hackfort** University of the Federal Armed Forces Munich, Germany

**Sven Hoffmann** Department of Performance Psychology, Institute of Psychology, German Sport University Cologne, Cologne, Germany

**Christos I. Ioannou** Institute of Music Physiology and Musicians' Medicine, Hanover University of Music, Drama and Media, Hanover, Germany

**Michael Kalicinski** Institute of Physiology and Anatomy, German Sport University Cologne, Cologne, Germany

**Christian Kennel** Department of Performance Psychology, Institute of Psychology, German Sport University Cologne, Cologne, Germany

**Sylvain Laborde** Department of Performance Psychology, Institute of Psychology, German Sport University, Cologne, Germany; UFR STAPS, EA 4260, University of Caen, Caen, France

**Franziska Lautenbach** Department of Performance Psychology, Institute of Psychology, German Sport University, Cologne, Germany

**Babett H. Lobinger** Institute of Psychology, German Sport University Cologne, Cologne, Germany

**Emma Mosley** Centre for Event and Sport Research, Bournemouth University, UK

**Jürgen R. Nitsch** Department of Performance Psychology, Institute of Psychology, German Sport University Cologne, Germany

**Alexandra Pizzera** Department of Performance Psychology, German Sport University Cologne, Cologne, Germany; Institute of Sports and Sports Sciences, Heidelberg University, Heidelberg, Germany

**Markus Raab** Department of Performance Psychology, Institute of Psychology, German Sport University Cologne, Cologne, Germany

**Daniel Schneider** Leibniz Research Centre for Working Environment and Human Factors, Dortmund, Germany

**Vassilis Sevdalis** Department of Performance Psychology, Institute of Psychology, German Sport University Cologne, Cologne, Germany

**Gloria B. Solomon** Department of Kinesiology, Texas Christian University, Fort Worth, TX, USA

**Gershon Tenenbaum** Department of Educational Psychology & Learning Systems, Florida State University, Tallahassee, FL, USA

**Monika Thomas** Institute of Physiology and Anatomy, German Sport University Cologne, Cologne, Germany

**K. Werner** Department of Performance Psychology, Institute of Psychology, German Sport University Cologne, Cologne, Germany

**Aaron Williamon** Centre for Performance Science, Royal College of Music, London, UK

**Clemens Wöllner** Institute of Systematic Musicology, University of Hamburg, Hamburg, Germany

# Preface

The scope of this book is to present a unique collective volume written by experts, with the aims of (1) providing a scientific guide to the field of performance psychology with a focus on research from multiple disciplines and domains and (2) synthesizing these perspectives to form a foundation for future theoretical, empirical, and applied developments.

## WHAT IS PERFORMANCE PSYCHOLOGY?

Performance is everywhere, and the word is often used as an umbrella term to describe the behavior of humans or animals or even larger entities, such as a country or an organization. In the following pages, the focus is on human performance in everyday life, often in relation to achieving specific goals, such as winning a sports competition or performing in music or performance arts, and on improving, stabilizing, or reestablishing performance after injury when preparing for such events.

Individual performance, such as making a shot in soccer or pressing the keys on a piano, can be studied in various disciplines and from many perspectives. For instance, the soccer shot can be analyzed within medicine in terms of its physiological or anatomical components, within mechanics in terms of kinetic and kinematic parameters, within psychology in terms of mental imagery or performance under pressure, or within sociology in terms of the societal impact of the shot producing a goal. The perspective from which such a performance is analyzed and what consequences are discussed naturally vary. In this book, we use the discipline of psychology as a starting point to understand performance, because for the majority of researchers who have contributed to this book, their basic research interests are based on psychological concepts, which are applied to sports, music, and other performance domains. We would like to stress that nonpsychological factors such as the brain activity when monitoring errors are also considered, as any complex performance will be better understood from an exchange between disciplines.

Studying performance from a psychological perspective is valid in all domains, including education, business, science, leisure, and many more, because common to performance in any of them are specific psychological building blocks required by individual and group performers alike. Here, we have restricted ourselves to a limited number of high-performance domains, but the evidence we report from sports, music, and the performing arts is relevant beyond the individual domains. The chapters focus on a variety of topics and cover both theoretical and applied research, making the book interesting for scientists and practitioners.

## WHAT COMPONENTS OF PERFORMANCE PSYCHOLOGY ARE CONSIDERED?

Without a doubt, any complex performance can be described, explained, improved, and maintained from a multitude of psychological perspectives, and many subdisciplines of psychology have been dividing the work. For instance, general psychology describes components of individual emotion, motivation, perception, action, and cognition, and within cognition aspects such as memory, problem solving, language processing, attention selection, and—without exaggerating—hundreds more. Clinical psychology may be interested in these factors if they are beyond normal functioning, and social psychology is interested in the influence of social dimensions. Many textbooks introduce these factors in isolation, but here we present pairs of factors that are linked in explaining performance. For instance, to describe the soccer shot, we look at how visual perception leads to a specific action, how a negative emotion can regulate dysfunctional cognitive thoughts, and how a person's own sensorimotor experiences change the perception and cognitive judgments of a foul. For simplicity, we refer to these factors—perception, action, cognition, and emotion—as core capacities.

The empirical studies included in this book that addressed these core capacities employed a variety of methods, including experimental, diagnostic, and intervention approaches. Researchers have developed theories to explain specific phenomena (e.g., Lobinger, Chapter 5, and Altenmüller & Ioannou, Chapter 7 for music and sport examples), a paradigm and model to measure performance more objectively (e.g., Tenenbaum & Filho, Chapter 3, and Hoffmann, Chapter, 10), and a new experimental design that may detect as-yet unreported factors (e.g., Lautenbach, Chapter 19, and Sevdalis & Wöllner, Chapter 14). In applied research, researchers have explored both domain-specific phenomena (e.g., Kalicinski, Thomas & Lobinger, Chapter 8) as well as phenomena that are present in performances in all domains (e.g., Willemon & Clark, Chapter 4, Mosley & Laborde, Chapter, 18), such as choking under pressure (e.g., Laborde, Chapter 17, and Werner, Chapter 20). Experiments and specific diagnostics can measure various aspects of observed or self-reported behavior at very different levels of description (e.g., Solomon & Lobinger, Chapter 6, Kennel & Pizzera, Chapter 15, and de Oliveira, Chapter 16). For instance, working memory can be tested in a working span test in an experiment or by using an intelligence core measure in a questionnaire or by measuring brain activation that is believed to be associated with the use of working memory or by measuring hormonal transmitters that are related to it, as well (e.g., Raab, Chapter 1, Nitsch & Hackfort, Chapter 2, Hoffmann, Chapter 9, Pizzera, Chapter 13). Finally, in addition to presenting theoretical and applied research, we include several chapters that demonstrate how the lessons learned can be applied in real environments (see Chapter 1, Table 1).

## HOW IS THE BOOK STRUCTURED?

The book is structured in five sections. Section A introduces the building blocks of performance—that is, the core capacities of cognition, perception, action, and emotion and their components (e.g., attention, memory, decision making)—that are necessary to performance, as well as the theories, methods, and applications of performance research that are addressed in the subsequent sections. Sections B through E investigate the links between the core capacities. Section B explores the link between action and cognition, Section C between perception and cognition, Section D between perception and action, and Section E between emotion and cognition. Throughout the book, chapters address these interactions in a number of disciplines (e.g., behavioral science, neuroscience) and domains (e.g., sports, music). A major advantage over domain-specific or application-specific approaches is that the structure allows readers from various backgrounds to integrate knowledge from multiple domains that is presented in a continuum from basic and applied research to concrete practice. In sum, 20 chapters provide a scientific guide to performance psychology through an exploration of the core capacities of perception, action, cognition, and emotion.

The nature of interdisciplinary research leads to summarizing and synthesizing. Most research laboratories engage in a "fast-forward" method of conducting empirical research and publishing, and there have been few attempts to summarize the accumulated knowledge in book format. Quite often, significant contributions of young researchers go unnoticed or are undervalued, due to the large volume of research output and the ease of access to information from the Internet. By promoting researchers and giving them a forum to express their views, one can foresee the developments of the performance psychology field in the future. We specifically invited young scholars to contribute to this book, synthesizing their voices with those of more senior researchers to document the present and future of performance psychology research. Our approach, which included a weekend for discussions and feedback on drafts of most of the chapters, produced a book that is a valid reference not only for the student or researcher beginning or advancing an academic career but also for the practitioner (e.g., the musician, the medical doctor, the manager, or the coach) and the clinician (e.g., the sport psychologist, the music therapist).

This book is the culmination of the efforts of many people we would like to thank. First and foremost are the members of the Department of Performance Psychology at the Institute of Psychology at the German Sport University Cologne. For editorial management and coordination Ellen Otte was very helpful in setting and monitoring deadlines and providing feedback on format. At a 2-day book retreat, Damian Jeraj, Christian Kennel, Sylvain Laborde, Franziska

Lautenbach, Babett Lobinger, Lisa Musculus, Ellen Otte, Alexandra Pizzera, Markus Raab, Vassilis Sevdalis, and Karsten Werner provided excellent discussions. We would also like to thank others who reviewed beyond editors and authors, including Ina-Marie Döring, Sebastian Heuer, Damian Jeraj, Oliver Kapner, Kristin Katschak, Jonna Löffler, Lisa Musculus, and Helena Stettner. Finally, thanks to Nikki Levy and Barbara Makinster at Elsevier for rapid, quality feedback and support of this book.

We hope that you will enjoy our road to performance psychology that is at the same time a tribute to our anniversary celebration of 50 years of performance psychology at the German Sport University's Institute of Psychology, Cologne, Germany.

**Markus Raab, Babett Lobinger, Sven Hoffmann,**
**Alexandra Pizzera, and Sylvain Laborde**
Cologne, March 2015

Section A

# What is Performance Psychology?

1. The Building Blocks of Performance: An Overview (Markus Raab)    3
2. Theoretical Framework of Performance Psychology: An Action Theory Perspective (Jürgen R. Nitsch and Dieter Hackfort)    11
3. Measurement Considerations in Performance Psychology (Gershon Tenenbaum and Edson Filho)    31
4. Applications within Performance Psychology (Terry Clark and Aaron Williamon)    45

## Overview

Performance is often used as an umbrella term when we describe behavior of humans or animals or even larger entities such as a country or an organization. In the following, we are interested in human performance in everyday life, often in relation to achieve specific goals such as winning a competition in sports, music, or the arts, improving, stabilizing, or re-establishing performance when preparing for such events that are important and meaningful for a group or an individual.

Individual performance, such as the shot in soccer or the keys pressed on a piano can be studied from various disciplines and perspectives. For instance, the soccer shot may be analyzed within medicine on its physiological or anatomical components, within mechanics on kinetic and kinematic, within psychology on the person's mental imagery or performance under pressure, within sociology on the societal impact of a potential goal outcome the shot produced and much more. The perspective from which such performance then is analyzed and what consequences are discussed would naturally vary. In this book, we use psychology as a starting point to understand performance because the majority of researchers merged in this volume use psychology as the discipline to drive their basic and applied research interests that are transferred to various domains such as sport, music, or other performance domains. However, we would like to stress that non-psychological factors will be considered as well, as any complex performance will be better understood from an exchange between the disciplines. Section A will provide an introduction into the field of performance psychology providing prerequisites (Chapter 1), theories (Chapter 2), methods (Chapter 3), and applications (Chapter 4).

Chapter 1

# The Building Blocks of Performance: An Overview

**Markus Raab**
*Department of Performance Psychology, Institute of Psychology, German Sport University Cologne, Cologne, Germany*

*Rio de Janeiro on a sunny Sunday afternoon, 2014. After 113 minutes running up and down on roughly 10,000 square meters of grass, Mario Götze, a 22-year-old man from a small town in southern Bavaria, received a pass from André Schürle. He stopped the soccer ball with his chest and, seemingly flying and gliding at the same time, used his left foot to score the only and final goal to win Germany the World Cup final. In a film about the team (Spiess, Wortman, Christ, Gronheid, & Voigt, 2014), Götze acknowledged that in soccer things happen "intuitively and by instinct" and there is a need to "act very fast and make a decision immediately."[1] How one performs in such split-second decisions can have lasting effects: For example, following Götze's personal moment of glory and the team's overall success in the tournament, Mario soon became the most popular name of newborn boys in Germany; average beer consumption in Germany went up 4% in comparison to the previous year; and the Adidas company's stock market value increased 2.7%, in part because they could sell World Cup winner jerseys.*

A number of factors influence performance and its short- and long-term consequences, from the technical and tactical skills of a player to basic information processing, including perception, memory, emotion, and cognition, among others. Understanding the interplay of these building blocks and how nature and nurture affect performance is the goal of this book. The chapters in this section explore this interplay from a theoretical (Nitsch & Hackfort, Chapter 2), methodological (Tenenbaum & Filho, Chapter 3), and applied (Clark and Williamon, Chapter 4) perspective. For instance, Götze (Spiess et al., 2014) spoke of his World Cup performance in terms of intuition, speeded action, and choice. In Chapter 2, Nitsch and Hackfort provide a theoretical framework that explains how these concepts are interrelated and how mind and motion work together. In Chapter 3, Tenenbaum and Filho discuss the various methods that can be used to explore the mechanisms of performance. For instance, Götze

---

1. Translated by M. Raab.

and others were interviewed after their performance. Interviewing performers or gathering their reflections on viewing footage of themselves in action is a useful tool for exploring a performer's hindsight. Other methods might involve laboratory simulations. For instance, Götze could be asked to view a number of videos and indicate how he would shoot or pass in the situation, while his gaze, heart, kinematics, and the ball speed are measured. Still other methods employ more ecologically valid environments. Researchers might study performance in the field or in sport performance databases, which allows them to look at the relations between specific variables, such as attacks from the left side of the field, successful goals, and the number of defenders behind the ball. In Chapter 4, Clark and Williamon address the practical level of performance analysis. Data need to be measured and interpreted and ultimately filtered and conveyed to coaches and players as specific recommendations for training or competition that they can easily grasp.

The building blocks previously mentioned—perception, memory, emotion, and cognition—can be seen as the *prerequisites* of performance, that is, the building blocks of performance. From a performance psychology standpoint, these building blocks are often described as constructs, and in fact, they are often the terms used for structuring sections in introductory textbooks in psychology. These constructs can be subdivided: For instance, cognition can be split into the categories of memory and attention, and memory can be differentiated by structure or process, such as working memory or executive function. Without a doubt, to perform in sports, art, or music, all of these constructs matter, and specific situational, task, or personal constraints can influence how their importance is weighted. These constructs also lend structure to this book. Sections B–E focus on the interactions and concrete applications in performance psychology. Specifically, Section B deals with cognition and action, Section C with perception and cognition, Section D with perception and action, and Section E with emotion and cognition. Table 1 lists the constructs and exemplifies related phenomena that have been studied in the remaining chapters. Next, I explore several constructs in more detail.

## PERCEPTION

Perception is part of the information-processing system. It is a complex phenomenon that provides the input for so-called higher-order processes such as a creative choice. From an ecological perspective, perception refers to perceiving a stimulus directly (Gibson, 1979); from a gestalt psychology perspective, it refers to perceiving a stimulus as more than the sum of its parts (Kanizsa, 1979); and in the computational approach of Marr (1982), perception needs to answer questions such as its function, how this function can be described in an algorithm of input and output, and how perception is implemented as neuronal activity. For performance, it is self-evident that perception matters, and often perceptual modalities such as visual or auditory information are one way to structure

**TABLE 1** Building Blocks and Phenomena of Performance Discussed in Chapters 5–20

| Building Blocks | Phenomena | Example | Chapter |
|---|---|---|---|
| Action | Focal dystonia | Pianist involuntary movements reduce performance and are cured | 06 |
| | Response monitoring | Tennis player's stroke errors need anticipative control | 09 |
| Cognition | Choking under pressure | Professional golfer putting at the 18th hole | 05 |
| | Expectancies of performance | Coaches' expectancies of athletes abilities influence coach–athlete interactions | 07 |
| | Mental imagery | Prevention of fall-related injuries among older people due to imagery training | 08 |
| | Performance judgments | Tennis judges are biased due to attention-related errors | 11 |
| | Embodied cognition | Gymnastic judges use their own sensorimotor experience for athletes, performance evaluations | 13 |
| | Agency detection | Dancers detect themselves in point-light displays due to their own motor system | 14 |
| | Emotional intelligence | Athletes, trait emotional intelligence predicts emotion regulation in stressful environments | 17 |
| | Embodied problem solving | A chief engineer creating a new car using his own sensorimotor system to influence his cognitive system | 20 |
| Emotion | Competitive trait anxiety | Ballet dancers with high anxiety use maladaptive coping strategies | 18 |
| | High ego involvement | Job interview candidates' response to be explained via a cortisol–performance relation | 19 |

*Continued*

**TABLE 1** Building Blocks and Phenomena of Performance Discussed in Chapters 5–20—cont'd

| Building Blocks | Phenomena | Example | Chapter |
| --- | --- | --- | --- |
| Perception | Error monitoring | A car driver controlling his attention of navigation system and traffic | 10 |
| | Response selection and error detection | Middle-aged assembly-line workers improve cognitive performance during multitask training | 11 |
| | Acoustic reafference | Track and field athletes use their own acoustic reafferences of the running sound to regulate their movements | 15 |
| | Optic flow | Basketball players use long fixations at the basket to control their actions | 16 |

building blocks. For example, visual perception is important for catching a fly ball, and acoustic perception for sensing the synchrony of a played instrument with a given beat.

Catching a fly ball is something children can learn, and after a few trials, it can even become a routine behavior. However, describing such a behavior is not easy, and robots have substantial problems performing this task. How does vision guide movements so a person can catch a ball? If a ball is moving straight toward a person's eyes, the size of the ball will be perceived as increasing, and the perceptual system will predict when the ball will arrive at eye height using the time-to-contact variable (Lee, 1980). If the ball is far away and flying in a curve, as when a baseball is hit to the outfield, keeping the angle of gaze constant by adjusting running speed will allow an outfielder to be at the place where the ball will land. Laboratory experiments using very simple tasks, such as grasping an object with constant speed and straight trajectory, have been quite helpful in understanding the mechanisms of grasping that allow such behavioral precision in time and space. In more complex tasks, such as perceiving the ball in basketball, players and opponents make use of these same principles but in a much more complex system of focusing attention and selecting specific information (see Chapter 16).

Multiple sensory channels allow humans to use information from the world and combine it with information available from the body or memory systems. In recent years, however, there has been a new appreciation of multisensory integration in complex behaviors. Chapter 14 focuses on visual information and its relation to action, and Chapter 15 considers acoustic information.

## ACTION (action u movement)

Actions have been described as intentional movements that serve a specific goal (Magill, 2011). In contrast to movements—reflexes or actions that are not intentional—actions can be structured by the intentions and the situations in which they occur. For instance, in many textbooks of movement learning and movement control (Magill, 2011; Schmidt & Lee, 2005), actions are often dichotomized into discrete (e.g., a golf putt) and continuous (e.g., riding a bike) movements. More complicated taxonomies separate actions into 16 groups defined by dimension (Gentile, 2000), for instance, whether the action happens in an unpredictable and changing environment, whether objects need to be manipulated, or whether the actor's own body needs to move in the environment. Soccer is an example of a complex environment, and a basketball free throw is an example of a stationary action in a quite stable environment, where the actor manipulates an object but does not move around.

From an action theory perspective, actions are regulated in multiples ways (see Chapter 2) and are more or less cognitively based. Prinz, Beisert, and Herwig (2013) separated actions into those parts that share representations with cognition and those that do not. This relation between action and cognition is discussed in Section B.

A theoretical challenge from an embodied cognition approach (see Werner, Chapter 20) is whether bodily information influences behavior for all cognitive processes (e.g., Wilson, 2002) or whether, in general, we need to assume a more radical form in which all cognitive processes are grounded in action (Chemero, 2009). A methodological challenge is the measurement of such interactions and whether we can exclude other regulatory mechanisms (Raab, Johnson, & Heekeren, 2009). Finally, the quantification and specification of the effects of action on cognition may have meaningful practical consequences, for instance, with respect to improving problem solving (Werner & Raab, 2013) or boosting creativity (Topolinski & Reber, 2010). One important step forward would be to differentiate the types of movements that have an effect on cognition (Tomporowski, 2009). Another would be to identify the specific movement directions or functions that alter the effect of actions, such as in numerical cognition (Fischer, 2012). Section B focuses on action theory concepts, which will be exemplified to general and applied aspects of complex action production in music and sports.

## COGNITION

Cognition is what allows us to solve problems, be creative, and use language in multiple forms (Goldstein, 2015). Research in this field is so extensive that today there are hundreds of theories even for specific concepts such as judgment and decision-making (Bar-Eli, Plessner, & Raab, 2011). Not all have been applied yet in performance psychology, but recent overviews have summarized how different theories describe performance in sports (Raab, 2012) or in other performance domains (see Clark and Williamon, Chapter 4, and Altenmüller & Ioannou, Chapter 7).

For some of these concepts, general theories have been applied to sports choices. For instance, general decision-making theories have been applied to ball-allocation decisions in volleyball (Raab, Gula, & Gigerenzer, 2011). Whereas in simple economic contexts such theories provide a normative solution, this is not the case in sports performance where people vary their choices and do not follow rational choice theorems. For instance, even when Michael Jordan is on the court, you will not find one playmaker who allocates every ball to him. Research areas such as the hot-hand phenomenon, that is, the belief that a player has a higher chance of scoring after two or three hits than after two or three misses, has shown that sequential choices are only partly dependent and sometimes fully independent (see Avugos, Köppen, Czienskowski, Raab, & Bar-Eli, 2012; Bar-Eli, Avugos, & Raab, 2006 for a meta-analytical and a narrative review, respectively). In addition, recent empirical studies have described and explained such behaviors in light of attack and defense strategies (Csapo, Avugos, Raab, & Bar-Eli, 2014, in press; Csapo & Raab, 2014) or depending on how the situations had been framed (MacMahon, Köppen, & Raab, 2014).

One specific theory that describes discrete and sequential choices is based on rules of thumb, or heuristics. The simple heuristics approach has been applied to many performance domains, including sports (Raab, 2012) and medicine (Wegwarth, Gaissmaier, & Gigerenzer, 2009). Researchers have been working to model such heuristics (e.g., Luan, Schooler, & Gigerenzer, 2014), to compare them to each other and to more traditional decision strategies, and to reveal their neurophysiological foundations (Volz et al., 2006). Whereas in some performance domains it has been shown that relying on heuristics produces better choices (such as in medicine; Wegwarth et al., 2009), for many fields, the success of heuristics is still being explored. In Section E, some of the heuristics are explained.

In Section C, the authors focus on the part of perception and cognitive processes that are related to the occurrence and monitoring of errors and the improvement of cognitive performance across the life span.

## EMOTION

Emotions are a crucial part of any performance and need to be differentiated from mood or general concepts such as affect (see Section E for definitions and concept descriptions). Classifications of emotions that focus on the attributes (e.g., positive or negative valence) or the intensity (high, low) of emotions are introduced for performance in different domains, as well as classifications that focus more on discrete emotions such as anxiety. Recent theoretical advances have clarified the conceptual relations of emotional and cognitive processes. Rather new developments in this area are theories that take advantage of neurophysiological research to measure brain and body responses (Daamen & Raab, 2012). These new trends are impressive in their ability to broaden our understanding, but nevertheless, it is not yet possible to measure and fully understand the dynamic changes in these processes in complex, performance-related behaviors such as a sports competition or a concert. Simulation of such emotionally laden environments has, however, been made possible with recent technological advances. Systems now exist that can, for instance, simulate a jury evaluation for a music academy student (Clark and Williamon, Chapter 4) or immerse a study participant in a three-dimensional display of a sport competition (Laborde & Raab, 2013).

Many of these research applications in different domains have been established without collaboration between disciplines. The exceptions are highlighted, for instance, in a project in which music research has benefited from instruction strategies used in sports science, and another in sports science that made use of music research regarding interventions that worked for music cramp and applied it to the yips, a cramp occurring in golf (see Lobinger, Chapter 5, and Altenmüller & Ioannou, Chapter 7).

In summary, emotions are omnipresent and often have beneficial and diametrical effects on performance, as will become evident throughout the remaining sections of this book. Nonetheless, Section E explains emotional processes in more detail, with a particular focus on the relation between emotion and cognition in performance under pressure. Mosley and Laborde, in Chapter 18, address the concepts of discrete emotions such as anxiety and focus on high-pressure situations in which stress occurs, providing evidence on general and individual differences linked to emotional processes.

## REFERENCES

Avugos, S., Köppen, J., Czienskowski, U., Raab, M., & Bar-Eli, M. (2012). The "hot hand" reconsidered: a meta-analytic approach. *Psychology of Sport and Exercise, 14*, 21–27.

Bar-Eli, M., Avugos, S., & Raab, M. (2006). Twenty years of "hot hand" research: review and critique. *Psychology of Sport and Exercise, 7*, 525–553.

Bar-Eli, M., Plessner, H., & Raab, M. (2011). *Judgement, decision making and success in sport.* Chichester, West Sussex: Wiley.

Chemero, A. (2009). *Radical embodied cognitive science.* Cambridge, MA: MIT Press.

Csapo, P., Avugos, S., Raab, M., & Bar-Eli, M. (2014). The effect of perceived streakiness on the shot-taking behaviour of basketball players. *European Journal of Sport Science, 27*, 1–8.

Csapo, P., Avugos, S., Raab, M., & Bar-Eli, M., in press, How should "hot" players in basketball be defended? The use of fast-and-frugal heuristics by basketball coaches and players in response to streakiness. *Journal of Sports Sciences, 8*, 1–9.

Csapo, P., & Raab, M. (2014). "Hand down, man down." Analysis of defensive adjustments in response to the hot hand in basketball using novel defense metrics. *PLoS One, 9*(12), e114184.

Daamen, M., & Raab, M. (2012). Psychological assessments in physical exercise. In H. Boecker, C. H. Hillmann, L. Scheef, & H. K. Strüder (Eds.), *Functional neuroimaging in exercise and sport sciences* (pp. 109–153). Cham, Switzerland: Springer.

Fischer, M. H. (2012). A hierarchical view of grounded, embodied and situated numerical cognition. *Cognitive Processing, 13*, 161–164.

Gentile, A. M. (2000). Skill acquisition: action, movement, and neuromotor processes. In J. H. Carr, & R. B. Shepard (Eds.), *Movement science: Foundations for physical therapy* (2nd ed.) (pp. 111–187). Rockville, MD: Aspen.

Gibson, J. J. (1979). *The ecological approach to visual perception*. Hillsdale, NJ: Erlbaum.

Goldstein, E. B. (2015). *Cognitive psychology: Connecting mind, research, and everyday experience* (4th ed.). Belmont, CA: Wadsworth/Cengage.

Kanizsa, G. (1979). *Organization in vision: Essays on gestalt perception*. New York: Praeger.

Laborde, S., & Raab, M. (2013). The tale of hearts and reason: the influence of moods on decision making. *Journal of Sport and Exercise Psychology, 35*, 339–357.

Lee, D. N. (1980). The optic flow field: the foundation of vision. *Philosophical Transactions of the Royal Society of London* B, *290*, 169–179.

Luan, S., Schooler, L. J., & Gigerenzer, G. (2014). From perception to preference and on to inference: an approach-avoidance analysis of thresholds. *Psychological Review, 121*, 501–525.

MacMahon, C., Köppen, J., & Raab, M. (2014). The hot hand belief and framing effects. *Research Quarterly for Exercise and Sport, 85*, 341–350.

Magill, R. A. (2011). *Motor learning and control: Concepts and applications* (9th ed.). New York: McGraw-Hill.

Marr, D. (1982). *Vision. A computational investigation into the human representation and processing of visual information*. New York: Freeman.

Prinz, W., Beisert, M., & Herwig, A. (2013). *Action science: Foundations of an emerging discipline*. Cambridge, MA: MIT Press.

Raab, M. (2012). Simple heuristics in sports. *International Review of Sport and Exercise Psychology, 5*, 104–120.

Raab, M., Gula, B., & Gigerenzer, G. (2011). The hot hand exists in volleyball and is used for allocation decisions. *Journal of Experimental Psychology: Applied, 18*, 81–94.

Raab, M., Johnson, J., & Heekeren, H. (Eds.). (2009). *Progress in brain research. Mind and motion: The bidirectional link between thought and action* (Vol. 174). Amsterdam, the Netherlands: Elsevier.

Schmidt, R. A., & Lee, T. D. (2005). *Motor learning: A behavioral emphasis* (4th ed.). Champaign, IL: Human Kinetics.

Spiess, T. (Producer), Wortman, S. (Coproducer), Christ, M. (Director), Gronheid, J. (Director), & Voigt, U. (Director). (2014). Die Mannschaft [The Team] [Motion picture]. Germany: Little Shark Entertainment.

Tomporowski, P. D. (2009). Methodological issues: research approaches, research design, and task selection. In T. McMorris, P. D. Tomporowski, & M. Audiffren (Eds.), *Exercise and cognitive function* (pp. 91–112). Chichester, England: Wiley.

Topolinski, S., & Reber, R. (2010). Gaining insight into the "Aha" experience. *Current Directions in Psychological Science, 19*, 402–405.

Volz, K. G., Schooler, L. J., Schubotz, R. I., Raab, M., Gigerenzer, G., & von Cramon, D. Y. (2006). Why you think milan is larger than modena: neural correlates of the recognition heuristic. *Journal of Cognitive Neuroscience, 18*, 1924–1936.

Wegwarth, O., Gaissmaier, W., & Gigerenzer, G. (2009). Smart strategies for doctors and doctors in training: heuristics in medicine. *Medical Education, 43*, 721–728.

Werner, K., & Raab, M. (2013). Moving to solution: effects of movement priming on problem solving. *Experimental Psychology, 60*, 403–409.

Wilson, M. (2002). Six views of embodied cognition. *Psychonomic Bulletin and Review, 9*, 625–636.

# Chapter 2

# Theoretical Framework of Performance Psychology: An Action Theory Perspective

Jürgen R. Nitsch[1], Dieter Hackfort[2]
[1]Department of Performance Psychology, Institute of Psychology, German Sport University Cologne, Germany; [2]University of the Federal Armed Forces Munich, Germany

Performance is a constituent element of human life and a particular objective of manifold everyday activities. Consequently, it is addressed from the perspective of different scientific disciplines ranging from philosophy to biochemistry. In psychology, performance became a traditional topic in various fields of fundamental and applied psychology, e.g., in educational psychology, occupational psychology, clinical psychology, and sport psychology. Aside from the test diagnostic assessment of "classic" performance variables (e.g., reaction time, concentration, intelligence), numerous empirical studies are focused on the *efficiency and vulnerability of mental functioning* on the one hand and on *social interaction in performance settings* on the other. Typical issues are learning and memory; problem solving; decision-making; movement control; time management; learning and achievement motivation; coping with stress, anxiety, and failure; error prevention; performance-related mental, emotional, and behavioral disorders; burnout and dropout; as well as team building; division of tasks; allocation of responsibilities; teamwork skills; conflict management; mobbing prevention; and leadership style. In applied sport psychology, "performance psychology" commonly covers a toolbox of intervention techniques related to "mental power," "mental strength," "mental toughness," "mental fitness," or more specifically to self-confidence and self-efficacy, for example, self-motivation, self-programming, goal-setting, self-talk, imagery, visualization and mental training, stress-inoculation, cognitive reframing, attention control, relaxation, and biofeedback (see, e.g., Dosil, 2006; Hackfort & Tenenbaum, 2006; Hardy, Jones, & Gould, 1996).

In spite of those multifarious aspects, there has been, however, neither a comprehensive and consensual definition of performance psychology until now nor an integrative theory that provides the potential to systematically guide research and application, thus making the dynamic complexity of human performance sufficiently understandable, controllable, and communicable. So, it is worth paying particular attention to these issues. This will happen in three theory-oriented steps: First, the psychological perspective on performance is characterized, providing a preliminary understanding of performance psychology and its subject area. The second and main step addresses the needed (meta-) theoretical foundation of performance psychology. Accordingly, the focus is not on listing various performance-related theoretical concepts (e.g., for team sports thoroughly carried out by Lebed & Bar-Eli, 2013) but on embedding and considering the performance issue within the overall context of human action organization. Therefore, essentials of the action theory perspective as developed by the authors are outlined and specified with regard to the issue at hand. Third, particular attention is given to the functional role of emotions in action organization. This will contribute to further illustrating action theoretical postulates and to a more proper theory-based understanding of emotional states and processes with special regard to both performance and in general.

## DEFINITION AND SCOPE OF PERFORMANCE PSYCHOLOGY

### Performance and Psychology

The general task of performance psychology is related to the description, explanation, prediction, and optimization of performance-oriented activities in accordance with general and domain-specific ethical standards. The psychological perspective on performance comprises three issues: (1) the *psychological fundamentals* of performance-oriented activities in various action domains such as labor, politics, arts, music, or sports; (2) *psychological transfer effects* of performance-oriented activities in particular with regard to personality development, self-esteem, time management, stress control, communication skills, etc.; and (3) optimization of the capability to achieve demanding *mental tasks*.

This understanding refers to *different agents*, for example, individuals, groups, and organizations, young and elderly, as well as people with or without disabilities. It covers *different motives, domains, and kinds of activity*, for example, school/academic education, the whole range of professional activities, health-oriented sport and exercise, and elite sports, housekeeping, and playing music, as well as strange and/or extraordinary performances documented in the *Guinness World Records*. Even health, well-being, youthfulness, beautifulness or life expectancy are increasingly considered to be products of more or less successfully self-managed activity for which the person is self-responsible. In addition, the preceding definition includes *different proficiency levels* (e.g., novices and experts, amateurs and professionals) as well as *different criteria of performance*, for example, primary criteria related to the action itself and its direct results (frequency, duration, speed, accuracy, novelty, required effort, and their combinations), and secondary criteria in the sense of external/extrinsic social evaluation and feedback. According to Bem's (1972) "self-perception theory," the latter follows a simple logic: If I (or someone else) receive recognition such as praise, awards, applause, or many scientific citations, then the corresponding performance must have been outstanding! (As we all know, that is often a misguiding conclusion!)

For a better understanding of the psychological perspective on performance, it is necessary to distinguish two functional aspects of performance: (1) *performance as a means to an end* with regard to the motives and interests that are intended to satisfy by the consequences of a performance action; (2) *performance as an end in itself*, that is, the accent is on the self-reinforcing performance activity itself and its progressive perfection. In this sense, striving for excellence more or less turns into functional autonomy.

Furthermore, we must be aware of the formally twofold usage of the term "performance" (1) as related to a *class of specific actions and outcomes* or (2) as a more or less marked *dimension of any kind of human action* (that is the position preferred here).

## Structure of Performance Orientation

The key features of any *performance orientation* can be summarized as follows (see Figure 1):

1. *Reference Standards*: Feeling challenged to set/raise and to meet/exceed demanding reference standards, which are considered as binding for the evaluation of the course and outcome of an action and specified by the habitual and/or actual aspiration level. According to well-known conceptions of achievement motivation, typical references are individual's prior performance (*Individual Reference Standard*; e.g., actual "handicap" of a golf player), the performance of relevant others (*Interindividual Reference*

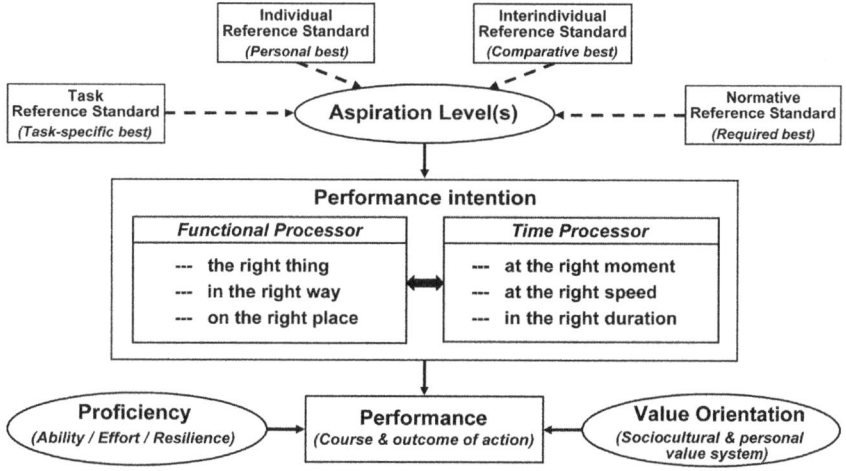

**FIGURE 1** Structure of performance orientation (broken lines = optional relations).

*Standard*; e.g., handicap or actual results of other golf players on a tour), the demands of a given task (*Task Reference Standard*; e.g., the "par" as the strokes calculated for one "hole" in golf), and/or a specific standard value that must be met (*Normative Reference Standard*; e.g., maximum handicap needed to be allowed to play on a golf course).

2. *Performance Intention*: Specification of both the *functional and the temporal components* of an action (see Thomas & Weaver, 1975; cit. Zakay, 1993, p. 64). Accordingly, performance orientations may differ with regard to the range and precision of the components concerned.
3. *Agent Causality*: Attributing the course and outcome of an action to the proficiency and responsibility of the agent (see Heckhausen, 1980, p. 112). Accordingly, the focus of performance attribution can be on ability and/or effort and/or resilience and/or outcome.
4. *Value Orientation*: Striving for excellence means intending/providing a valuable contribution related to the sociocultural and/or personal value system. That is, the course and outcome of an action is evaluated from the perspective of meeting, adding, or creating relevant values.

These four aspects provide essential performance markers and the target points for the analysis and the compensation of performance deficits as well. "Performance" is a *relational* concept that depends on the applied reference standard and aspiration level in relation to the individual's proficiency and the personal and sociocultural value systems: Objectively different courses and results of an action may indicate similar *relative* performance values. Attaining a goal is not identical with success, success is not identical with high performance, and high performance is psychologically not identical with objectively best performance ever. Thus, the psychological meaning of performance can be neither reduced to absolute excellence nor assessed without regarding the kind, regularity, and comparability of performance preconditions and task demands.

## Characteristics of Peak Performance

As mentioned previously, the objective of applied performance psychology is optimization of performance-related activity. The term *optimization* refers to a *maximin principle* of intervention with respect to a single action as well as to the individual's long-term performance orientation. Specifically, that means *maximizing* the efficiency of performance behavior and *minimizing* unfavorable side effects and undesired long-term consequences.

The *maximizing intention* refers to developing, enhancing, maintaining, and reestablishing the habitual and actual *motivation, competence*, and *resilience* for efficiently dealing with high-demanding tasks. Usually addressed is a broad range of target qualifications such as setting a conclusive series of clear, realistic, and challenging goals; willingness to invest time, resources, and effort needed to achieve the goals; overcoming obstacles and injuries, resistance against temptations and distractions; bearing deprivations, discomfort, and setbacks; coping with stress, failure, fatigue, monotony, and satiation; and acquiring and realizing efficient action strategies and skills. *The minimizing intention* includes, for instance, strategies and measures for efficient recovery, injury prevention, and health care, psychological crisis management, career counseling, and career transition.

Striving for peak performance implies much more than attaining qualitatively and/or quantitatively absolutely outstanding action results achieved under regular conditions. In particular, three aspects are worth being added:

1. *Performing on top implies acting at one's limits.* This results in a very vulnerable balance of high-performance goals, increased risk-taking, and low tolerance for errors. Even minimal fluctuations in concentration can result in errors followed by fatal consequences with regard to a broad range of potential personal, social, and economic disadvantages or damages.
2. *Performing on top requires more or less neglecting other orientations and domains of human life and activity.* It includes allocation of all temporal, personal, social, and economic resources needed for the achievement of one superior goal: enhancing and maintaining performance. Thus, striving for excellence means focusing attention on performing a given task and focusing life on improving performance.
3. *Peak performance is principally public performance.* It is an event that attracts the attention of spectators, media, public figures, organizations, and institutions. Thus, success is publicly known success, and failure is publicly known failure.

In this threefold sense, the preconditions of peak performance are also potential causes of performance crises.

## The Action Paradigm—A Meta-Theoretical Perspective on Performance

### The Primacy of Action

As it has been illustrated, human performance is (1) a highly complex phenomenon that is (2) attributed to an active agent who (3) processes a task within (4) the given environmental setting. This dynamic complexity cannot be sufficiently analyzed by specialized investigations of isolated if–then relations nor from the perspective of a single scientific discipline. What is needed is a guiding conception providing a high potential for intra- and interdisciplinary integration focused on the key element of the performance issue, that is, *human action*.

Action theoretical conceptions have been developed since the 1940s at the latest in different countries and in different scientific disciplines, that is, philosophy, linguistics, sociology, economy, pedagogy, and especially in various subdisciplines of psychology (see Nitsch, 2004 for an overview and references). Since the 1970s, the action perspective increasingly became a leading idea for the theoretical foundation of sport psychology (see, for an overview of various approaches, e.g., Hackfort, Munzert, & Seiler, 2000a; Kaminski, 2009; Kunath & Schellenberger, 1991; Nitsch, 2004; Volpert, 1974). Specific theoretical aspects were additionally conceptualized, for instance, by Allmer (1997), Munzert (1997), Quinten (1994), Samulski (1986), Schack (2010), Seiler (1995), and Wiskow (1992).

In summary, the action approach is based on a long-time and widespread tradition of integrative thinking that is in continuous progress. From our point of view, the action theoretical perspective can be characterized by three fundamental assumptions:

1. *The basic nature of humans is substantiated by the necessity and capability of organizing life by actions.* Thus, action is considered as the key reference for theory building, research, and intervention in the human sciences (see, e.g., Cranach, Kalbermatten, Indermühle, & Gugler, 1980, p. 279; Gehlen, 1971, p. 23; Groeben, 1986, p. 59ff.; Hauser, 1948; Nitsch, 1975; Rubinstein, 1984, p. 229). Action is understood as intentionally organized behavior within a meaningfully structured situational context (see also Hackfort, Munzert, & Seiler, 2000b). This includes both doing and omitting something deliberately. In particular, with respect to the performance issue, it is important to have in mind that any action implies a quadruple function in varying accentuation: (1) *exploration function* in the sense of gathering new information and experiences; (2) *construction function* in the sense of actively solving present problems and tasks; (3) *protection function* in the sense of guarding against threats and disturbances; (4) *presentation function* in the sense of demonstrating personal characteristics as a means of impression management.

2. *Action is a system process, that is, the integrated response of an agent to his or her present situation in the world.* Constitutive for any action is (1) the dynamic interrelation of person and environment; (2) the coordinated interaction of principally all intrapersonal functions (see, e.g., Lersch, 1962, p. 461); and (3) the temporal and functional embedding within the action continuum with regard to the individual's action biography and future perspective. The present action is at the same time the endpoint of the previous and the starting point of the future development. The system perspective has a very challenging methodological consequence, as illustrated in Figure 2, for psychological interrelations in traditional terminology. The active attributes of the functional components of a dynamic self-optimizing system are *mutually interdependent*: Variations in the state of one component more or less results in adaptive alterations in the state of the others. Thus, the usual empirical investigation of unidirectional if–then relations, e.g., the impact of a certain emotion on cognitive performance, is inevitably insufficient.

3. *Psychological processes, states, and traits are considered as fundamentally related to action.* On the one hand, the analysis and optimization of an action must specify the role played by the different psychological functions in the regulation of the action under study. This implies that the particular psychological orientation, activation, and control function will be specified. On the other hand, the impact of the course and outcome of an action on the short- and long-term modification of psychological functions is to be taken

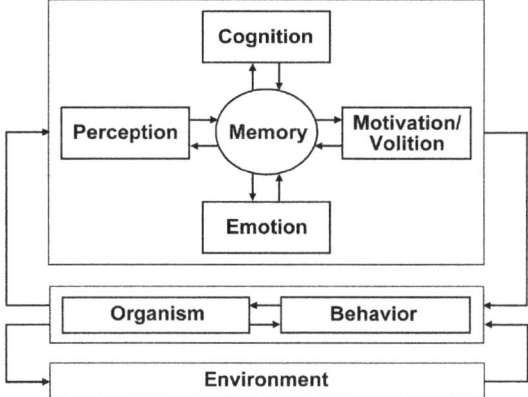

**FIGURE 2** Illustration of the functional network of an action system by using traditional terminology (based on the first author's former lectures on introduction to sport psychology at the German Sport University Cologne, 1993ff.).

into account as, for instance, illustrated by the potential interrelation of sport activity and mental fitness.

The implications of this perspective cannot be outlined here at full length. On the basis of a series of the authors' previous publications on action theory, the focus will be on summarizing some selected aspects that may contribute to a more differentiated and comprehensive understanding of the performance issue.

## Intention—The Organizing Principle of Action

Actions are actively organized (1) with regard to their anticipated and intended consequences and (2) based on internal representations. The meaning of intention is often reduced to setting a particular goal to be achieved. From an action theoretical point of view, intention is to be considered more adequately as a set of operational definitions with respect to satisfying personal needs within a framework of individual values. This understanding implies the differentiation of four *intention components* that may be differently accentuated and elaborated on a case-by-case basis (see Nitsch, 1996, 2005):

1. *Value Intention*: Intention formation happens within a framework of superior values. Values—like fair play in sport—may be understood as landmarks in the action space that define the limits for acceptable action options. Positively defined, the values we refer to constitute the particular *moral significance* of an action.
2. *Outcome Intention*: It refers to the needs to be satisfied, on the one hand, and the anticipated, desired consequences of action, that is, its *purpose*, on the other (e.g., performance enhancement, self-affirmation, social recognition, health improvement, welfare, and professional career).

3. *Goal Intention*: An action goal is defined as an intended and operationally defined action result (e.g., performance score in sports) that should be instrumental and thus required for fulfilling the purpose of an action by attaining the desired action consequences. Achieving the goal of an action, however, does not necessarily imply achieving the action purpose.
4. *Implementation Intention*: To actually achieve the goal, appropriate *means* as well as situational *cues* for the initiation of an action are to be defined. This is where implementation intention comes in (e.g., "On next Tuesday, 18:00h, I will start with training X"). Cues are considered as if conditions in an if–then relationship. They are very important for the more or less automatic activation of intended behavior, as was demonstrated by the studies of Bargh and colleagues in the early 1990s (see, e.g., Bargh & Gollwitzer, 1994).

Beyond defining values, purposes, goals, means, and cues for initiating a certain action, intention formation also involves the definition of *interrupting* and *goal-terminating mechanisms* related to specific cues (see Simon, 1967): If, and only if, particular cues occur, then the ongoing activity will be interrupted for a certain period of time to respond to unexpected urgent events, or the activity will be terminated if the goal is achieved or perceived as not attainable. Accordingly, we have to complete the intention structure with two additional aspects that are of high importance for an efficient organizing of one's actions: *interruption intention* and *goal-terminating intention*.

Emphasizing the intentional organization of actions implies a fundamental methodological consequence, that is, shifting the focus from causal explanation to *intentional explanation* related to a pragmatic syllogism. The key question now changes from "Caused by what?" to "*For what reasons does somebody do or omit something?*" That is, we explain an action by identifying the subjective premises from which the action is deduced as a subjective-logical conclusion. The practical consequence is that a central approach to action modification refers to modifying the individual's *action logic*. In this sense, intentions can be understood as complex, more or less complete, and conclusive *argumentation patterns* that subjectively constitute the execution of an action. The strength of a performance intention then depends on how far the reasons for an action are subjectively perceived as (1) acceptable, (2) convincing, and (3) sufficiently operationally defined.

## The Action Space and Its Situational Configuration

Actions are multifaceted events within a multidimensional *action space*. The action space is defined by the principal options and limitations of an agents' activity; the present constellation specifies the *action situation* (Nitsch & Hackfort, 1981; see also Hackfort, 1986; Nitsch, 1997, 2004, 2009). The structural characteristics can be summarized as follows (see Figure 3):

1. *What we are actually doing or omitting as well as the kind and degree of our well-being depend on the attributes and interrelations of three components:*

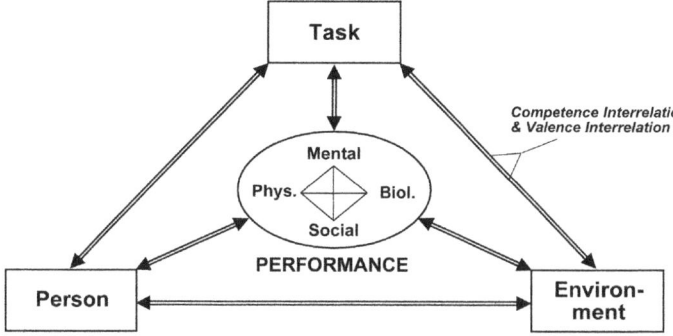

**FIGURE 3** The situational context of performance-oriented action.

*person, environment, and task*. It is important to be aware of the transactional nature of the person–environment–task interrelation: An objective or subjective change of one component triggers a change of the others. For example, the perception of one's capabilities will change depending on the given task; changing personal properties, for example, due to increasing fatigue, will result in a change in the perception of task difficulty and environmental conditions. In this sense, for example, an injury or other performance restrictions are more than changes in a person's abilities and reduced resilience but imply at the same time changes of the objective and subjective personal world.
2. *We act as physical, biological, mental, and social beings in a physically, biologically, mentally, and socially structured world while performing more or less equally structured tasks*. Changes of one of these aspects will potentially change the precondition of an action directly and/or indirectly by modifying the functioning of the others. We will return to that point later.
3. The subjective definition of one's own situation, that is, the perceived attributes and configuration of the situation components, establishes the psychological basis of intentional behavior. The entire situation and the specific relevance of its components is individually appraised in terms of two aspects: (1) the *action competence*, that is, the perceived degree of controllability of the situation by the subject dependent on the expected effort and the probability of success; and (2) the *action valence*, that is, the perceived urgency and importance of optimizing a situation by one's own action. Then, deciding on acting or not acting depends on the perceived degree of valence and competence, and the valence–competence relation. Consensual situation definitions are essential prerequisites of efficient social interaction and communication. Dissonant definitions are one important cause of social conflicts and performance decrements in and of a group or team (Nitsch & Hackfort, 1984). Optimizing these situational features is the general objective of both the individual's action as well as the interindividual coordination of actions and the general intention of practical interventions as it is characteristic for performance management.

## The Functional Architecture of Actions

Actions are characterized by a multimodal functional architecture that is important to be taken into account in analyzing and optimizing performance activities: Integrated by the intentionality principle mentioned previously, the entire organization of an action is particularly based on (1) different organization levels, (2) a triadic phase structure, and (3) three functional systems of action control. Some of the essential features are summarized next (for more details see, e.g., Nitsch, 2004, 2009; see also Hackfort, 2006).

### System Levels of Action Organization

The organization of action involves the interplay of four personal system levels: the *physical*, the *biological*, the *mental*, and the *social organization level*, which are characterized as follows (see also Figure 3):

1. *Each level of action organization is governed by specific rules of functioning and establishes specific sets of objective constraints and subjective options of actions*, that is, *physical* ones (in the sense of anthropometric properties, e.g., body height, leg length, weight, volume etc.; cinematic and dynamic preconditions), *biological* ones (especially referred to the neurological, neuromuscular, endocrine, and metabolic functioning of the organism; motor skills), *mental* ones (e.g., intentions, cognitions, feelings; mental skills), and *social* ones (especially internalized social role expectations; social skills). We do or omit something at the same time under the influence of physical laws, biological preconditions, mental processes and representations, and social values, norms, attitudes, and rules.
2. *The organization levels are considered to be functionally interrelated.* Each level can trigger the functioning of the other ones. In particular, anthropometric properties (physical level) have potential impact on energy expenditure during action (biological level) or may even prime or shape mental processes (see, e.g., the embodiment issue); internalized social values and expectations (social level) potentially constrain the individual's decision-making and intention formation (mental level).
3. *Different personal disposition levels specify different relations to a given task and the environmental context* (see the situation concept mentioned earlier).

Keeping these aspects in mind, the main intention of performance management is optimizing the functioning of a *system* based on empirical findings related to the *system's dynamics*.

### The Phase Structure of Actions

Actions cannot be reduced to their overt behavioral part of execution. From a psychological point of view an action principally passes through a sequence of three phases performing anticipation, realization, and interpretation functions in the course of an action (Nitsch, 1975, 2004; see Figure 4). The highlighted processes represent key factors for the mental foundation of performance on the one hand and essential links for mental performance enhancement on the other.

1. *Anticipation Phase*: In the first phase, an action is psychologically conceptualized. Accordingly, the functional focus is on situation analysis and subjective situation definition, action planning, and intention formation associated, for instance, with processes of decision-making.
2. *Realization Phase*: In confrontation with the reality and depending on the actual circumstances is the execution of the intended behavior initiated or cancelled. The functional focus is on the automatic, emotional, and cognitive processing of an action (see below).
3. *Interpretation Phase*: Finally, the course and outcome of an action is retrospectively analyzed and evaluated with regard to the predeterminations made in the Anticipation Phase. Particularly, the functional focus is on outcome assessment, attribution of causes to success or failure, and the reevaluation of the situation as a starting point for subsequent actions.

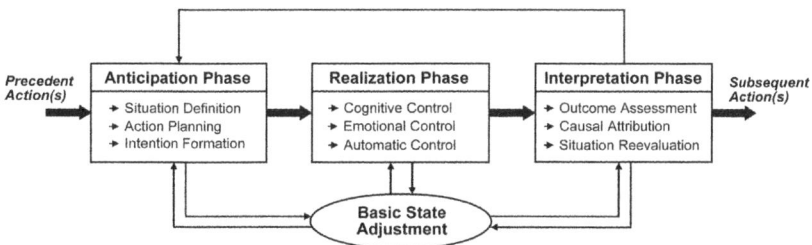

FIGURE 4   Triadic phase structure of performance-oriented action.

In each of these phases, special deficits or disturbances that impair the entire course of an action and threaten its intended outcome may occur. This may happen, for instance, in the case of reducing the available time for anticipation processes under time pressure, distracting spectator reactions in the Realization Phase, or time-consuming overattention to action evaluation in the Interpretation Phase. Furthermore, each of the phases mentioned above (as well as the included functions) can become a separate three-phased action on its own in the sense of *functional autonomy*.

In addition to that time sequence of action organizing functions, a second general aspect has to be taken into account: the specific *adjustment of the agent's basic state* as a necessary precondition of the optimal processing of the action phases (see Figure 4). Accordingly, psychological performance enhancement includes both the optimization of the phase-specific functions (e.g., goal setting, mental training) and the optimization of the agent's state (e.g., relaxation training, self-motivation).

In summary, to understand sufficiently what happens in the course of an action, we have to take into account what happens in each of the three phases:

anticipation, realization, and interpretation. In practical consequence, psychological improvement of performance includes the optimization of both the various processes and the entire phase structure of an action at hand.

## Functional Systems of Action Control

According to empirical action analyses and stimulated by evolution theory, particularly with regard to the gradual differentiation of functional systems of behavior control throughout the development of humankind, human intentional action organization is considered to include three interrelated and specifically operating functional control systems: *the automatic, emotional,* and *cognitive systems of action control* (Nitsch, 1985, 2004; see also Hackfort, 2006). Each of them contributes specifically to the overall organization of an action with respect to orientation, activation and regulation functions, and may become dominant in the case of habitual, emotional, or voluntary action.

1. *Automatic Action Control System*: It provides immediate and quick adaptation to relatively simple and stereotypic conditions of the present situation. The basic principle is the automatic reaction to specific stimuli or cues by preestablished modes of behavior. The underlying learning processes include, in particular, respondent and operant conditioning as well as the automatization of behavioral or mental processes.
2. *Emotional Action Control System*: In dissent from some other emotion concepts, emotion is understood as a basic function in the orientation, activation, and regulation of actions. The functional focus (and advantage) is on the holistic orientation and quick synchronization of complex cognitive and psychomotor processes. Reactions are not triggered directly by specific stimuli but indirectly triggered and shaped by the individual's emotional labeling of the present situation, for example, as threatening or joyful.
3. *Cognitive Action Control System*: Its functional focus is the long-term adaptation to complex, variable, novel, and above all future conditions. With this system, the relation of situational conditions and action is differently established, that is, by cognitive processes. Thus, new types of action organization come into play: conscious situation analysis, anticipation, and evaluation of action situations as well as learning, goal setting, planning, and action monitoring based on symbolic, especially verbally encoded mental representations, and the insight in structural and functional relations.

These three functional systems of action control are principally designed for an integrated contribution to the overall organization of an action. Beyond their specific functions, they can not only mutually influence and support, but also disturb or impair the optimal functioning of each other. Thus, the situation-specific tuning and synchronization of the three systems' activity becomes an essential task of performance management.

The central message to performance management with general respect to the functional action architecture can be summarized as follows: The guiding question in research and intervention is not asking *whether* physical, biological, mental, or social aspects, anticipation or interpretation processes, automatic, emotional, or certain cognitive conditions play a role at all. The question is *how* they come into play and *what* is their special—favorable or unfavorable—contribution to the phenomenon under study. This will be exemplarily portrayed next with special respect to emotions.

## Functionality of Emotional Processes with Special Reference to Performance

While analyzing the emotion–performance relationship, the functional link to the biological and social (sub-)systems is essential and especially the interrelationship with cognitive processes in action regulation as it is stipulated above. Emotions are not action-decoupled units or detached entities interacting with actions as described by Hanin and Ekkekakis (2014). Emotions are generated in the course and by the result of an action, and in turn, they are influencing the action process (see also, e.g., Rubinstein, 1984, p. 582). In this spirit already, Piaget (1954; see also Piaget & Inhelder, 1972) pointed out that affective and cognitive processes are complementary, irreducible to each other, functionally linked, and not in a cause–effect relationship. Speaking about a differential or basic emotion (see e.g., Ekman, 1992; Izard, 1972; Plutchik, 1970; Zajonc, 1980), a reduction to a single symptom like arousal is an inappropriate simplification, and it is necessary to consider that emotions are organized like syndromes "consisting of cognitive appraisals, action impulses, and patterned somatic reactions" (Lazarus, Kanner, & Folkman, 1980, p. 198f.). The outlined action theory approach assimilates these insights and provides a functional reference system for a sufficient understanding of emotions, affective processes, and the emotion–performance relationship and enables for a differentiated view of the functionality of emotions.

## Multifacetedness and Functional Complexity

While it is the unique combination of cognitive and physiological processes and a certain action tendency that is characteristic for the experience and expression of a single emotion, the different components refer to various regulation processes in special action-regulation systems realizing specific functional purposes in the process of acting. Research on emotions in addition to this complex interplay is facing the difficulty that most of the time we are involved in various actions and these actions are in different phases of the action process. Thus, usually an affective state in which partially various emotions are included is experienced in daily life as well as in competitive or performance situations. Considering this functional approach and understanding, it is less surprising to find only low or moderate correlations in studies to examine, for example, the relation between a single emotion like anxiety and performance. Hence, the functionality of emotions is twofold: (1) emotions are generated and modified by acting, and (2) emotions are influencing action regulation (Hackfort, 1991; Nitsch, 1985).

In the process of action regulation, various functions and experiences of emotions can be differentiated (Hackfort & Birkner, 2006), as is summarized in Figure 5.

The functionality of emotions in action regulation can be summarized by experiencing an emotion to be pleasant or unpleasant and to be functional or dysfunctional. What is felt to be pleasant not always is beneficial, and what is experienced to be unpleasant not always is dysfunctional. To clarify the emotion–performance relationship, the temporary extension, the intensity, and the emotional pattern at that period of time have to be considered. Furthermore, in the analysis of the emotional impact for performance, it is important to recognize that the components of an emotion are functioning on different levels with various time courses and with respect to the action phase(s) and its regulation processes. Neurophysiological and cognitive processes are not running parallel in time, for example, in the actual genesis of emotions, hormone secretion needs more time than cognitions like worries to come up. When a composite of emotions is experienced when acting in a given situation, several emotions, each of them with a certain pattern of activation, and cognitions are varying in the time course. At the same point in time all of them are differently influential in action regulation, which has to be considered in the emotion–performance analysis. For such an approach, it is essential to refer to the complex dynamics, and consequently, assumptions like the inverted U-shaped relationship or an optimal zone are obviously insufficient conceptions to cover corresponding dynamics with respect to the performance execution.

## Functional Disturbances

In addition, when discussing phenomena like what is described by choking, a system approach in the action analysis may be beneficial to uncover the functional role and develop a better understanding. First, choking has been shown to correlate with anxiety, fear of negative evaluation, and ineffective coping

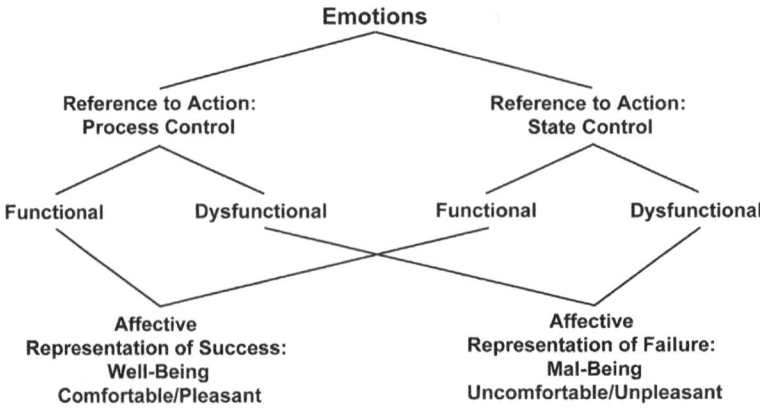

**FIGURE 5** Functional relevance of emotions in action regulation.

(Marchant, Maher, & Wang, 2014). The authors also point out (p. 449) that it is a tautology to add "under pressure" as pressure is explained consistently to be a necessary condition for the phenomenon. All this appears logical following the first/original definition of choking as "performance decrements under pressure circumstances" by Baumeister (1984, p. 610). In the course of 25 years of research, a more sophisticated understanding turned out, and Hill, Hanton, Fleming, and Matthews (2009, p. 206) emphasized choking to be a "process whereby the individual perceives that their resources are insufficient to meet the demands of the situation, and concludes with a significant drop in performance." The approach to analyze the process embedded in the perception of the demands of the situation, obviously the action situation is addressed, is a significant step forward toward the functional role with regard to the action at hand.

On the basis of the indicated calculation of demands and resources, it absolutely makes sense to "conclude" a drop in performance or to reduce or stop further investment. Following this understanding, choking serves the preservation of the functionality of the system in order to protect the system against serious injury or damage. Furthermore, when looking from a system perspective and with reference to the subsystems of action regulation, it would be possible to hypothesize that choking is developed by and indicates a conflict between processes in the subsystems and, thus, disturbing a proper action regulation and performance decrement is the effect. From this point of view, the performance decrement is not astonishing, surprising, inexplicable, or surmountable. The appropriate and relevant approach to enable, ensure, and enhance performance is to uncover the conflict and to manage the improvement of the tuning in the action regulation processes. This strategy would also be in line with studies providing proofs for performance improvement with experts in dual task designs.

## Options in Emotional Processing

Furthermore, it has been well known for a long time from research on stress, resilience, and coping (for an overview see Nitsch, 1981) that the definition of the situation and the appraisal of resources and competencies in proportion to demands and the significance of the situation are essential for the development of (emotional) stress and its consequences. Quite similar to the threat or challenge definition of the situation, choking is not the only option, but concentration of attention, increase of effort, and so on, are the alternatives to reacting to pressure. If choking is coming up, it is associated with one of two tendencies and an increase of the probability to result in one of two completely different programs of the behavior system, which we know very well from the analysis of psychological crises: (1) playing dead reflex and (2) a storm of movement. The functional sense of such programs, as has been proved to be suitable in the process of evolution, with respect to performance is realized either by a search for orientation, in case it is assumed that there (still) are possibilities to achieve performance (enhancement), or by a reduction of (any) activity as a protective response in case no (further) effort makes any sense. Sometimes, it is difficult to detect the intention of a separate action, but in a broader perspective, the meaningfulness for the system and suitability in the situation turns out, especially considering long-term achievement potential.

## CONCLUSION

Human performance, in particular peak performance, would be insufficiently understood as being generated by single internal or external conditions. It is the end product of the dynamics of a complex system with regard to the intentional organization of task-related behavior within a meaningful environmental context. In this sense, the outlined action theory perspective provides a metatheoretical frame of reference for efficiently dealing with the complexity of the performance issue:

1. It enables the continuous and constructive integration of concepts and findings from various fields of research to an overarching picture.
2. It provides general theoretical coordinates with respect to theory-building, the development of investigation strategies and methods, and theory-based intervention. In this sense, the action theoretical frame of reference may be considered as a *cognitive map* that helps to systematically structure research and intervention.
3. As a holistic conception of human action, it offers the theoretical links urgently needed for intra- and interdisciplinary collaboration in the performance domain.

## REFERENCES

Allmer, H. (1997). Intention und Volition [Intention and volition]. In R. Schwarzer (Ed.), *Gesundheitspsychologie: Ein Lehrbuch* [Health psychology: A textbook] (pp. 67–89). Göttingen, Germany: Hogrefe.

Bargh, J. A., & Gollwitzer, P. M. (1994). Environmental control of goal-directed action: Automatic and strategic contingencies between situations and behavior. *Nebraska Symposium on Motivation, 41*, 71–124.

Baumeister, R. F. (1984). Choking under pressure: Self-consciousness and paradoxical effects of incentives on skilful performance. *Journal of Personality and Social Psychology, 46*, 610–620.

Bem, D. J. (1972). Self-perception theory. In L. Berkowitz (Ed.), *Advances in experimental social psychology* (Vol. 6) (pp. 1–62). New York: Academic Press.

Cranach, M. von, Kalbermatten, U., Indermühle, K., & Gugler, B. (1980). *Zielgerichtetes Handeln* [Goal-directed action]. Bern, Switzerland: Huber.

Dosil, J. (Ed.). (2006). *The sport psychologist's handbook: A guide for sport specific performance enhancement*. Wiley: Chichester, WS.

Ekman, P. (1992). An argument for basic emotions. *Cognition and Emotion, 6*, 169–200.

Gehlen, A. (1971). *Der Mensch. Seine Natur und seine Stellung in der Welt* [Man: His nature and place in the world] (9th ed.). Frankfurt am Main, Germany: Athenäum.

Groeben, N. (1986). *Handeln, Tun, Verhalten als Einheiten einer verstehend-erklärenden Psychologie. Wissenschaftstheoretischer Überblick und Programmentwurf zur Integration von Hermeneutik und Empirismus* [Action, doing, and behavior as units of a descriptive-explanatory psychology: Epistemological overview and program design for the integration of hermeneutics and empiricism]. Tübingen, Germany: Francke.

Hackfort, D. (1986). *Theorie und Analyse sportbezogener Ängstlichkeit* [Theory and analysis of sport-related trait anxiety]. Schorndorf, Germany: Hofmann.

Hackfort, D. (1991). Emotion in sports: An action theoretical analysis. In C. D. Spielberger & I. G. Sarason (Eds.), *Stress and emotion* (pp. 65–73). New York, NY: Hemisphere.

Hackfort, D. (2006). A conceptual framework and fundamental issues for investigating the development of peak performance in sports. In D. Hackfort & G. Tenenbaum (Eds.), *Essential processes for attaining peak performance* (pp. 10–23). Aachen, Germany: Meyer & Meyer.

Hackfort, D., & Birkner, H.-A. (2006). Funktionen von Emotionen [Functions of emotions]. In M. Tietjens & B. Strauß (Eds.), *Handbuch Sportpsychologie* [Handbook of sport psychology] (pp. 165–177). Schorndorf, Germany: Hofmann.

Hackfort, D., Munzert, J., & Seiler, R. (Eds.). (2000a). *Handeln im Sport als handlungspsychologisches Modell* [Action in sport as an action-psychological model]. Heidelberg: Germany: Asanger.

Hackfort, D., Munzert, J., & Seiler, R. (2000b). Handlungstheoretische Perspektiven für die Ausarbeitung eines handlungspsychologischen Ansatzes [Action-theory perspectives for the elaboration of an action-psychology approach]. In D. Hackfort, J. Munzert, & R. Seiler (Eds.), *Handeln im Sport als handlungspsychologisches Modell* [Action in sport as an action-psychological model] (pp. 31–46). Heidelberg: Germany: Asanger (2000).

Hackfort, D., & Tenenbaum, G. (Eds.). (2006). *Essential processes for attaining peak performance*. Aachen: Germany: Meyer & Meyer Sport (UK).

Hanin, Y. L., & Ekkekakis, P. (2014). Emotions in sport and exercise settings. In A. Papaioannou & D. Hackfort (Eds.), *Routledge companion to sport and exercise psychology* (pp. 83–104). London: Routledge.

Hardy, L., Jones, G., & Gould, D. (1996). *Understanding psychological preparation for sport. Theory and practice of elite performers*. Chichester, WS: Wiley.

Hauser, R. (1948). *Psychologie als Lehre vom menschlichen Handeln* [Psychology as science of human action]. Wien, Austria: Herder.

Heckhausen, H. (1980). *Motivation und Handeln* [Motivation and action]. Berlin, Germany: Springer.

Hill, D. M., Hanton, S., Fleming, S., & Matthews, N. (2009). A re-examination of choking in sport. *European Journal of Sport Science*, 9(4), 203–212.

Izard, C. E. (1972). *Patterns of emotion: A new analysis of anxiety and behavior*. New York, NY: Academic Press.

Kaminski, G. (2009). Sport in the perspective of Barkerian psychological ecology. *International Journal of Sport Psychology*, 40, 50–78.

Kunath, P., & Schellenberger, H. (Eds.). (1991). *Tätigkeitsorientierte Sportpsychologie* [Activity-oriented sport psychology]. Frankfurt am Main: Germany: Harri Deutsch.

Lazarus, R. S., Kanner, A. D., & Folkman, S. (1980). Emotions: A cognitive phenomenological analysis. In R. Plutchik & H. Kellermann (Eds.), *Theories of emotions* (pp. 189–217). New York, NY: Academic Press.

Lebed, F., & Bar-Eli, M. (2013). *Complexity and control in team sports: Dialectics in contesting human systems*. Abingdon, OX: Routledge.

Lersch, P. (1962). *Aufbau der Person* [The structure of personality] (8th ed.). München, Germany: Barth.

Marchant, D., Maher, R., & Wang, J. (2014). Perspectives on choking in sport. In A. G. Papaioannou & D. Hackfort (Eds.), *Routledge companion to sport and exercise psychology* (pp. 446–459). London: Routledge.

Munzert, J. (1997). *Sprache und Bewegungsorganisation. Untersuchungen zur Selbstinstruktion beim Bewegungslernen* [Language and movement organization. Investigations on self-instruction in motor learning]. Schorndorf, Germany: Hofmann.

Nitsch, J. R. (1975). Sportliches Handeln als Handlungsmodell [Action in sport as action model]. *Sportwissenschaft, 5*(1), 39–55.

Nitsch, J. R. (Ed.). (1981). *Stress: Theorien, Untersuchungen, Maßnahmen* [Stress: Theories, investigations, interventions]. Bern, Switzerland: Huber.

Nitsch, J. R. (1985). Emotionen und Handlungsregulationen [Emotions and action regulation]. In G. Schilling & K. Herren (Eds.), *Angst, Freude und Leistung im Sport* [Anxiety, joy, and performance in sport] *Proceedings of the VIth FEPSAC Congress in Magglingen, 1983* (Vol. 1) (pp. 37–60). Magglingen, Switzerland: FEPSAC and Eidgenössische Turn- und Sportschule.

Nitsch, J. R. (1996). Intention und Handlungsregulation [Intention and action regulation]. In R. Daugs, K. Blischke, F. Marschall, & H. Müller (Eds.), *Kognition und Motorik* [Cognition and motor function] (pp. 69–86). Hamburg, Germany: Czwalina.

Nitsch, J. R. (1997). Situative Handlungsorganisation [Situated action organization]. In H. Ilg (Ed.), *Gesundheitsförderung – Konzepte, Erfahrungen, Ergebnisse aus sportpsychologischer und sportpädagogischer Sicht* [Health promotion: Concepts, experiences and findings from a sportpsychological and sportpedagogical point of view] (pp. 351–363). Köln, Germany: bps.

Nitsch, J. R. (2004). Handlungstheoretische Grundlagen der Sportpsychologie [Action-theoretical foundations of sport psychology]. In H. Gabler, J. R. Nitsch, & R. Singer (Eds.), *Einführung in die Sportpsychologie: Teil 1. Grundthemen* [Introduction to sport psychology: Part 1. Basic issues] (4th ed.) (pp. 43–164). Schorndorf, Germany: Hofmann.

Nitsch, J. R. (2005). Motivation reconsidered – an action-logical approach. In R. Stelter & K. K. Roessler (Eds.), *New approaches to sport and exercise psychology* (pp. 55–82). Oxford: Meyer & Meyer Sport (UK).

Nitsch, J. R. (2009). Ecological approaches to sport activity: A commentary from an action-theoretical point of view. *International Journal of Sport Psychology, 40*, 152–176.

Nitsch, J. R., & Hackfort, D. (1981). Stress in Schule und Hochschule: Eine handlungspsychologische Funktionsanalyse [Stress at school and university: A functional analysis based on action theory]. In J. R. Nitsch (Ed.), *Stress: Theorien, Untersuchungen, Maßnahmen* [Stress: Theories, investigations, interventions] (pp. 263–311). Bern, Switzerland: Huber.

Nitsch, J. R., & Hackfort, D. (1984). Basisregulation interpersonalen Handelns im Sport [Basic state regulation of interpersonal action in sport]. In E. Hahn & H. Rieder (Eds.), *Sensumotorisches Lernen und Sportspielforschung* [Sensorimotor learning and research on sports games] (pp. 148–166). Köln, Germany: bps.

Piaget, J. (1954). *Les relations entre l'affectivité et l'intelligence dans le development mental de l'enfant* [The relation between affectivity and intelligence in the mental development of the child]. Paris, France: Centre de documentation universitaire.

Piaget, J., & Inhelder, B. (1972). *Die Psychologie des Kindes* [The psychology of the child]. Freiburg, Germany: Walter.

Plutchik, R. (1970). Emotions, evolution, and adaptive processes. In M. B. Arnold (Ed.), *Feelings and emotions. The Loyola Symposium* (pp. 3–23). New York, NY: Academic Press.

Quinten, S. (1994). *Das Bewegungsselbstkonzept und seine handlungsregulierenden Funktionen. Eine theoretische und empirische Studie am Beispiel Bewegungslernen im Tanz* [The movement-related self-concept and its action-regulative functions. A theoretical and empirical study on motor learning in dancing]. Köln, Germany: bps.

Rubinstein, S. L. (1984). *Grundlagen der Allgemeinen Psychologie* [Fundamentals of general psychology] (10th ed.). Berlin, Germany: Volk und Wissen.

Samulski, D. (1986). *Selbstmotivierung im Sportunterricht* [Self-motivation in physical education]. Köln, Germany: bps.

Schack, T. (2010). *Die Kognitive Architektur menschlicher Bewegungen* [The cognitive architecture of human movement]. Aachen, Germany: Meyer & Meyer.

Seiler, R. (1995). *Kognitive Organisation von Bewegungshandlungen. Empirische Untersuchungen mit dem Inversionsprinzip* [Cognitive organization of movement actions. Empircal investigations using the inversion principle]. Sankt Augustin, Germany: Academia.

Simon, H. A. (1967). Motivational and emotional controls of cognition. *Psychological Review, 74*(1), 29–39.

Thomas, E. A., & Weaver, W. B. (1975). Cognitive processing and time perception. *Perception and Psychophysics, 17*, 363–367.

Volpert, W. (1974). *Handlungsstrukturanalyse als Beitrag zur Qualifikationsforschung* [Structural analysis of action as a contribution to qualification research]. Köln, Germany: Pahl-Rugenstein.

Wiskow, M. (1992). *Konkreatives Handeln. Theoretische und empirische Ansätze zur Umorientierung in der Kreativitätsforschung* [Con-creative action. Theoretical and empirical approaches to a reorientation of creativity research]. Köln, Germany: bps.

Zajonc, R. B. (1980). Feeling and thinking. Preferences need no inferences. *American Psychologist, 35*, 151–175.

Zakay, D. (1993). The impact of time perception processes on decision making under time stress. In O. Svenson, & A. J. Maule (Eds.), *Time pressure and stress in human judgment and decision making* (pp. 59–72). New York, NY: Plenum Press.

# Chapter 3

# Measurement Considerations in Performance Psychology

Gershon Tenenbaum[1], Edson Filho[2]
[1]*Department of Educational Psychology & Learning Systems, Florida State University, Tallahassee, FL, USA;* [2]*School of Psychology, University of Central Lancashire, Lancashire, UK*

## MEASUREMENT CONSIDERATIONS IN PERFORMANCE PSYCHOLOGY

Psychology is a scientific domain aimed at understanding human behavior. Psychology also aims at helping people to meet their genetic potential and reach self-fulfilment. Performance psychology is a scientific and applied discipline, which concentrates on meeting these goals. Performance can be visual (artistic-creative), motor (athletic), cognitive (language acquisition), or any other professional activity (e.g., engineering, medicine, architecture, astronautics, etc.), and thus must be quantifiable in any kind of judgmental form. For example, athletic performance is evaluated via statistical features and competitive outcomes; medical surgeries are evaluated through the rate of successful patients' recovery and their functionality; and artistic performance is evaluated through experts' reports and the audience reactions. These examples illustrate that performance can be evaluated through different types of measures that must be trustworthy to make scientific judgment about one's performance quality.

In this chapter, we refrain from describing specific performance measures, as there are numerous measures addressing both individual- and group-level performance that have been thoroughly detailed elsewhere (for a review, see Daamen & Raab, 2012; Fernandez-Ballesteros, 2003; Tenenbaum, Eklund, & Kamata, 2012). Rather, our focus is to discuss core measurement concepts applied to performance psychology. Foremost, we introduce several aspects inherent to the development of trustworthy quantitative and qualitative performance measures. Moreover, we comment on the importance of parsimonious theoretical integration in performance psychology (see Chapter 2). We expand on this argument by introducing an integrative two-parameter measurement model aimed at capturing the cognitive–affective–behavioral linkage in sports performance. We conclude the chapter by introducing new measurement trends in neuroscience and genetics.

## TYPES OF MEASURES

Measures are broadly classified as primarily quantitative or qualitative. If the goal is to run faster and jump longer, then the measures are generally *quantitative and objective*. However, if the goal is to perform a concert or a dance in front of an audience, then the measures are usually *qualitative and expressive*. In this case, the goal is to follow the notes and rhythm while adding personal interpretation and expression to the performance.

### Quantitative Measures

Quantitative and objective measures use physical and psychometrical tools and represent *manifest variables*. By definition, manifest variables are those that can be observed and measured, such as heart rate, lactic acid, visual span, landing characteristics, and rhythmic reproduction. Responses to items in a questionnaire are also considered manifest variables because it is assumed that they represent a latent variable (e.g., personality dimension or psychological state). In other words, several of the items interactively may represent a latent variable, which has not been directly measured.

A quantitative positivistic, "hard-natural science" approach to measurement of any entity requires that the measures will share an *origin, equal units of measurement*, and *linearity*. When these measurement features are evident, the measure is considered sample-free. In this case, any measureable difference in one location of a given scale is equal to any identical difference along the scale. In other words, people vary freely on a scale, which remains consistent and does not need any modifications—it stays stable across samples and thus is sample-free. A sample-free measurement tool (e.g., a yard stick) allows objectivity and generalizability to be inferred with much confidence (Wright & Douglas, 1977; Wright & Masters, 1982).

Noteworthy, the call for quantitative objective measures in the realm of the "soft sciences" had been made at the beginning of the twentieth century by Thurstone (1925, 1927, 1928), Torgerson (1958), Angoff (1960), and Guttman (1944, 1974). Guttman argued that if a sample of attributes given to a sizable random sample is scalable, generalizability can be inferred and a linear continuum can be formed. When these conditions are satisfied, then two individuals with identical scores are almost similar quantitatively on the measurable variable. During these early times, the need for linearity via an "item calibration" procedure was developed—all for meeting the requirements of objective measurement. However, because people are not always behaving in a physical manner, valid responses on introspective measures have been tested via reliability and validity procedures. We refrain from thoroughly describing reliability and validity in this chapter because there is a substantial literature, which details these concepts in depth. Rather, we refer to them as essential requirements in the domain of performance psychology.

## The Concept of Reliability

Performance measures must be reliable—meaning that they must be as accurate and error-free as possible. Performance scores that ought to be stable must show consistency over trials. Conversely, measures that ought to fluctuate across varied events must show inconsistency and variance within the performers. Consider the following:

$$S = T + E$$

Where $S$ is the performance score, $T$ is the true score, and $E$ is the measurement error, we strive for $E=0$, so that $S=T$, meaning 100% reliability and 0% error. Any deviation from this value indicates some untrustworthiness in the measure. The sources of measurement errors are internal and external to the examinees. These must be carefully controlled if we wish to rely on the measure outcome and make secure decisions.

## The Concept of Validity

Validity refers to the confidence we have that a given measure provides us with information about the variable on which we wish to obtain information. To gain validity, one must design a strategic plan, which takes into account the goals and the situations under which the test must be performed. A test, which aims at supporting a certain decision, may be completely irrelevant for supporting another decision. Thus, each measure is both *decision-directed* and *decision-dependent* in the sense that satisfying both decisions must be done in a valid manner. There are several concepts of validity references (Vaughn & Daniel, 2012). The most common ones are (a) content-related validity (relevance and representativeness), (b) criterion-related validity (construct and predictive), and (c) construct-related validity (convergent–correlational, contrasted groups, experimental, and discriminant).

## Statistical Power and Sample Size

Statistical power influences the reliability and validity of a given measure. McDonald and Marsh (1990) have long noted that both low and high power have an effect on goodness-of-fit indices. Specifically, low power may lead to "underfit," whereas high power may lead to "overfit" measurement models. Hence, assuming all factors equal, researchers relying on different statistical power are likely to select different measurement models. For this reason, scholars concur that an a priori power analysis is paramount in measurement research and practice. Performance measures should be developed based on appropriate power, and thus the sample size should be "large enough to detect the hypothesized differences (if present) but not significantly larger" (Jennings & Gianaros, 2007, p. 814). It is important to note that statistical power is greater when error variance is low. Therefore, power can be augmented not only by increasing the sample size but also by testing more homogeneous populations, and improving data acquisition methods (e.g., neurological measures).

## Administering and Interpreting Performance Measures

Prior to administering a measurement, researchers and practitioners must consider its specific goals, theoretical and practical grounds, and reliability and validity evidence. The measure must also share uniformity in relation to the test-taker, tester, procedures, instructions, equipment, and facilities—all aimed at minimizing the measurement error. Moreover, one should take into account the measurement length and number of repetitions to avoid boredom and fatigue. To increase performance tests trustworthiness, one should address the following:

1. Consider ethical standards pertaining to research and provision of services.
2. Allow for warm-up and familiarity, especially when the test is complex and/or relies on memory load.
3. Allow sufficient time interval for recovery when the test is long or when several tests are administered sequentially.
4. Scorers and judges must be fully familiar with the testing procedure.
5. Consider the number of test takers and the testing time estimate with respect to the age, developmental stage, and other relevant mental and physical features of the test takers.
6. Pilot the test and the procedure prior to the main administration.
7. Ensure efficient test organization, particularly in complex, sequential, and/or demanding conditions.
8. Consider the cost of the test and/or technology required to measure the variables of interest.
9. Tests must be calibrated to avoid constant error to be included in the performance scores.
10. Set standards for evaluation, assessment, and measurement of the performance outcomes to avoid confusion and inconsistencies.
11. Document all of the developmental processes and procedure of the test.

After data collection, the measurement scores should be interpreted through norm and/or criterion-related reference. A norm-referenced performance score is relative to the norm group against which the score is compared. Criterion-related referenced performance consists of comparing a performer's score to an external criterion (e.g., qualifying standards for the Olympic Games).

## Qualitative Measures

Positivistic, "hard science" measurement requirements are of less, if at all, value in the evaluative process of performances that are expressive in nature and/or those that must not have the rigidity of the scientific measurement assumptions, as well as the need to be generalizable. However, such measures must have sampling, theoretic-methodological, and transferability grounds.

## Sampling: The Importance of Case Selection

Qualitative assessment is not about large sample sizes or generalizability power. Rather, the selection of "the case" or "the participant" is based on strong theoretical reasoning (deductive approach), empirical data, and/or follows a reasonable logical path (inductive approach). Therefore, one must know why a given case or participant is of interest, either based on theory or exploratory logic. Numerous sampling strategies are described in the literature (e.g., criterion sampling, snow-ball sampling; for a review, see Patton, 2002) and the underlying idea is similar: to recruit "information-rich cases," specifically those cases or subjects able to contribute to the understanding of a given phenomenon. In measuring performance through qualitative lenses, scholars and practitioners should avoid remote and marginal cases, and rather focus on cases and subjects of marked importance.

## Methodological Triangulation and Interpretative Pluralism

A mono-method assessment is not appropriate in designing qualitative measures or conducting qualitative research (Parker, 2004). This is because qualitative measures inevitably carry the researchers' "subjective" biases. Given the subjectivity associated with single measures, qualitative researchers recommend the use of multiple methods to describe performance measures. For instance, in judging a dancers' performance, a judge may qualitatively evaluate one's bodily expression, while considering the audience reaction and the music melody. In essence, qualitative measures are designed to orient intensive case studies, and as such, must be validated in more than one methodological route.

In addition to methodological triangulation, researchers interested in the development of qualitative measures should engage in "interpretative pluralism" (Coyle, 2010). Whereas methodological triangulation involves the use of multiple sources and methods, interpretative pluralism rests on applying different theoretical perspectives to the same data set. In other words, interpretative pluralism consists of "learning from the juxtaposition of divergent ideas and ways of seeing" (Kincheloe, 2005, p. 344). Thus, to develop trustworthy qualitative performance measures, one should apply numerous theoretical, epistemological, and ontological approaches to the same data set.

## Transferability of Measures and Research Outcomes

As discussed herein, qualitative measures and research findings are not to be generalized. Thus, in qualitative research, the concept of "transferability" does not imply generalization to universal laws. However, properly triangulated performance measures may indeed be transferable to similar cases, contexts, and subjects (see Kelle, 2006). Transferability is important in advancing applied research in the human sciences. For example, in performance psychology, the well-established individual zones of optimal functioning (IZOF) framework was also coined through qualitative performance measures (Hanin, 2007). Although the qualitative results of the IZOF framework are not generalizable across different performers, the key theoretical tenet of the framework (that different performers have idiosyncratic bio-psycho-social states) applies to different performers and is transferable to different sports. Altogether, properly designed qualitative measures are key in developing "grounded theories" that can be transferable to similar cases depending on their recognizable theoretical strength. In effect, theory development and integration is the ultimate goal of performance measurement.

## MEASUREMENT AND THEORY DEVELOPMENT

Researchers and practitioners rely on theory to develop trustworthy measures to advance theoretical knowledge. In this continuous process of "theory-measurement development-advanced theory," it is essential to consider alternative models and strive for parsimonious solutions.

### Alternative Models

Science evolves when alternative solutions (thesis and antithesis) are proposed and integrated. Therefore, scholars should constantly evaluate alternative measurement models. Alternative models are models with different specifications but yielding a similar fit (Hershberger, 2006). The literature on performance psychology includes important examples of well-established measurements that have been redesigned and improved (see Eys, Carron, Bray, & Brawley, 2007; Kellmann & Kallus, 2001). In this regard, psychometric theory describes two specific alternative models: *equivalent* and *non-equivalent*. Equivalent models have different path links but yield the same fit. Non-equivalent models have different path links and/or factorial configurations and yield different fit indices. Alternative models in qualitative research are advanced through "member checking" and the use of "critical friends" and "external judges." In practice, the participants in a qualitative study, or trained scholars not directly involved with the research, offer feedback on alternative ways of understanding and drawing conclusions from the raw data.

### The Principle of Parsimony

In addition to proposing alternative explanations for a given phenomenon, the goal of science is to develop parsimonious models that capture, explain, and predict natural, humanistic, and social occurrences (Gigerenzer, 2010). As opposed to being developed in a theoretical vacuum, any measurement tool, whether qualitative or quantitative, should be based on theory and used to advance new theories. It follows that the *principle of parsimony* is at the core of both measurement and theory development. The principle of parsimony reflects the notion that researchers should strive for simple measurement models that use the minimum number of parameters needed to explain a given phenomenon (Raykov & Marcoulides, 1999).

Parsimonious modeling is particularly important for the development of integrated theoretical frameworks. Various scholars concur that integrated theoretical frameworks should be simple (i.e., "less is more" effective; for a review, see Goldstein & Gigerenzer, 2011), as simple models are more easily interpreted than overly parameterized models. Although the principle of parsimony has primarily been used in the definition of quantitative measures, qualitative scholars also speak of interpretative pluralism in the sense of creating integrated theoretical models that have a clear focus. Next, we illustrate the concept of parsimonious modeling by introducing a framework for capturing the cognitive–affective–behavioral linkage in performance psychology.

## TWO-PARAMETER MODEL FOR CAPTURING THE COGNITIVE–AFFECTIVE–BEHAVIORAL LINKAGE IN PERFORMANCE PSYCHOLOGY

Performance psychologists are interested in modeling the process of adaptation through the use of reliable, valid, and parsimonious modeling. Specifically, in this context is the question of how performers cope with stressors that necessitate fast reaction, as well as environmental constraints, which allow for more processing time. Adaptation is in essence a dynamic process of seeking a better, richer, more complete, and in-depth format of adaptation, founded upon constructive interpretations of earlier experiences (Webb, Schweiger Gallo, Miles, Gollwitzer, & Sheeran, 2012). To conceptualize the adaptation process, Tenenbaum, Lane, Razon, Lidor, and Schinke (2015) operationalized the degree of adaptation (A) as a function of the difference between how one perceives the environment/task (e.g., challenging, threatening, etc.), $\delta_i$, and how one perceives his or her own ability/capacity (i.e., self-efficacy) to interact and cope with the environmental/task demands, $\beta_v$, to achieve a given goal (Function 1).

$$\text{Adaptation (A)} = f(\beta_V - \delta_i) \quad (1)$$

Perceived ability that exceeds the perception of task demands, $\beta_v - \delta_i > 0$, that is, $\beta_v > \delta_i$, results in an adaptation process that has a higher probability of resulting in a desirable adaptation state. However, when one perceives a task as being too difficult for his/her own capacity, $\beta_v - \delta_i < 0$, that is, $\beta_v < \delta_i$, the probability of investing effort and energy to accomplish the task is reduced, and the state of adaptation declines. When $\beta_v - \delta_i = 0$, that is, $\beta_v = \delta_i$, the task may be perceived as a challenge, which also may result in a positive adaptation, or alternatively, in a failed adaptation. In motivational terms, when $\beta_v < \delta_i$, the performer will most likely exhibit an *avoidance behavior*, and when $\beta_v > \delta_i$, an *approach behavior* is more probable. The adaptation process, therefore, is probabilistic in nature, and is presented by Eqn (2) as follows:

$$P(A - 0 - 1; \beta_V, \delta_i) = \exp(\beta_V - \delta_i) / [1 + \exp(\beta_V - \delta_i)] \qquad (2)$$

where $P$ is the probability of poor-to-excellent adaptation (say, ranging from 0 to 1), $\beta_V - \delta_i$ is the self-perception of ability and self-perception of the environmental/task demand difference, and exp is the exponent of this difference that makes the function linear. This concept was presented by Rasch (1960) pertaining to item response theory (IRT), but more recently Tenenbaum et al. (2015) expanded it to the realm of adaptation and performance, as well as emotions, cognitions, and behavioral outcomes. This simple two-perceptual concept is presented in Figure 1.

At any stage of the adaptation process one should measure the performer's perception about oneself and the task (may it be a dance, surgery, or athletic) along with measures of the environmental and task appraisals, and the resulting emotions, cognitions, and behaviors. Different emotions and thoughts are associated with the $\beta_V - \delta_i$ quantum before, during, and after task completion. Task-related evoked emotions are considered here as neither helpful nor harmful, but rather as a source of information for eliciting a designated action (Baumeister, Tice, & Zell, 2007). Tenenbaum et al. (2015) have argued that this simple measurement concept can account for all performance-related environments.

The concept applies to fast and slow adaptation processes. Fast adaptation depends on the effectiveness of the perceptual-cognitive system to direct and focus attention while eliminating noise, anticipating upcoming events, and selecting a response (Tenenbaum et al., 2009). The effectiveness of these processes consists of the richness and variety of perceptions processed at a

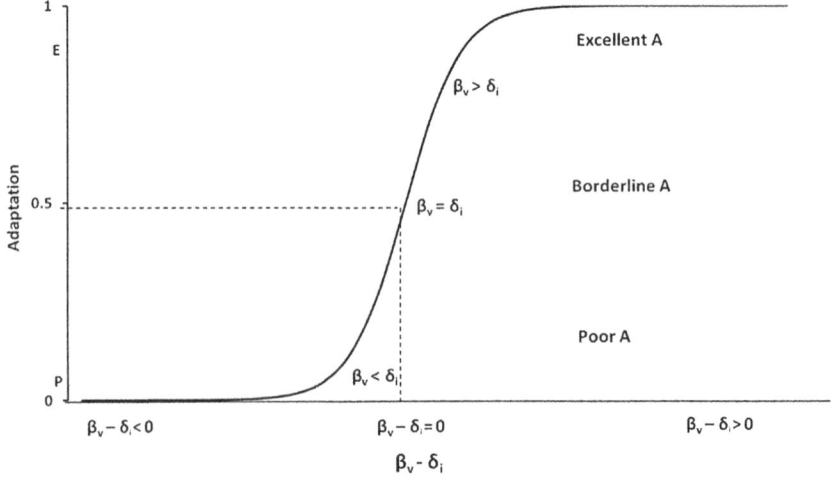

**FIGURE 1** Probabilistic function describing the adaptation process as a function of $\beta_V - \delta_i$.

given time; that is, the system's capacity to encode (store and represent) and access (retrieve) information relevant to the task being performed (Tenenbaum & Bar-Eli, 1993). Once the performer acquires a vast number of well-developed, efficient coping strategies, these can be retrieved and implemented under any circumstances. Thus, a knowledge base in the form of neural mental representations controls the retrieval of response pathways and makes the action easy to execute, $\beta_v - \delta_i > 0$. Once the task is too complex and the knowledge base is lacking, the adaptation process suffers, and it subsequently affects the motivation and emotional states because $\beta_v - \delta_i < 0$.

Skilled performers were found to maintain a consistent set of physical and psychological routines during the preparation time for a given closed self-paced act (Lidor, 2007). That is, maintaining a consistent sequence of pre-performance routines can result in improved levels of proficiency (i.e., increasing $\beta_v$; see Lidor, 2007). Moreover, imposing pre-performance routines, such as external focusing attention, imagery, and relaxation, was found to facilitate the accuracy of closed self-paced tasks. In other words, the task, $\delta_i$, is perceived as one, which is achievable. In essence, performance routines calibrate all competitive components to a mental level that enables the achievement of optimal preparation and adaptation prior to and during the execution of a closed self-paced task, where $\beta_v - \delta_i > 0$, that is, $\beta_v > \delta_i$. When the performer perceives the task as being too difficult to perform, namely $\beta_v - \delta_i < 0$, that is, $\beta_v < \delta_i$, the probability for maladaptation increases.

An additional example of this concept is presented via the Reversal Theory (RT; Apter, 1984) where the adaptation (e.g., the extent one feels balance as one's state of mind) depends on how one perceives his or her meta-motivational dominance ($\beta_v$, e.g., capability to cope with environmental demands) and how challenging and demanding the environment is ($\delta_i$). Accordingly, telic-dominant people seek a safe environment where $\beta_v = \delta_i$, that is, $\beta_v > \delta_i$, in order to feel comfortable and avoid anxiety. Paratelic people seek challenging, and sometimes dangerous, environments to challenge their abilities in order to feel excitement. Maintaining oneself in a comfort zone brings boredom and a desire for a reversal. Thus, the adaptation processes depend on the amount of time people remain stable in their preferred meta-motivational dominance (e.g., positive adaptation), and the time they maintain stability in the less preferred meta-motivational state (negative adaptation) and wish to revert back to their dominant meta-motivational state. Thus, reversal into the preferred state of mind happens when $\beta_v < \delta_i$.

The two-parameter measurement concept can relate to the phenomenon of burnout, a slow process accompanied by emotional and physical exhaustion, sport devaluation, and a reduced sense of accomplishment (Raedeke, 1997; Raedeke & Smith, 2001). Other stressors may cause similar symptoms in workers and performance (see Eklund, Smith, Raedeke, & Creswell, 2012). When burnout occurs, environmental demands (e.g., extensive practice with limited recovery time, family obligations, work, or any other stress-evoking sources) accumulate to the point where they exceed the capability of the person to cope efficiently with the tasks at hand, and the result is maladaptive behaviors, typical of the burnout syndrome. In other words, when the stressors are perceived as too demanding for the person to cope with, $\beta_v < \delta_i$, the person has a higher probability of experiencing personal difficulties and performance decline, along with limited coping capabilities. However, when $\beta_v > \delta_i$ is prevalent, a positive and comfortable state of adaptation is expected to be experienced.

Another slow process is known as the acculturation phenomenon—how do people and performers alike adapt to a culture that is unfamiliar to them. The acculturation process is not linear, unlike a career trajectory (Chirkov, 2009). High satisfaction with the personal and environmental demands might move the performer toward a positive state of adaptation and contextual engagement when $\beta_v - \delta_i > 0$. Low satisfaction can result in maladaptation and its possible manifestations of alienation when $\beta_v - \delta_i < 0$ (Schinke et al., 2006). The more inclusive the environment, the more likely the relocated athletes become and then remain engaged within their sport contexts (e.g., $\beta_v - \delta_i > 0$). However, acculturation is influenced by cognitions as much as social context. In essence, the current state of acculturation depends on the depth of cognitive skills necessary to function within the sport environment, along with the social support bestowed on the athlete. Once the environmental demands equate or exceed personal demands, $\beta_v - \delta_i < 0$, the performer is expected to feel stress (and related emotions)—a state of maladaptation that may result in performance decline.

This two-parameter model is an alternative approach to overly parameterized models in performance psychology. Performance psychologists interested in developing measurement models must strive for parsimonious, reliable (i.e., error-free), and valid solutions. Foremost, performance psychologists must bear in mind that measurement development is a continuous (theory-measurement development-advanced theory) and multidimensional process. In some cases, measures ought to be generalizable, whereas in other causes, the measures are specific. Furthermore, in some cases, measures must be culture-free, while in other cases, measures must be culture-dependent. Additionally, measures may be constrained by gender, skill level, and age cohort. These distinctions depend on the context, domain, and environmental features within which the performance is measured. Researchers and practitioners must have the motivation to implement new trends in performance measurement, including neuroscience and genetic measures.

## NEW TRENDS IN PERFORMANCE MEASUREMENT

Performance analysts have increasingly used neuroscience measures, particularly over the last two decades. It is likely that the use of these measures will continue to gain momentum within psychology research (Cacioppo, Tassinary, & Berntson, 2007). It is important that researchers and practitioners remember that neurophysiological measurement involves the mappings of either *peripheral* (e.g., skin conductance; breathing and heart rates) or *central* (e.g., brain systems) processes. The mapping of peripheral or central processes carry distinct methodological challenges and lead to different theoretical implications. In developing

**TABLE 1** Measurement Tools for Capturing Brain Activity

| Measurement Tool | Description |
| --- | --- |
| Electroencephalogram (EEG) | EEG is a noninvasive tool used to measure brain wave activity. EEG has a high temporal resolution (in the order of milliseconds). It has been used to measure motor learning and control, and skilled human performance through coherence analysis across brain cortices, event-related potentials, and motor-related cortical potentials. |
| Magnetoencephalography (MEG) | MEG is a noninvasive technique that allows for spatial and temporal measurement (on a millisecond basis) of ongoing brain activity. In the performance psychology domain, MEG has been commonly used to study attention, awareness, cognitive control, and memory. |
| Transcranial Magnetic Stimulation (TMS) Transcranial Direct-Current Stimulation (tDCS) | TMS and tDCS are noninvasive tools used to alter the activity of the human brain through the application of low-current magnetic pulses over the scalp. It is been used in studies on motor control, neurofeedback, and neuropsychiatry. |
| Positron Emission Tomography (PET) Functional Magnetic Resonance Imaging (fMRI) | PET and fMRI are noninvasive measurements of brain activity used to spatially map changes in blood flow within the brain. PET captures the rate of blood flow, whereas fMRI is sensitive to the level of oxygenation in the brain. These tools been used to measure motor, social, and cognitive performance in a variety of tasks (e.g., memory, language, and numeracy). |
| Near-Infrared Spectroscopy (NIRS) | NIRS is a noninvasive method of monitoring brain activity that involves measuring the absorption of near-infrared light, thus the evaluation of hemodynamic activity. The main uses of NIRS in performance psychology include research efforts on movement control and neuro-rehabilitation research. |

measurement and theory, performance psychologists must pay attention to the fact the psychological implications stemming from neurophysiological measurement are not to be drawn on a *one-to-one* basis. Rather, the implications of neurophysiological research exist on a *one-to-many* relation, in the sense that a psychological variable (e.g., anxiety) is usually associated with various physiological markers (e.g., increased heart rate, decreased body temperature, changes in brain synchrony; see Cassioppo et al., 2007, Chapter 1).

In this context, performance psychologists must continue to expand the use of neurophysiological measures to understand multilayered human behavior and performance. In particular, performance psychologists are recommended to focus on measuring brain processes through a variety of tools including, among others (see Table 1), electroencephalogram, transcranial magnetic stimulation, and functional magnetic resonance imaging (for a review, see Nakata, Yoshie, Miura, & Kudo, 2010). Although all these psychophysiological methods are available, breakthrough research in performance psychology may remain contingent on the development of mobile neurophysiological instruments, particularly electroencephalogram devices that are the most commonly used instruments in brain research in performance psychology.

In addition to neurophysiological measurement, advancement in *forward genetics* (behavioral phenotype→gene of interest) and *reverse genetics* (gene of interest→behavioral phenotype) shall place gene research at the epicenter of human performance measurement (Anokhin, 2014). Rather than using latent variables to build measurement models, researchers may rely on human genome databases (e.g., Human Genome Resource) to test relationships among different genes and performance outcomes. Experimental research on gene-mediated differences and genetic polymorphisms may reignite the long-standing nature versus nurture debate and revolutionize measurement in performance psychology. While it is not easy to forecast what genetic research may lead us to, the importance of the core measurement concepts discussed herein (e.g., *reliability, validity, methodological triangulation, interpretative pluralism, principle of parsimony*) shall remain unaltered.

## SUMMARY

There are numerous established measures of performance psychology described in the literature (see Daamen & Raab, 2012; Fernandez-Ballesteros, 2003; Tenenbaum et al., 2012), and all of them share some "fundamental features." In this chapter, we focused on these fundamental features by discussing the importance of *reliability, validity, power,* and *interpretative norms* in the development of quantitative measures of performance. We also discussed the importance of *sampling, methodological triangulation, interpretative pluralism,* and *transferability* in the development of trustworthy qualitative measures. We introduced neuropsychophysiological and genetic trends in performance measurement and highlighted the importance of developing measures with the primary aim of advancing theory in performance psychology.

## REFERENCES

Angoff, W. H. (1960). Measurement and scaling. *Encyclopedia of Educational Research, 3,* 807–817.

Anokhin, A. P. (2014). Genetic psychophysiology: advances, problems, and future directions. *International Journal of Psychophysiology, 93,* 173–197. http://dx.doi.org/10.1016/j.ijpsycho.2014.04.003.

Apter, M. J. (1984). Reversal theory and personality: a review. *Journal of Research in Personality, 18,* 265–288.

Baumeister, R. F., Tice, D. M., & Zell, A. L. (2007). How emotions facilitate and impair selfregulation. In J. J. Gross (Ed.), *Handbook of emotion-regulation* (pp. 408–426). New York, NY: Guilford Press.

Cacioppo, J. T., Tassinary, L. G., & Berntson, G. G. (2007). Psychophysiological science: Interdisciplinary approaches to classic questions about the mind. In J. T. Cacioppo, L. G. Tassinary, & G. Berntson (Eds.), *Handbook of psychophysiology* (3rd ed.) (pp. 1–16). New York, NY: Cambridge University Press.

Chirkov, V. (2009). Critical psychology of acculturation: what do we study and how do we study it, when we investigate acculturation? *International Journal of Intercultural Relations, 33*, 94–105.

Coyle, A. (2010). Qualitative research and anomalous experience: a call for interpretative pluralism. *Qualitative Research in Psychology, 7*, 79–83. http://dx.doi.org/10.1080/14780880903304600.

Daamen, M., & Raab, M. (2012). Psychological assessments in physical exercise. In H. Boecker, C. H. Hillman, L. Scheef, & H. K. Strüder (Eds.), *Functional neuroimaging in exercise and sport sciences*. New York, NY: Springer.

Eklund, R. C., Smith, A. L., Raedeke, T. D., & Creswell, S. (2012). Measurement of athlete burnout. In G. Tenenbaum, R. C. Eklund, & A. Kamata (Eds.), *Handbook of measurement in sport and exercise psychology* (pp. 359–366). Champaign, IL: Human Kinetics.

Eys, M. A., Carron, A. V., Bray, S. R., & Brawley, L. R. (2007). Item wording and internal consistency of a measure of cohesion: the group environment questionnaire. *Journal of Sport & Exercise Psychology, 29*, 395–402.

Fernandez-Ballesteros, R. (2003). *Encyclopedia of psychological assessment*. Thousand Oaks, CA.

Gigerenzer, G. (2010). Personal reflections on theory and psychology. *Theory & Psychology, 20*, 733–743. http://dx.doi.org/10.1177/0959354310378184.

Goldstein, D. G., & Gigerenzer, G. (2011). The beauty of simple models: themes in recognition heuristic research. *Judgment and Decision Making, 6*, 392–395.

Guttman, L. (1944). A basis for scaling qualitative data. *American Sociological Review, 9*, 139–150.

Guttman, L. (1947). The Cornell technique for scale and intensity analysis. *Educational and Psychological Measurement, 7*, 247–279.

Hanin, Y. L. (2007). Emotions in sport: current issues and perspectives. In G. Tenenbaum, & R. C. Eklund (Eds.), *Handbook of sport psychology* (3rd ed.) (pp. 31–58). Hoboken, NJ: Wiley & Sons.

Hershberger, S. L. (2006). The problem of equivalent structural models. In G. R. Hancock, & R. O. Mueller (Eds.), *Structural equation modeling: A second course* (pp. 13–41). Charlotte, NC: Information Age Publishing.

Jennings, J. R., & Gianaros, P. J. (2007). Methodology. In J. T. Cacioppo, L. G. Tassinary, & G. Berntson (Eds.), *Handbook of psychophysiology* (pp. 812–833). New York, NY: Cambridge University Press.

Kelle, U. (2006). Combining qualitative and quantitative methods in research practice: purposes and advantages. *Qualitative Research in Psychology, 3*, 293–311.

Kellmann, M., & Kallus, K. W. (2001). *Recovery-stress questionnaire for athletes*. Champaign, IL: Human Kinetics.

Kincheloe, J. (2005). On to the next level: continuing the conceptualization of the bricolage. *Qualitative Inquiry, 11*, 323–350.

Lidor, R. (2007). Preparatory routines in self-paced events: do they benefit the skilled athletes? Can they help the beginners? In G. Tenenbaum, & R. C. Eklund (Eds.), *Handbook of sport psychology* (3rd ed.) (pp. 445–465). New York, NY: Wiley.

McDonald, R. P., & Marsh, H. W. (1990). Choosing a multivariate model: noncentrality and goodness of fit. *Psychological Bulletin, 107*, 247–255. http://dx.doi.org/10.1037/0033-2909.107.2.247.

Nakata, H., Yoshie, M., Miura, A., & Kudo, K. (2010). Characteristics of the athletes' brain: evidence from neurophysiology and neuroimaging. *Brain Research Reviews, 62*, 197–211. http://dx.doi.org/10.1016/j.brainresrev.2009.11.006.

Parker, I. (2004). Criteria for qualitative research in psychology. *Qualitative Research in Psychology, 1,* 95–106. http://dx.doi.org/10.1191/1478088704qp010oa.

Patton, M. Q. (2002). *Qualitative research and evaluation methods.* Thousand Oaks, CA: SAGE Publications.

Raedeke, T. D. (1997). Is athlete burnout more than just stress? A sport commitment perspective. *Journal of Sport and Exercise Psychology, 19,* 396–417.

Raedeke, T. D., & Smith, A. (2001). Development and preliminary validation of a measure of athlete burnout. *Journal of Sport and Exercise Psychology, 23,* 281–306.

Rasch, G. (1960). *Probabilistic models for some intelligence and attainment tests* (Copenhagen, Danish Institute for Educational Research). Chicago: University of Chicago Press.

Raykov, T., & Marcoulides, G. A. (1999). On desirability of parsimony in structural equation model selection. *Structural Equation Modeling, 6,* 292–300.

Schinke, R. J., Michel, G., Gauthier, A. P., Pickard, P., Danielson, R., Peltier, D., et al. (2006). The adaptation to elite sport: a Canadian Aboriginal perspective. *The Sport Psychologist, 20,* 435–448.

Tenenbaum, G., & Bar-Eli, M. (1993). Decision making in sport: a cognitive perspective. In R. N. Singer, M. Murphy, & L. K. Tennant (Eds.), *Handbook of research on sport psychology* (pp. 171–192). New York, NY: Macmillan.

Tenenbaum, G., Eklund, R., & Kamata, A. (2012). *Measurement in sport and exercise psychology.* Champaign, IL: Human Kinetics.

Tenenbaum, G., Hatfield, B. D., Eklund, R. C., Land, W. M., Calmeiro, L., Razon, S., et al. (2009). A conceptual framework for studying emotions-cognitions-performance linkage under conditions that vary in perceived pressure. *Progress in Brain Research, 174,* 159–178. http://dx.doi.org/10.1016/S0079-6123(09)01314-4.

Tenenbaum, G., Lane, A., Razon, S., Lidor, R., & Schinke, R. (2015). Adaptation: a two-perception probabilistic conceptual framework. *Journal of Clinical Sport Psychology, 9,* 1–23.

Thurstone, L. L. (1925). A method of scaling psychological and educational tests. *Journal of Educational Psychology, 16,* 433–451.

Thurstone, L. L. (1927). The unit of measurement in educational scales. *Journal of Educational Psychology, 18,* 505–524.

Thurstone, L. L. (1928). The absolute zero in intelligence measurement. *Psychological Review, 35,* 175–197.

Torgerson, W. S. (1958). *Theory and methods of scaling.* Oxford, England: Wiley.

Vaughn, B. K., & Daniel, S. R. (2012). Conceptualizing validity. In G. Tenenbaum, R. C. Eklund, & A. Kamata (Eds.), *Measurement in sport and exercise psychology* (pp. 33–40). Champaign, IL: Human Kinetics.

Webb, T. L., Schweiger Gallo, I., Miles, E., Gollwitzer, P. M., & Sheeran, P. (2012). Effective regulation of affect: an action control perspective on emotion regulation. *European Review of Social Psychology, 23,* 143–186.

Wright, B. D., & Douglas, G. A. (1977). Best procedures for sample-free item analysis. *Applied Psychological Measurement, 1,* 281–295.

Wright, B. D., & Masters, G. N. (1982). *Rating scale analysis.* Chicago, IL: MESA Press.

# Chapter 4

# Applications within Performance Psychology

Terry Clark, Aaron Williamon
*Centre for Performance Science, Royal College of Music, London, UK*

---

*I used to be very worried about my memory backstage before performing. So, to strengthen my memory when learning a piece of music I work on different aspects of my memory separately. I imagine the sound of the piece in my head; I've sometimes notated the whole piece before a performance too. I find it helps my memory if I sit and notate the score because then I have a visual representation to fall back upon during the performance if need be. For another aspect, I'll do a lot of imagery when sitting at the piano. I'll look at my hands a lot, or I'll even just sit at the piano and, without touching the keyboard, imagine seeing my hands playing all of the notes. I've also done things like play both hands but then stop playing one, holding it above the keys without actually sounding the notes. Then I'll switch and do that with the other hand. I think it helps my ear because it requires me to imagine what that hand sounds like.*

*I also try to imagine the perfect performance as I want it to go. I listen and look at how I want to perform. I think about all the technical difficulties that I might struggle with, and I imagine that everything just works and try to make my body feel that it works. I also imagine that I'm onstage because I find it gives a little bit of extra nerves, a taste of what it's going to feel like. It also gives me something to aim for when I'm performing the piece. I always feel that if I know what I want from a performance then I've got a much better chance than if I go into the performance not knowing. I find if I give myself an exact image of how I want it to go then I feel a lot more confident going on stage (Clark, 2011, p. 93).*

Performance psychology is commonly employed within sport to facilitate skill development and training. Numerous reports document the psychological "success factors" (Orlick & Partington, 1998) that distinguish successful athletes from those who are less successful (Durand-Bush & Salmela, 2002; Gould, Diffenbach, & Moffett, 2002; Krane & Williams, 2006). Moreover, athlete development and support programs, such as those based upon the Long-term Athlete Development model (LTAD; Balyi, Way, Higgs, Norris, & Cardinal, 2005), are now commonly adopted by sporting bodies throughout the world, and performance enhancement teams in place at Olympic and high-performance training centers and university athletic programs (Radlo, 2012) identify performance psychology as a key component of their work. While performance psychology has become an integral part of training in sport, its implications and applications are by no means limited to this one area of pursuit. Above are a series of quotes from a concert pianist describing his use of multisensory imagery to learn and memorize a piece of music and then to perform that piece in a public context. This chapter begins by discussing functions of applied performance psychology across a range of diverse domains. Next, the relevance of performance psychology to the development of performance excellence is considered. Finally, a research program recently conducted within music that sought to develop and evaluate an ecologically appropriate performance psychology program is presented to facilitate discussion of the wide-ranging and powerful implications of applied performance psychology for human performance in general.

## FUNCTIONS OF APPLIED PERFORMANCE PSYCHOLOGY

As performance psychology sees application within a growing number of domains, so too do the functions for which performance psychology can be applied. While some of these are specific to certain areas, others are common across multiple fields.

The most common and widespread function for applied performance psychology is performance enhancement. From the initial acquisition and refinement of motor skills to the seemingly effortless presentation of those skills in a competitive or performance environment, there is a wealth of literature on how performance psychology can help those working in various domains learn, train, and perform better. Typically comprising mental or psychological skills training, this application of performance psychology has been referred to as the educational approach to sport psychology practice (Zizzi, Zaichkowsky, & Perna, 2002).

While optimal performance will always be a key concern for performers and performance psychologists who work with them, ensuring that performers are able to cope with the physical and mental demands associated with their domains so as to sustain healthy, long-term involvement in the performance arena is also of considerable concern. One model that seeks to explain the interaction between the experience of stressful situations and injury occurrence is the multicomponent theoretical model of stress and injury (Andersen & Williams, 1988). When faced with a stressful situation, Andersen and Williams (1988) propose that the likelihood of an injury occurring will be moderated by a person's personality, history of stressors, and coping resources. Consequently, someone with problematic or debilitating personality characteristics, such as high levels of anxiety or perfectionism, that exacerbate the stress response, a lengthy history of stressors, and few coping resources would be more likely to appraise the situation as anxiety provoking and have a higher likelihood of suffering an injury in response to the situation (Williams & Andersen, 1998).

In support of the stress and injury model, research has shown that athletes and dancers who report high life-stress incur more injuries that those with lower life-stress (Williams & Roepke, 1993; Noh, Morris, & Andersen, 2005), and that assessment of coping resources can distinguish between athletes who get injured and those who do not (Hanson, McCullagh, & Tonymon, 1992). By increasing coping resources, the stress-injury model implies that a person can limit the physiological activation and attentional disruptions resulting in response to a stressful situation experienced during participation in an activity, thereby reducing the risk of injury. Should an injury occur, performance psychology can also play a role in minimizing the negative affect resulting from an injury and assist the performer to return to work as quickly as possible (Rose & Jevne, 1993; Taylor & Taylor, 1997). In addition to ensuring that performers are physically ready to undertake full training and performance following an injury, it is essential to ensure that they are also psychologically ready to do so (Rotella & Heyman, 1993).

While a "more practice is better" mentality still exists within many performance domains, heightened training loads are now recognized as having the potential to impair physical and psychological function. Deleterious practice behaviors have been linked to burnout among athletes and dancers (Goodger, Gorely, Lavallee, & Harwood, 2007; Koutedakis, 2000; Quested & Duda, 2011) and physical and mental ill-being among athletes across all age groups (Brenner, 2007; DiFiori et al., 2014). In response, techniques including periodization are being applied to help performers manage training intensities. Periodization refers to the purposeful sequencing of a training season into smaller, distinct training periods and units, each with their own characteristics and foci, so as to reduce the potential for overtraining and facilitate optimal performance at a given competition (Matveyev, 1966). Periodization considers physiological, psychological, and biomechanical skill elements and seeks to improve the performer as a whole person (Verkhoshansky, 1998), with an increasing number of studies demonstrating the effectiveness of periodization-based training programs within sports (e.g., Issurin, 2008; Lorenz, Reiman, & Walker, 2010; Turner, 2011). Support for the use of such training programs can also be found within dance (Clark, Gupta, & Ho, 2014; Wyon, 2010).

## THE ROAD TO EXCELLENCE

Scientists have long sought to understand and explain exactly how high-level skill and expertise emerge. One of the most prominent explanations surrounding talent development is that engagement in and accumulation of many hours of deliberate training and practice are required. This is based on findings from a wide range of domains where research has suggested that a minimum of 10 years of goal-directed, hard work is required for an individual to reach a level of expert proficiency (Ericsson, Krampe, & Tesch-Romer, 1993). Support for this has come from fields as diverse as chess (Simon & Chase, 1973), music

(Sosniak, 1985), mathematics (Gustin, 1985), tennis (Monsaas, 1985), swimming (Kalinowski, 1985), and long-distance running (Wallingford, 1975).

Ericsson et al. (1993) define deliberate practice as follows (1993; p. 368):

> ...*a highly structured activity, the explicit goal of which is to improve performance. Specific tasks are invented to overcome weaknesses, and performance is carefully monitored to provide clues for ways to improve it further. We claim that deliberate practice requires effort and is not inherently enjoyable. Individuals are motivated to practice because practice improves performance.*

Subsequent discussion on deliberate practice has given rise to the 10,000-h or 10-year rule, with research considering the particular characteristics of high-quality practice (Ginsborg, 2002; Ginsborg & Chaffin, 2007; Ginsborg, Chaffin, & Nicholson, 2006; Miklaszewski, 1989; Williamon, Lehmann, & McClure, 2005; Williamon & Valentine, 2000, 2002) as well as the potentially moderating role of metacognition, self-regulated learning, and other personal traits in the development of musical expertise (e.g., Hallam, 2001; Jørgensen, 2004; McPherson & Williamon, 2006; McPherson & Zimmerman, 2002; Zimmerman, 1989). Ericsson and Lehmann (1999) have offered a revised version of the definition of deliberate practice that is more encompassing, recognizing that not all hours of practice or training are equal. Deliberate practice, Ericsson and Lehmann (1999, p. 695) stated, is as follows:

> *[S]tructured activity, often designed by teachers or coaches with the explicit goal of increasing an individual's current level of performance. In contrast to work and play, it requires the generation of specific goals for improvement and the monitoring of various aspects of performance. Furthermore, deliberate practice involves trying to exceed one's previous limits, which required full concentration and effort. Consequently, it is only possible to engage in these activities for a limited amount of time until rest and recuperation are needed.*

In addition to practice that leads performers to "achieve the desired end product, in as short a time as possible, without interfering with long term goals" (Hallam, 1998, p. 142), what else contributes to the development of expert performance? Within the arts, many performers and scholars now point to the importance of strong mental skills necessary to facilitate effective practice and to deliver consistently successful performances, as well as assist with the management of performance-related demands and stressors (Connolly & Williamon, 2004; Hays, 2002; Hays & Brown, 2004; Smith, Maragos, & Van Dyke, 2000; Talbot-Honeck & Orlick, 1998). Resembling early investigations conducted within sport (e.g., Highlen & Bennett, 1979; Mahoney & Avener, 1977; Ravizza, 1977), interviews with performing artists have also shed light upon apparent psychological prerequisites of expertise. Interviewing professional orchestral musicians, Talbot-Honeck and Orlick (1998) compiled a list of characteristics and skills that were present in high-level musicians and that the participants deemed to be essential for achieving musical excellence. These included a high commitment to music making; use of goal setting both in terms of self-development and musical learning; a positive, healthy self-understanding; a feeling of being in control of one's own life or destiny; strong concentration and focusing abilities; and effective use of mental rehearsal and visualization skills. Based upon their work with music students at the Royal College of Music, Connolly and Williamon (2004) identified the following mental skills as being of benefit to musicians: mental and physical relaxation, imagery and mental rehearsal, pre-performance routines, focus and concentration, ideal performance states and simulation, and goal identification and setting.

In addition to psychological skills and characteristics distinguishing successful from less successful performers, researchers now also examine the role of psychological skills to facilitate performers' movement along developmental pathways. This has stemmed, in part, from the recognition that talent identification and development models can be limited in their ability to assess an individual's capacity to develop (Abbott & Collins, 2004). MacNamara, Holmes, and Collins (2006) proposed that Psychological Characteristics for Developing Excellence (PCDEs) are also important for the developing musician, in addition to learning how to engage in the requisite quantity and quality of practice. Examining progression through Bloom's (1985) three stages of development (early, middle, and late years) in eight professional classical musicians, MacNamara et al. (2006) identified the personal and psychological characteristics prevalent at each stage. The early years comprised the PCDEs of natural ability, dedication, and planning. The middle years comprised self-belief, planning, ability to adapt and learn, social and multi-skills, and dedication and determination. Lastly, the later years included the PCDEs of versatility and adaptability, self-belief, planning, determination and dedication, discipline, and drive. PCDEs have also been found to facilitate athletes' progression along talent development pathways. Interviewing seven elite athletes and a parent of each athlete, MacNamara, Button, and Collins (2010) identified the following skills and attributes perceived to facilitate the development of elite performance: competitiveness, commitment to excelling in sport, understanding of the relevance of physical attributes to performance, appreciation of what is required to excel, coping with pressure, self-belief, imagery, and staying there. PCDEs, MacNamara, and colleagues propose, appear to play a crucial role in the successful development of performance skills and are particularly evident at key transition points. Given the increased use of curriculum-based approaches to developing PCDEs in aspiring athletes (i.e., Developing the Potential of Young People in Sport; SportScotland), MacNamara et al. (2006) and MacNamara, Holmes, and Collins (2007) suggest that similar training provided to performers in domains other than sport could be of particular benefit in the development of their performance skills.

Research is casting greater light on the psychological skills and characteristics required by performers to attain and sustain elite performance. They also demonstrate that performance psychology can be of interest and concern to diverse groups of performers. The delivery of mental skills training, and performance psychology more generally, within sport has been well discussed in terms of content and application. There is little suggestion in the literature, however, on how other types of performers are to learn about such techniques in a manner that will facilitate their integration into practice and performance. What follows, therefore, is discussion of a research program that sought to develop, implement, and evaluate an ecologically appropriate mental skills training program for developing musicians. In particular, this program aimed to examine the question: how can musicians best employ mental skills to enhance their performance experiences as well as assist them to prepare effectively and manage the demands associated with performing?

## DEVELOPING AN ECOLOGICALLY APPROPRIATE PERFORMANCE PSYCHOLOGY PROGRAM IN MUSIC

The first phase of research employed semi-structured interviews to gain greater understanding into advanced-level student ($n=18$) and professional ($n=11$) musicians' performance preparation routines and performance experiences (Clark, Lisboa, & Williamon, 2014a, b). It has been noted that skilled musicians employ a variety of highly developed preparation routines (Partington, 1995). A similar usage of such routines has also been found within sport (Hanton & Jones, 1999; Jones & Hanton, 2001) and surgery (Wetzel, 2006). It was felt that developing an awareness of the types of pre-performance activities in which advanced musicians engage would be of benefit to other musicians and to those who train them.

The results of the interviews highlighted a number of key differences between experienced and less experienced musicians, both in terms of their practice and performance routines and experiences but also in how they view music performance as an occupation. For instance, the more experienced musicians discussed engaging in well-thought-out activities and behaviors with the aim of explicitly preparing themselves musically, physically, and mentally to perform. Additionally, they expressed greater clarity and understanding of their pre-performance preparation routines, likely resulting from experience and reflective learning (Clark et al., 2014a). This was strongly evident when the musicians discussed factors that they felt contributed to successful and less successful performances, suggesting that a musician's response to a situation, rather than the situation itself, may have a stronger influence on the resulting quality of the performance (Clark et al., 2014b).

Both experienced and less experienced musicians were exposed to imagery and developed their own use of it through a range of means, but almost never through any structured manner. Given the practice and learning benefits that can be obtained through imagery use (e.g., Driskell, Copper, & Moran, 1994), providing structured imagery training to music performance students could be advantageous for the development of performance preparation skills (Clark, Williamon, & Aksentijevic, 2012). However, there was at the time (and largely still today) scant indication in the literature regarding the most efficacious means of doing so.

These findings expanded earlier work by involving a variety of instrumental groups while also comparing musicians with lesser and greater amounts of performance experience. As a result, they provided insight into the impact of accrued experience. This raised the question: to what extent is it possible to teach less experienced musicians using the perceptions and understanding of seasoned performers, and in particular to engage in greater amounts of reflective learning, or must younger musicians come to such insight through their own means?

Using the interview findings, a nine-week musician-specific mental skills training program was developed (Clark, 2011; Clark & Williamon, 2011). Despite being commonplace within sport and some other elite performance domains, mental skills training programs at present are rarely applied within the performing arts (Hays, 2002). A number of anecdotal reports of mental skills use for musicians exist, yet empirical investigations into their efficacy are largely lacking.

The primary objective of the mental skills training program was to foster increased awareness in the participants concerning their own practice behaviors and performance preparation activities, how they respond to various performance-related situations, and how such behaviors and responses may impact their ability to learn and perform. To this end, systematic training was provided in a number of key mental skills that have particular relevance for performance preparation. The topics covered fell under three main categories: (1) motivation and effective practice, which included goal setting, peak performance awareness, and effective practice and time management; (2) relaxation and arousal control, which included relaxation strategies, arousal control through cognitive restructuring, and self-talk; and (3) performance preparation and enhancement, which included mental rehearsal and imagery, focus and concentration, and performance preparation and analysis.

Each of the three categories of topics included three 60-min group sessions and three 30-min individual sessions spread over three weeks. This amounted to 13.5 h of contact time for each participant. The group sessions included a mix of the theoretical and empirical background information on the topics as well as group and individual exercises designed to assist the participants in developing their basic ability with each of the skills and then integrate the skills into their daily musical activities. For example, when discussing imagery, the participants were first taken through exercises designed to enhance their imagery vividness and controllability. Subsequent exercises then introduced the participants to different functions of imagery such as enhancing practice effectiveness, memorization, and performance preparation. The individual sessions offered time for the participants to discuss their personal experiences (previous and current) with the topics and exercises and provided an opportunity for the program to be tailored more specifically to their wants and needs.

The mental skills training program was delivered to 14 undergraduate and postgraduate performance majors at the Royal College of Music. A further nine students were recruited to form a control group. A range of quantitative and qualitative assessment tools were used to explore the potential effects of the program on the musicians' behaviors and perceptions surrounding practicing and performing. When comparing the experimental group with the control group, a number of positive effects of the training program were found. Specifically, significant differences emerged in the experimental group's attitudes and behaviors surrounding practice and performance preparation when compared with the control group. For instance, the experimental group participants reported a greater belief in the importance of employing self-regulated learning behaviors as well as increased self-reported quantity of practice and technical proficiency. A significant change was also noted in imagery ability when comparing the experimental and the control groups, as was a significant increase in self-efficacy for performing for the experimental group. These quantitative findings were further substantiated and expanded upon by comments provided by the experimental group participants during and following the training phase. Broadly, their comments grouped into four main themes: (1) increased self-awareness of effective performance preparation; (2) improved practice efficiency; (3) shift towards a more positive view of feelings usually associated with performance anxiety; and (4) positive impact on attitudes toward music making.

Investigations have demonstrated that mental skills can have a positive impact upon musicians' performance anxiety. The skills investigated have included attention training and behavioral rehearsal (Kendrick, Craig, Lawson, & Davidson, 1982), self-talk with relaxation delivered under hypnosis (Stanton, 1994), visualization and guided imagery (Esplen & Hodnett, 1999; Gratto, 1998), breathing and relaxation exercises (Gratto, 1998), and a combination of cognitive and behavioral techniques (Osborne, Greene, & Immel, 2014). In addition to this, studies have demonstrated the positive practice and performance benefits that can be obtained through incorporating mental rehearsal or imagery into a musician's practice activities (e.g., Coffman, 1990; Driskell et al., 1994; Highben & Palmer, 2004; Ross, 1985). The training program described here expanded upon these earlier mental skills investigations by adopting a multifaceted approach, ensuring that participants had adequate time to practice and develop proficiency with the skills prior to testing, and striving to guarantee that testing occurred in a naturalistic performance environment. It also demonstrated the efficacy of mental skills for functions beyond controlling debilitating anxiety symptoms and supplementing physical practice.

In addition, qualitative feedback was collected concerning the participants' thoughts on the actual topics and exercises employed within the program. As was expected, each of the participants responded idiosyncratically to the topics covered. Despite this, the topics of self-talk, imagery, and performance preparation sparked the greatest interest, and personal exploration among the participants, as well as being discussed most often in a focus group conducted at the conclusion of the program. By contrast, their opinions on relaxation, cognitive restructuring, and focus and concentration exercises varied more widely. Whether this was due to the particular combination of topics covered, their delivery by the program facilitator, or the skill strengths and weaknesses of the participants themselves is uncertain.

## CONSIDERATIONS FOR IMPLEMENTING PERFORMANCE PSYCHOLOGY

Throughout the music performance psychology research program presented here, several concepts of relevance to the implementation of performance psychology to performers emerged that are worth bearing in mind by researchers and practitioners wishing to create and deliver applied performance psychology initiatives. Many of these can be related to Thomas' seven phases of the performance enhancement processes (unpublished manuscript):

1. Orientation
2. Activity analysis
3. Individual/team assessment
4. Conceptualization
5. Psychological skills training
6. Implementation
7. Evaluation

Expanding upon Martens' (1987) three phases of psychological skills training (education, acquisition, and practice), Thomas' model highlights the activities and responsibilities required of the athlete/performer, practitioner, and relevant coaches and teachers as appropriate. Concerns have been expressed regarding the appropriateness of Thomas' (unpublished manuscript) seven phases (e.g., necessity of formal assessments, generalizability, limited flexibility within certain contexts; Morris & Thomas, 2005) with some practitioners advocating a more intuitive and individualized counseling approach (e.g., Andersen, 2000, 2005; Tenenbaum, 2001). Nonetheless, Thomas' model offers helpful guidelines for those wishing to apply performance psychology concepts with performers and provides a useful framework within which to discuss the music performance psychology research program outlined here.

### Phase 1—Orientation

A necessary first step when delivering performance psychology programs is for the practitioner to determine the interests and commitment level of the performers and to clarify the purpose of the program (Thomas, unpublished manuscript). Prior to providing training on novel skills, it is worth seeking to develop awareness or understanding of where the performers sit in terms of their readiness or openness to new ideas. This issue is typified by theories such as the

Transtheoretical Model that proposes that people progress through a series of stages of change when modifying their behavior (Prochaska & DiClemente, 1984). This model identifies five stages: pre-contemplation, contemplation, preparation, action, and maintenance. There were a number of instances during the music performance psychology training program in which the participants' receptiveness to the skills discussed could have been moderated, at least in part, by their position on such a framework. No doubt, by signing up to take part in such a program as discussed here, the participants were already demonstrating an interest and willingness to engage in new approaches. It is recognized, of course, that practitioners need not formally assess a performer's placement upon something such as the Transtheoretical Model in order to deliver an effective intervention. Indeed, a number of other models and theories are available that can help explain a performer's willingness to engage in a novel training activity. Nonetheless, by striving to develop a clearer understanding of the participants' receptiveness and openness to new ideas, the program could be delivered in such a way so as to take those aspects into consideration.

## Phase 2—Activity Analysis

Prior to developing and implementing a performance psychology training program, Thomas (unpublished manuscript) suggests that the practitioner gain a detailed analysis or understanding of the specific domain within which the program will run. Such an analysis should draw upon psychological, biomechanical, and physiological components. Boutcher and Rotella (1987) propose that this places the practitioner in a stronger position from which to determine the source of training or performance impairments needed to be addressed in the performer. Doing so also ensures that the practitioner possesses the requisite levels of knowledge to be able to meet the unique needs of performers within a particular domain. In the music research program, the person conducting the research was also an experienced performer and had a high level of first-hand, domain-specific knowledge. However, conducting the interviews prior to developing and delivering the program ensured that a domain-specific language particular to the content delivered was obtained.

## Phase 3—Individual/Team Assessment

A challenge, or reality perhaps, associated with delivering multifaceted training programs to groups of performers is that not all participants will respond to each topic or skill with similar interest, or experience similar success with each of the exercises. This issue emerged within the participant evaluation of the music research program. Such idiosyncrasy of topic and exercise preference lends further support for the use of multifaceted programs (e.g., Smith et al., 2000; Weinberg & Williams, 2006). It is important to acknowledge as well that not only will there be individual differences in terms of skill and exercise preference but that certain areas will be more relevant for certain facets of practice and performance for different people. Again, this reaffirms the importance that when providing training in mental skills use, emphasis should be on discussing and demonstrating the broad range of functions for which each mental skill can be employed. Concurrent with this, when teaching mental skills, it needs to be stressed that students should aim to develop proficiency with each of the skills so as to ensure effective application, rather than presuming from the outset that certain skills are inherently paired with specific outcomes or functions.

## Phase 4—Conceptualization

Linked to participants' possible progression through stages of behavior change when introduced to novel training methods, the relevance of another stage model of development was raised by the findings of the music interview study (Clark et al., 2014a, 2014b). Investigating the learning and development trajectories of new teachers, Fuller and Brown (1975) proposed that teachers progress through a series of stages regarding their primary concerns related to teaching. In the first stage, teachers' concerns are often vague, with teachers demonstrating low involvement in teaching. In the second stage, their concerns are more linked to survival. These include classroom control, mastery of content, and their own adequacy as teachers. By the third stage, they are more focused on the task or, in other words, on teaching performance. Finally, Fuller and Brown (1975) posited that teachers ultimately reach a fourth stage in which their primary concern is the impact of their teaching on their students. At this stage, teachers demonstrate concern for their pupils' social, academic, and emotional needs.

When comparing comments from the experienced and less experienced musicians within the interview study reported here, a distinction between areas of concern appeared to emerge. Most commonly, the less experienced musicians expressed concern for audience evaluation, considering a perfect performance to be of upmost importance when performing. The more experienced musicians, meanwhile, expressed concerns regarding the development of audience–performer connections and enjoyment of the performance event as a whole, while still conscious of the quality of their performance. This relates to findings by Hallam (2001) who noted that novice and expert musicians possess different objectives when practicing. While novice musicians were primarily concerned with note accuracy, Hallam found that expert musicians demonstrated a broader range of concerns encompassing aspects such as planning, monitoring, and evaluation.

Teaching and performance—whether in music, sport, or any other domain—are no doubt rather different activities, but it would seem likely that young and developing performers do progress in a similar manner as they develop proficiency and confidence as performers. Training programs such as those associated with PCDEs (MacNamara et al., 2006) and Long-Term Athlete Development (Balyi et al., 2005) have this concept at their core and encourage a focus on different skills and strategies depending on the level of proficiency of the performer. Ensuring developmental specificity within the content of training programs would help ensure that the type of training performers receive is appropriate for their stage of development, ultimately assisting them to progress more smoothly and successfully.

### Phase 5—Psychological Skills Training

The context within which performance psychology training is provided exerts perhaps the greatest amount of influence on the efficacy of the training program. Beyond an individual performer's own interests, influence can be exerted on a performer from an enormous range of dimensions, impacting the extent to which that performer chooses to engage with novel or nontraditional training methods. These different dimensions can be thought of as the contexts or cultures within which a performer exists. Within music, for instance, at the most microscopic level is the individual musician. An instrumental or vocal music student works primarily with one principal study professor who will have considerable influence over the musical activities and behaviors of all of the students in his or her studio. Beyond this is the instrumental group to which the musician belongs. Many instrument groups can have particular characteristics to which a member of that group can associate. This can also influence the choices a musician makes. Wider still is the culture of the institution in which the musician studies, as well as the profession as it stands locally and nationally.

The impact that a culture or environment can have upon an individual has been examined extensively by Bronfenbrenner in the development of his ecological systems theory (1979). Bronfenbrenner proposes that, in order for an individual to be fully understood, an understanding of their greater ecological environment is essential, which he likens to a series of concentric circles with the individual placed at the center. Commonly held beliefs or behaviors at any of these levels can have a significant impact upon a young performer's own beliefs and behaviors. Not only is it important to develop an understanding of the beliefs of an individual performer prior to the development and delivery of a novel training program, an awareness of the context within which the training will take place is also essential for effective delivery. Indeed, the development of a multifaceted, holistic perspective in order to understand how learning occurs has been advocated (Welch, 2007).

### Phase 6—Implementation

Unlike at universities, where students can expect to spend the majority of tuition time in lecture-type settings, music conservatoire training is primarily practical and hands-on, with comparatively little time spent in lectures. The same can be said of performers from domains other than music: participants are drawn to those activities because they enjoy *performing*. Given this, any sort of supplementary training offered to students or developing performers should strive to be practical and applicable. This issue was raised during the qualitative evaluation of the mental skills training program in which the participants requested a greater number of examples and case studies of musicians employing their skills, a more substantial amount of practical, precise application of the skills discussed, and a further integration of the program into the conservatoire's overall timetable together with more mock auditions and performances. It is well recognized that psychological skills require extensive practice prior to implementation within the performance environment. The development and use of simulation and virtual reality training labs within music (Williamon, Aufegger, & Eiholzer, 2014) and surgery (Bashankaev, Baido, & Wexner, 2011; Stefanidis et al., 2014) offer exciting new opportunities for performers to learn and practice performance psychology skills in contextually rich, yet low-risk, performance environments.

## Phase 7—Evaluation

When providing instruction on novel training methods, such as mental skills, change and the acquisition of proficiency will likely be slow. It is possible that this very issue may have hindered, at least in part, earlier mental skills investigations in which participants were not allowed time to practice skills before being tested on them. Within any training program, let alone novel mental skills training programs, participants will need to spend much time working with the skills in order to derive benefits fully from them. In line with this, each participant will need to experiment with a range of exercises to determine what will be most appropriate for them as everyone has different wants and needs. This again lends further support for the use of providing multifaceted programs. Indeed, integration of novel skills and training methods into a performer's regular practice and performance routines can take a long time. Discussing new skills once or even twice will likely be insufficient to enable change. It is strongly believed that one-shot or occasional psychological skills training, no matter how competently delivered, cannot be as effective as that provided on a continuous basis over an extended period of time (Weinberg & Williams, 2006). In fact, it has been recommended that an average of three to six months be devoted to the implementation of a mental skills training program. This is due to the fact that it takes time for the participants to learn mental skills with which they may not have had previous experience, together with integrating them effectively into practice and performance situations (Weinberg & Williams, 2006).

Alongside the development and delivery of applied psychology programs, employing a robust, mixed-methods evaluation protocol is essential to generating a greater evidence base of applied concepts and programs particular to various groups of performers. Traditional formative and summative methods of evaluation are useful for improving and evaluating the effectiveness of established models and programs. When established models or programs do not exist, however, or when attempting to apply existing concepts to new populations, evaluation methods that focus on the developmental process can prove useful (e.g., Clark et al., 2014; Fagan et al., 2011; Patton, 2011).

## CONCLUSION

As the number of domains within which performance psychology is applied continues to rise, so too will the number of functions for which performance psychology can be of use to performers. For instance, medicine is one such domain within which performance psychology is seeing greater use. Psychological skills—and imagery in particular—are being employed to facilitate the development and enhancement of surgical technical skills (e.g., Arora et al., 2011, 2010; Rogers, 2006; Sevdalis, Moran, & Arora, 2013). Such avenues offer exciting opportunities for performance psychology practitioners to expand their work down novel paths. Still, more empirical investigations, particularly in domains other than sport, are needed in order to provide practitioners with a better understanding of the most appropriate content for applied performance psychology interventions and programs, as well as the most effective means of delivering that content to performers.

Further to ensuring the rigor of novel performance psychology interventions will be efforts to maintain the professional standards of performance psychology practitioners. Many sports psychology associations have done well in developing professional standards that their members must meet in order to qualify for accreditation or affiliation. As the field of performance psychology continues to grow, and as an increasing number of practitioners from diverse backgrounds engage in performance psychology-type training, it would be worthwhile developing professional guidelines and standards of practice for those working as practitioners in the field.

Ongoing dialog and collaboration between domains of performance will facilitate the exchange of ideas and methods. No doubt, such investigations will uncover new topics of interest to researchers and practitioners, helping to develop further the field of performance psychology and hold considerable promise for all involved.

## REFERENCES

Abbott, A., & Collins, D. (2004). Eliminating the dichotomy between theory and practice in talent identification and development: considering the role of psychology. *Journal of Sports Sciences*, 22, 395–408.

Andersen, M. B. (Ed.). (2000). *Doing sport psychology*. Champaign, IL: Human Kinetics.

Andersen, M. B. (Ed.). (2005). *Sport psychology in practice*. Champaign, IL: Human Kinetics.

Andersen, M. B., & Williams, J. M. (1988). A model of stress and athletic injury: prediction and prevention. *Journal of Sport and Exercise Psychology*, 10, 294–306.

Arora, S., Aggarwal, R., Sevdalis, N., Moran, A., Sirimanna, P., Kneebone, R., et al. (2010). Development and validation of mental practice as a training strategy for laparoscopic surgery. *Surgical Endoscopy*, 24, 179–187.

Arora, S., Aggarwal, R., Moran, A., Sirimanna, P., et al. (2011). Mental practice: effective stress management training for novice surgeons. *Journal of the American College of Surgeons, 212,* 225–233.

Balyi, I., Way, R., Higgs, C., Norris, S., & Cardinal, C. (2005). *Canadian sport for life: Long-term athlete development* (Resource paper). Vancouver, Canada: Canadian Sport Centres.

Bashankaev, B., Baido, S., & Wexner, S. D. (2011). Review of available methods of simulation training to facilitate surgical education. *Surgical Endoscopy, 25,* 28–35.

Bloom, B. S. (Ed.). (1985). *Developing talent in young people.* New York: Ballantine.

Boutcher, S. H., & Rotella, R. J. (1987). A psychological skills educational program for closed-skill performance enhancement. *The Sport Psychologist, 1,* 127–137.

Brenner, J. S. (2007). Overuse injuries, overtraining, and burnout in child and adolescent athletes. *Pediatrics, 119,* 1242–1245.

Bronfenbrenner, U. (1979). *The ecology of human development.* Cambridge, MA: Harvard University Press.

Clark, T. (2011). *Mental skills in music: Investigating use, ability, and training.* Unpublished doctoral thesis. Royal College of Music.

Clark, T. & Williamon, A. (2011). Evaluation of a mental skills training program for musicians. *Journal of Applied Sport Psychology, 23,* 342–359.

Clark, T., Gupta, A., & Ho, C. H. (2014). Developing a dancer wellness program employing developmental evaluation. *Frontiers in Psychology, 5,* 731. http://dx.doi.org/10.3389/fpsyg.2014.00731.

Clark, T., Lisboa, T., & Williamon, A. (2014a). Learning to be an instrumental musician. In I. Papageorgi, & G. Welch (Eds.), *Advanced musical Performance: Investigations in higher education learning* (pp. 287–300). Farnham, UK: Ashgate.

Clark, T., Lisboa, T., & Williamon, A. (2014b). An investigation into musicians' thoughts and perceptions during performance. *Research Studies in Music Education, 36,* 17–34. http://dx.doi.org/10.1177/1321103X14523531.

Clark, T., Williamon, A., & Aksentijevic, A. (2012). Musical imagery and imagination: the function, measurement and application of imagery skills for performance. In D. Hargreaves, D. Miell, & R. MacDonald (Eds.), *Musical Imaginations: Multidisciplinary perspectives on creativity, performance, and perception* (pp. 351–365). Oxford: Oxford University Press.

Coffman, D. (1990). Effects of mental practice, physical practice, and knowledge of results on piano performance. *Journal of Research in Music Education, 38,* 187–196.

Connolly, C., & Williamon, A. (2004). Mental skills training. In A. Williamon (Ed.), *Musical Excellence: Strategies and techniques to enhance performance* (pp. 221–245). Oxford: Oxford University Press.

DiFiori, J. P., Benjamin, H. J., Brenner, J. S., Gregory, A., Jayanthi, N., Landry, G. L., et al. (2014). Overuse injuries and burnout in youth sports: a position statement from the American Medical Society for Sports Medicine. *British Journal of Sports Medicine, 48,* 287–288.

Driskell, J. E., Copper, C., & Moran, A. (1994). Does mental practice enhance music performance? *Journal of Applied Psychology, 79,* 481–492.

Durand-Bush, N., & Salmela, J. H. (2002). The development and maintenance of expert athletic performance: perceptions of world and Olympic champions. *Journal of Applied Sport Psychology, 14,* 154–171.

Ericsson, K. A., Krampe, R. T., & Tesch-Romer, C. (1993). The role of deliberate practice in the acquisition of expert performance. *Psychological Review, 100,* 363–406.

Ericsson, K. A., & Lehmann, A. C. (1999). Expertise. In M. A. Runco, & S. Pritzker (Eds.), *Encyclopaedia of creativity* (Vol. 1) (pp. 695–707). New York: Academic Press.

Esplen, M. J., & Hodnett, E. (1999). A pilot study investigating student musicians' experiences of guided imagery as a technique to manage performance anxiety. *Medical Problems of Performing Artists, 14,* 127–132.

Fagan, M. C., Redman, S. D., Stacks, J., Barrett, V., Thullen, B., Altenor, S., et al. (2011). Development evaluation: building innovations in complex environments. *Health Promotion Practice, 12,* 645–650.

Fuller, F. F., & Brown, O. H. (1975). Becoming a teacher. In K. Ryan (Ed.), *Teacher education (74th Yearbook of the National Society for the study of education, Pt. II* (pp. 25–52). Chicago: University of Chicago Press.

Ginsborg, J. (2002). Classical singers learning and memorising a new song: an observational study. *Psychology of Music, 30,* 58–101.

Ginsborg, J., & Chaffin, R. (2007). The effect of retrieval cues developed during practice and rehearsal on an expert singer's long-term recall for words and melody. In A. Williamon, & D. Coimbra (Eds.), *Proceedings of the International Symposium on Performance Science 2007* (pp. 23–28). European Association of Conservatoires (AEC).

Ginsborg, J., Chaffin, R., & Nicholson, G. (2006). Shared performance cues in singing and conducting: a content analysis of talk during practice. *Psychology of Music, 34,* 167–194.

Goodger, K., Gorely, T., Lavallee, D., & Harwood, C. (2007). Burnout in sport: a systematic review. *Sport Psychologist, 21,* 127–151.

Gould, D., Diffenbach, K., & Moffett, A. (2002). Psychological characteristics and their development in Olympic champions. *Journal of Applied Sport Psychology, 14,* 172–204.

Gratto, S. (1998). The effectiveness of an audition anxiety workshop in reducing stress. *Medical Problems of Performing Artists, 13,* 29–34.

Gustin, W. (1985). The development of exceptional research mathematicians. In B. S. Bloom (Ed.), *Developing talent in young people* (pp. 270–331). New York: Ballantine Books.

Hallam, S. (1998). *Instrumental teaching: A practical guide to better teaching and learning.* London: Heinemann.

Hallam, S. (2001). The development of metacognition in musicians: implications for education. *British Journal of Music Education, 18,* 27–39.

Hanson, S. J., McCullagh, P., & Tonymon, P. (1992). The relationship of personality characteristics, life stress, and coping resources to athletic injury. *Journal of Sport and Exercise Psychology, 14,* 262–272.

Hanton, S., & Jones, G. (1999). The acquisition and development of cognitive skills and strategies: I. Making the butterflies fly in formation. *The Sport Psychologist, 13,* 1–21.

Hays, K. (2002). The enhancement of performance excellence among performing artists. *Journal of Applied Sport Psychology, 14,* 299–312.

Hays, K., & Brown, C. H. (2004). *You're On: Consulting for peak performance.* Washington: American Psychological Association.

Highben, Z., & Palmer, C. (2004). Effects of auditory and motor mental practice in memorized piano performance. *Bulletin of the Council for Research in Music Education, 158,* 58–67.

Highlen, P. S., & Bennett, B. B. (1979). Psychological characteristics of successful and non-successful elite wrestlers: an exploratory study. *Journal of Sport Psychology, 1,* 123–137.

Issurin, V. (2008). Block periodization versus traditional training theory: a review. *The Journal of Sports Medicine and Physical Fitness, 48,* 65–75.

Jones, G., & Hanton, S. (2001). Pre-competitive feeling states and directional anxiety interpretations. *Journal of Sports Sciences, 19,* 385–395.

Jørgensen, H. (2004). Strategies for individual practice. In A. Williamon (Ed.), *Musical Excellence: Strategies and techniques to enhance performance* (pp. 85–104). Oxford: Oxford University Press.

Kalinowski, A. (1985). The development of Olympic swimmers. In B. S. Bloom (Ed.), *Developing Talent in young people* (pp. 139–192). New York: Ballantine Books.

Kendrick, M. J., Craig, K. D., Lawson, D. M., & Davidson, P. O. (1982). Cognitive and behavioural therapy for musical performance anxiety. *Journal of Consulting and Clinical Psychology, 50*, 353–362.

Koutedakis, Y. (2000). Burnout in dance. *Journal of Dance Medicine & Science, 4*, 122–127.

Krane, V., & Williams, J. M. (2006). Psychological characteristics of peak performances. In J. M. Williams (Ed.), *Applied sport psychology: Personal growth to peak performance* (5th ed.) (pp. 207–227). New York: McGraw-Hill.

Lorenz, D. S., Reiman, M. P., & Walker, J. C. (2010). Periodization: current review and suggested implementation for athletic rehabilitation. *Sports Health: A Multidisciplinary Approach, 2*, 509–518.

MacNamara, Á., Holmes, P., & Collins, D. (2006). The pathway to excellence: the role of psychological characteristics in negotiating the challenges of musical development. *British Journal of Music Education, 23*, 285–302.

MacNamara, Á., Holmes, P., & Collins, D. (2007). Negotiating transitions in musical development: the role of psychological characteristics of developing excellence. *Psychology of Music, 36*, 335–352.

MacNamara, Á., Button, A., & Collins, D. (2010). The role of psychological characteristics in facilitating the pathway to elite performance-Part 1: Identifying mental skills and behaviors. *Sport psychologist, 24*(1), 52–73.

Mahoney, M. J., & Avener, M. (1977). Psychology of the elite athlete: an exploratory study. *Cognitive Therapy and Research, 1*, 135–141.

Martens, R. (1987). *Coaches guide to sport psychology*. Champaign, IL: Human Kinetics.

Matveyev, L. P. (1966). *Periodization of sports training*. Moscow: Fiscultura I Sport.

McPherson, G., & Williamon, A. (2006). Giftedness and talent. In G. McPherson (Ed.), *The child as musician* (pp. 239–256). Oxford: Oxford University Press.

McPherson, G., & Zimmerman, B. (2002). Self-regulation of musical learning: a social cognitive perspective. In R. Colwell, & C. Richardson (Eds.), *The new handbook of research on music teaching and learning* (pp. 327–347). Oxford: Oxford University Press.

Miklaszewski, K. (1989). A case study of a pianist preparing a musical performance. *Psychology of Music, 17*, 95–109.

Monsaas, J. (1985). Learning to be a world-class tennis player. In B. S. Bloom (Ed.), *Developing talent in young people* (pp. 211–269). New York: Ballantine Books.

Morris, T., & Thomas, P. (2005). Applied sport psychology. In T. Morris, & J. Summers (Eds.), *Sport psychology: Theory, applications, and issues* (pp. 236–277). Milton, Australia: John Wiley & Sons Australia, Ltd.

Noh, Y. E., Morris, T., & Andersen, M. B. (2005). Psychosocial factors and ballet injuries. *International Journal of Sport and Exercise Psychology, 3*, 7–25.

Orlick, T., & Partington, J. (1998). Mental links to excellence. *The Sport Psychologist, 2*, 105–130.

Osborne, M. S., Greene, D. J., & Immel, D. T. (2014). Managing performance anxiety and improving mental skills in conservatoire students through performance psychology training: a pilot study. *Psychology of Well-Being: Theory, Research, and Practice, 4*. http://dx.doi.org/10.1186/s13612-014-0018-3.

Partington, J. (1995). *Making music*. Ottawa: Carlton University Press.

Patton, M. Q. (2011). *Developmental evaluation: Applying complexity concepts to enhance innovation and use*. New York: The Guilford Press.

Prochaska, J. O., & DiClemente, C. C. (1984). *The transtheoretical approach: Crossing the traditional boundaries of change*. Homewood, IL: Irwin.

Quested, E., & Duda, J. L. (2011). Antecedents of burnout among elite dancers: a longitudinal test of basic needs theory. *Psychology of Sport and Exercise, 12*, 159–167.

Radlo, S. J. (2012). "Mindless athletes: a need for holistic university sport performance enhancement programs. *Journal of Athletic Enhancement, 1*. http://dx.doi.org/10.4172/2324-9080.1000e106.

Ravizza, K. (1977). Peak experiences in sport. *Journal of Humanistic Psychology, 17*, 35–40.

Rogers, R. G. (2006). Mental practice and acquisition of motor skills: examples from sports training and surgical education. *Obstetrics and Gynecology Clinics of North America, 33*, 297–304.

Rose, J., & Jevne, R. F. J. (1993). Psychosocial processes associated with athletic injuries. *The Sport Psychologist, 7*, 309–328.

Ross, S. (1985). The effectiveness of mental training practice on improving performance of college trombonists. *Journal of Research in Music Education, 33*, 221–230.

Rotella, R. J., & Heyman, S. R. (1993). Stress, injury, and the psychological rehabilitation of athletes. In J. M. Williams (Ed.), *Applied sport psychology* (2nd ed.) (pp. 338–355). Mountain View, CA: Mayfield.

Sevdalis, N., Moran, A., & Arora, S. (2013). Mental imagery and mental practice applications in surgery: state of the art and future directions. In S. Lacey, & R. Lawson (Eds.), *Multisensory imagery* (pp. 343–363). New York: Springer.

Simon, H., & Chase, W. (1973). Skill in chess. *American Scientist, 61*, 394–403.

Smith, A., Maragos, A., & Van Dyke, A. (2000). Psychology of the musician. In R. Tubiana, & P. Amadio (Eds.), *Medical problems of the instrumentalist musician* (pp. 135–170). London: Martin Dunitz, Ltd.

Sosniak, L. (1985). Learning to be a concert pianist. In B. S. Bloom (Ed.), *Developing talent in young people* (pp. 19–67). New York: Ballantine Books.

Stanton, H. (1994). Reduction of performance anxiety in music student. *Australian Psychologist, 29*, 124–127.

Stefanidis, D., Sevdalis, N., Paige, J., Zevin, B., Aggarwal, R., Grantcharov, et al. (2014). Simulation in surgery: what's needed next? *Annals of Surgery*. http://dx.doi.org/10.1097/SLA.0000000000000826.

Talbot-Honeck, C., & Orlick, T. (1998). The essence of excellence: mental links of top classical musicians. *Journal of Excellence, 1*, 61–75.

Taylor, J., & Taylor, S. (1997). *Psychological approaches to sports injury rehabilitation*. Lippincott, Williams, & Wilkins.

Tenenbaum, G. (2001). *The practice of sport psychology*. Morgantown, WV: Fitness Information Technology.

Thomas, P. R. *An overview of performance enhancement processes in applied sport psychology*. Colorado Springs: United States Olympic Training Centre, Unpublished manuscript.

Turner, A. (2011). The science and practice of periodization: a brief review. *Strength & Conditioning Journal, 33*, 34–46.

Verkhoshansky, Y. (1998). Main features of modern scientific sports training theory. *New Studies in Athletics, 13*(3), 9–20.

Wallingford, R. (1975). Long distance running. In A. W. Tayler, & F. Landry (Eds.), *The scientific aspects of sports training* (pp. 118–130). Springfield, IL: Charles C. Thomas.

Weinberg, R., & Williams, J. (2006). Integrating and implementing a psychological skills training program. In J. M. Williams (Ed.), *Applied sport psychology: Personal growth to peak performance* (5th ed.) (pp. 425–457). New York: McGraw-Hill.

Welch, G. F. (2007). Addressing the multifaceted nature of music education: an activity theory research perspective. *Research Studies in Music Education, 28*, 23–37.

Wetzel, C. M., Kneebone, R. L., Woloshynowych, M., Nestel, D., Moorthy, K., Kidd, J., & Darzi, A. ( 2006). The effects of stress on surgical performance. *The American Journal of Surgery, 191*(1), 5–10.

Williamon, A., Aufegger, L., & Eiholzer, H. (2014). Simulating and stimulating performance: introducing distributed simulation to enhance musical learning and performance. *Frontiers in Psychology, 5*. http://dx.doi.org/10.3389/fpsyg.2014.00025.

Williamon, A., Lehmann, A., & McClure, K. (2005). Studying practice quantitatively. In R. Kopiez, A. C. Lehmann, I. Wolther, & C. Wolf (Eds.), *Proceedings of the Fifth Triennial ESCOM Conference* (pp. 182–185). Hanover, Germany: Hanover University of Music and Drama.

Williamon, A., & Valentine, E. (2000). Quantity and quality of musical practice as predictors of performance quality. *British Journal of Psychology, 91*, 353–376.

Williamon, A., & Valentine, E. (2002). The role of retrieval structures in memorizing music. *Cognitive Psychology, 44*, 1–32.

Williams, J. M., & Andersen, M. B. (1998). Psychosocial antecedents of sport injury: review and critique of the stress and injury model. *Journal of Applied Sport Psychology, 10*, 5–25.

Williams, J. M., & Roepke, N. (1993). Psychology of injury and injury rehabilitation. In R. N. Singer, M. Murphey, & L. K. Tennant (Eds.), *Handbook of research on sport psychology* (pp. 815–839). New York: Macmillan.

Wyon, M. (2010). Preparing to perform: periodization and dance. *Journal of Dance Medicine and Science, 14*, 67–72.

Zimmerman, B. J. (1989). A social cognitive view of self-regulated learning. *Journal of Educational Psychology, 81*, 329–339.

Zizzi, S., Zaichkowsky, L., & Perna, F. M. (2002). Certification in sport and exercise psychology. In J. L. Van Raalte, & B. Brewer (Eds.), *Exploring sport and exercise psychology* (2nd ed.) (pp. 459–478). Washington, DC: American Psychological Association.

# Section B

# Performance Phenomena of Cognitive–Action Interaction

5. Bridging the Gap between Action and Cognition: An Overview (Babett H. Lobinger) 67
6. Improving Performance by Means of Action–Cognition Coupling in Athletes and Coaches (Gloria B. Solomon) 87
7. Music Performance: Expectations, Failures, and Prevention (Eckart Altenmüller and Christos I. Ioannou) 103
8. Motor Imagery and Mental Training in Older Adults (Michael Kalicinski, Monika Thomas and Babett H. Lobinger) 121

## Overview

Understanding how skilled actions in the field of sports are planned and controlled, learned and taught is an important enterprise. This section, therefore, addresses expert performance in different domains (e.g., sport and music), for different actors (e.g., athletes, coaches, and elderly) and different performance outcomes—trying to uncover/reveal underlying mechanisms of action and cognition coupling beyond physical activity/sports. Analyzing paradoxical performance deepens the understanding of how intention, action planning, execution, and performance outcomes are linked (Chapter 5). Musicians and top athletes face similar challenges in practicing or training and performing on stage or in a tournament. Chapters 6 and 7, therefore, discuss prospects and constraints of comparing the link between action and cognition in golfers and pianists (suffering from golf cramp and musician's cramp). Expert performance also is a matter in old age, including prevention, but also optimization of action. Chapter 8 addresses motor and mental imagery in old age in terms of diagnostics and intervention.

Chapter 5

# Bridging the Gap between Action and Cognition: An Overview

### Babett H. Lobinger
*Institute of Psychology, German Sport University Cologne, Cologne, Germany*

*Imagine a golf professional is preparing to putt. He is at the green of the 18th hole and in the leading position with the opportunity to win the tournament with a successful putt. The golf ball is lying at a distance of 3 feet from the hole. He is one last stroke away from triumph. What do you think will come to his mind? Will he imagine his success and a standing ovation by the spectators? Or will he start to think about the consequences of a possible failure? What will he focus on? Will he talk to himself to encourage and self-instruct? Will he stick to his routines? Or will he be overwhelmed by negative thoughts and overcontrol his movements? How will he attribute his performance outcome to his skills and abilities?*

To understand what is happening in such moments and to prepare individuals so that they can successfully meet the demands of such situations, and perform at their very best and according to their personal aims and well-being, it is important to understand how action and cognition influence each other when it comes to (psychomotor) performance. Although the preceding example comes from the world of sports, the issues raised are relevant to other areas of performance, too, as chapters in this section illustrate.

To reach peak performance, that is, to become an expert, it is necessary to develop outstanding and specific skills. One way the route to expertise can be understood is as an interactive process of acquiring specific, situation-related heuristics that can be used to meet the demands of the situation (de Oliviera, Lobinger, & Raab, 2014). Along this route in sports, coaches play a crucial role in talent detection and development. Solomon (Chapter 6) provides an exploration of the development of coaching expertise and reviews the impact of coaches on athlete development and performance.

Especially in high-performance settings, similarities between musicians and athletes can be found with regard to hours of (deliberate) practice (Ericsson, Krampe, & Tesch-Römer, 1993) and the need to perform under pressure (Altenmüller, Ioannou, Raab, & Lobinger, 2014). Altenmüller and

Ioannou (Chapter 7) discuss music performance in terms of expectations, failures, and prevention.

The link between action and cognition is also an issue in old age. For example, facing the psychomotor demands of a situation and calibrating personal resources, skills, and abilities is quite important for older adults to prevent falls (Lobinger, 2005). Kalicinski, Thomas, and Lobinger (Chapter 8) address the potential of motor and mental imagery to improve mobility in older adults.

To explore the state of the art of research on the action–cognition link, in this overview, I provide an example in the world of sports. I also address the more traditional understanding of cognitive processes and their impact on actions (Eklund & Tenenbaum, 2014) as well as the latest pragmatic turn in cognitive science toward a new paradigm that considers the "enactivity" of cognition (Engel, Maye, Kurthen, & König, 2013). I provide extended examples of research programs that focus on different phenomena of performance, such as expert performance or paradoxical performance, and on different target groups, such as experts, novices, top coaches, and talent (elite performers).

The research projects presented in this chapter were inspired by problems and challenges in the world of sports and motivated by the aim of improving performance. The diagnostics and interventions described, therefore, are characterized by finality,[1] which means that methods, settings, and procedures are applied to answer an initial research question. Trying to answer the research questions goes beyond solely describing a current state or a condition. It leads to analyzing performance by focusing on the agent(s) as well as on the characteristics of the environment and specific tasks. Some of the models presented are heuristic ones, which try to deepen our understanding of the different factors influencing behavior in different settings. Additionally, we prefer multitrait–multimethod paradigms and mixed methods (Creswell, 2009). Resulting interventions aim at improving skills as well as environmental conditions or specific tasks. For illustration purposes, this chapter presents three research lines and projects that demonstrate some of these principles and demands.

## COGNITION AND ACTION

Traditionally, cognition in sports has most often been addressed in terms of cognitive capabilities or cognitive function (Eklund & Tenenbaum, 2014). The focus has been on the steps of information processing that underlie problem solving and decision-making, with emphasis on how stimuli are perceived, processed, and then translated into a plan of action, and therefore the link between the perceptual and motor systems is of high interest. To explore this area researchers have, for example, focused on the cognitive capabilities of experts and novices and on cognitive styles (Tenenbaum & Filho, 2014).

---

1. "Finalitätscharakter"; Amelang and Zielinsky (2002).

Operationalizing cognitive capabilities at a time when neuroscientific methods were not that elaborated or common led to experimental paradigms that involved, for instance, temporal and spatial occlusion and to more behaviorally oriented research paradigms. Examples of more qualitative paradigms are retrospective and concurrent verbal reports and the critical incident method or technique (see Ward, 2014). The shared aim was to gain insight into information processing while the body or even parts of the body are in motion.

Action, therefore, within this paradigm seems to be primarily understood as a kind of dependent performance outcome. Perception leads to cognition, which leads to action (see Figure 1, inner square showing a linear relationship). Cognition has—until recently—often been understood as simply any kind of mental operation related to information processing (Raab, Chapter 1). Action seems to be a sort of by-product, or, according to Chang (2014), cognitive function is understood as the mental process of thinking, which involves action.

In contrast, more recent research in cognitive science has taken a pragmatic turn, stressing the enactivity of cognition (Engel et al., 2013). Organisms do much more than simply receive information. They interact with their environment. Individuals match their actions to their environment, and the environment influences actions. Such "enactivism" (Protevi, 2006) is closely related to the general idea of embodied cognition (Pizzera, Chapter 13). As a consequence, research has to consider the complexity of cognition–action relatedness in special situations and environmental settings, as shown in Figure 1 by the curved arrows. Emotions, although I do not discuss them here, also play a crucial role in cognitive functions such as decision-making (Laborde, Chapter 17). Research topics such as flow and choking have been the focus of explorations of the relationship between thinking or specific mental states and action or performance (Moran, 2012). Researchers, for example, have become interested in "what to think to improve performance."

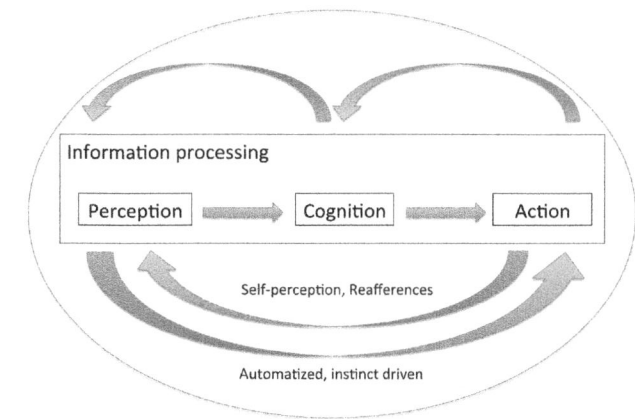

**FIGURE 1** Link between action and cognition.

For intervention purposes and mental training technique, focus of attention can be manipulated by instruction or self-instruction. Cognitive restructuring, for example, a technique in sports psychology, aims at reducing self-defending thoughts and negative self-talk while performing (Meyers, Whelan, & Murphy, 1996; Thomas, Maynard, & Hanton, 2007). The principal idea behind this seems to be to promote inner speech and self-instruction as a fruitful technique to control movements while performing complex motor tasks. Exploring mental processes in the field of sports indeed has revealed that experts have a special way of thinking and making decisions (Raab, 2007) and has shown that expertise in sports is knowledge based.

Expert coaches' knowledge and thinking are vital for their actions in coaching an athlete or a team. To give an example, my colleagues and I recently analyzed common coaches' beliefs about the 7-m throw in handball, trying to determine if relying on coaching instructions improved athletes' success rates of 7-m throws (Lobinger et al., 2014a). When we analyzed 7-m throws of the European Championship in 2010, we found no evidence for the common belief that a player who has been fouled should not take the penalty throw. Coaches' instructions, therefore, should be evaluated with empirical data, and beliefs and expert knowledge should be discussed and reflected upon within the coaching team.

To sum up, along with the pragmatic turn in cognitive science, sports seems to be regarded as a dynamic natural laboratory to examine the link between cognitive processes and action (Moran, 2012). In the following sections, I present three research programs. The first one addresses a movement disorder in golf called the yips, which can be classified as paradoxical performance (Lobinger et al., 2014b). To explain the yips, different models and theories are presented and interventions are discussed. The second research program presented is on coach education in soccer. This program serves as an example of applying mixed methods in analyzing expert performance. Top-level coaches from the German *Bundesliga* (national league) were interviewed and asked to reflect on the content of their coaching education to evaluate whether it suits in situ affordances of being a head coach in the Bundesliga. The presented research line therefore links cognitive aspects such as educational content and acquired knowledge to experts' need to perform on a high level. The third research project outlines a theoretical position on talent assessment in high-performance sports. Talent models are reviewed, and an extended model of stages in talent development is presented. Taking into account the enactivity of cognition, I present a heuristic approach to talent development (de Oliviera et al., 2014).

## THE YIPS IN GOLF

Let us revisit the example at the beginning of this Chapter. It is the 18th hole and our golf professional is preparing to putt. All of a sudden, right at the impact, that is, before his club touches the ball, the wrist of his dominant arm jerks involuntarily, the club rotates, and the ball misses the hole. From this moment on wards, the golf professional might suffer from what is called "the golfer's worst curse": the yips. What are the reasons for this movement disorder and who is prone to getting it? How can this be overcome or cured?

The term the yips was introduced by golf player and triple-Major winner Tommy Armour. Due to the sudden appearance of involuntary and jerky twitches in his lower arm, his career came to an abrupt end in 1935, when he was 41 years old. He called this phenomenon "yips," by which it is still known today. The yips can be defined as sudden jerks, tremors, or even spasms that predominantly affect the distal upper extremity (Smith et al., 2000). The yips occurs mainly during putting, as this stroke requires the finest motor skills and highest precision. Yet it can also arise in other strokes in golf, such as the chip, pitch, or drive (McDaniel, Cummings, & Shain, 1989; Smith et al., 2000). The symptoms appear shortly before hitting the ball, preventing a controlled putt and usually resulting in a miss of the hole (McDaniel et al., 1989; Smith et al., 2000). Being affected increases the rate of strokes for 18 holes by an average of about 4.7 strokes (Sachdev, 1992).

Although earlier research pointed out that professional golfers with a low handicap (3 or less) are more prone to developing the yips (Smith et al., 2000), recent studies have shown that these movement disorders can affect golfers at different levels of expertise and at different ages (Klämpfl, Lobinger, & Raab, 2013b; Klämpfl, Philippen, & Lobinger, 2014), from total beginners up to professionals at any adult age. The yips is a common phenomenon in golf. The estimated prevalence rates range from 17% to 35% (Klämpfl et al., 2013b), although Smith et al. (2000) suggested that it could be as much as 48%, which very likely is an overestimation.

In looking for profiles of affected golfers, research revealed psychological risk factors. A typical yips-affected golfer, for example, seems to be prone to perfectionism and to have a more anxious personality. There has been a lack of consensus on the etiology of the yips. It has been classified as either a neurological or a psychological movement disorder or a mixture of both, located on a continuum between neurological and psychological origin (Smith et al., 2000).

## Neurological Origin: Focal Dystonia

A condition known as task-specific focal dystonia (TSFD) occurs during the execution of a repetitive, specific task or action requiring fine motor skills, such as writing or playing a musical instrument (e.g., the piano or the violin; Altenmüller & Jabusch, 2006). TSFD can be described as involuntary muscle contractions leading to twisting or repetitive movements or abnormal postures of a body part (Pont-Sunyer, Martí, & Tolosa, 2010). An exact mechanism has not been discovered yet, but a loss of inhibitory function at the spinal, brainstem, and cortical levels may be one underlying neurological explanation. This loss of inhibitory function leads to a loss of selectivity and overflow, followed by an unwanted muscle spasm (Hallett, 2011). Stinear, Fleming, and Byblow (2006) discovered, for example, a reduced ability to inhibit muscle activity during the execution of putts. The prevalence rate is above 1% in professional musicians (Adler, Crews, Hentz, Smith, & Caviness, 2005; Altenmüller & Jabusch, 2006, 2008). The yips does not match the characteristics of TSFD entirely, however, which leads to the assumption of further causes for the yips.

## Psychological Origin: Choking

From a psychological perspective, the yips and similar phenomena are thought to be stress related and akin to choking under pressure (Smith et al., 2003; Stinear et al., 2006). Choking under pressure is defined as "a process, whereby the individual perceives that their resources are insufficient to meet the demands of the situation, and concludes with a significant drop in performance—a choke" (Hill, Hanton, Fleming, & Matthews, 2009, p. 206). As this phenomenon appears in different sports, several theories have emerged to elucidate the mechanism behind choking (for an overview, see Lobinger et al., 2014): Attention theories include distraction theories and self-focus theories. The latter are considered an especially valuable approach to explaining the phenomenon of the yips. "Self-focus theories propose that performance anxiety increases an athlete's level of self-consciousness, which will cause them to focus their attention inwardly, consciously monitor and/or control their skill execution, and choke as a result" (Hill, Hanton, Fleming, & Matthews, 2010, p. 27; see also Baumeister, 1984). In other words, inward attention impedes a previously learned automated movement, which leads to a decrease in performance (Wulf & Su, 2007). In this context, Masters and Maxwell (2008) introduced the theory of reinvestment, which talks about the detrimental effect of the "manipulation of conscious, explicit, rule based knowledge, by working memory, to control the mechanics of one's movements during motor output" (Masters, 1992, p. 208). Yips-affected golfers indeed were found to be more prone to reinvestment (Klämpfl, Lobinger, & Raab, 2013a; Laborde et al., 2015).

## Motor Origin: Dynamic Stereotype

Recently, examining the gray zone of the continuum between a neurological and a psychological origin, researchers have proposed a motor origin (Lobinger et al., 2014). This is in line with Marquardt (2009), which described the yips as a contextual movement disorder.

Affected golfers seem to be trapped in a vicious circle that includes anxiety, overcontrol, interference, and perception. Anxiety and fear of the yips might lead to a reinforcement of the movement disorder. Trying to consciously control the movement might have a negative impact on the stroke. The yips could be reinforced by negative transfer or proactive interference of similar movements. Fatal compensation strategies can become automated. Perception of the jerking or cramping might be a precondition of a pathologic putting problem. The yips, therefore, can be seen as a dynamic stereotype (Windholz, 1996), that is, as learned movement routines that are performed automatically and do not necessarily need to be consciously controlled (Altenmüller & Ioannou, Chapter 7; Lobinger, Klämpfl, & Altenmüller, 2014).

## Diagnosing and Treating the Yips

Given that there may be multiple origins of the yips, a multimethod diagnostic approach that considers neurological, physiological, psychometric, and kinematic variables should be applied. The research line at the Institute of Psychology at the German Sport University (GSU) started with pilot studies that led in three directions, focusing on diagnostics, etiology, and intervention. To explore the yips phenomenon, the first study aimed at classifying the different types of yips and drawing a more precise picture of it. Seventeen semi-structured interviews with golfers revealed negative cognitive as well as emotional associations with putts when the golfers were affected by the yips (Philippen & Lobinger, 2012). The usual solution strategy was to change technique or change equipment. In a prevalence study, the GSU questioned 1306 golfer to detect the yips and possible correlations regarding the golfers' backgrounds, such as sports biography (Klämpfl et al., 2014). The results showed that 22.4% of the participating golfers were affected by the yips. This estimated rate was much higher than that for a focal dystonia, which is 0.01–0.03% in musicians (Altenmüller & Jabusch, 2006). Interestingly, the study indicated that 61.7% of the yips-affected golfers played another racket sport besides golf.

The results inspired further research aimed at gaining a better understanding of the phenomenon in the area of diagnostics, etiology, and interventions by comparing yips-affected golfers with pianists affected by TSFD[2] (also see Chapter 7). The first of three studies in the program addressed the diagnostics of the yips by comparing yips-affected and non-affected golfers via an extensive diagnostic battery including measuring psychological variables as well as physiological parameters and gathering kinematic data (Klämpfl et al., 2013a). The multimethod approach included administering several online questionnaires to assess personality and gathering behavioral data generated by lab experiments. Forty golfers (20 affected, 20 non-affected) filled in questionnaires about trait anxiety or perfectionism, among other topics, and were asked to perform 15 putts from a 1.5-m distance in each of five different conditions on an artificial putting surface. The putting conditions were selected based on etiological assumptions and included normal putting, putting under pressure, putting only with the dominant hand, putting with a field hockey stick (context change),

---

2. This was a cooperative project with the group of Prof. Eckart Altenmüller of the Hanover University of Music, Drama, and Media and was supported by the German Research Foundation (DFG grant Al 278/9).

and putting with latex gloves (sensory trick). The researchers collected kinematic data (using SAM-Putlab) and measured muscle activity, heart rate, and putting accuracy. The most sensitive method for detecting the yips was having the participants putt one-handed while performance and kinematics were measured. Yips-affected golfers had problems performing the putts with only their dominant hand, because they could not compensate for the yips symptoms by applying the second hand. Surprisingly, other measures, especially the personality questionnaires, did not differentiate between yips-affected and non-affected golfers.

The second study, therefore, focused on the etiology of the yips and aimed at testing the validity of the reinvestment theory of Masters and Maxwell (2008) for the occurrence of the yips (Klämpfl et al., 2013b). The underlying hypothesis was that yips-affected golfers would improve their putting skills if they were prevented from consciously controlling their putting. Therefore, yips-affected participants ($N=19$) had to randomly putt in a skill-focus condition and a distraction condition generated by tone-based dual tasks. The results did not support the hypothesis: the performance of yips-affected golfers did not improve when they were distracted while putting.

Finally, the third study dealt with intervention methods and their effectiveness in treating the yips (Klämpfl, 2014). Seventeen yips-affected golfers participated in an intervention study that lasted seven months, including nine weeks of intervention. Two groups received etiology-based interventions, either sensorimotor retraining, which targeted task-specific dystonia, or a performance routine addressing the psychological origin of the yips. Sensorimotor retraining consisted of putting exercises based on variability of practice by variation of the affected skill with respect to swing, ball, racket, and surface. Performance routines covered learning progressive muscle relaxation, external focus of attention, and self-talk. Klämpfl (2014) hypothesized that both interventions would lead to a decrease in yips behavior.

Yips behavior was measured via the one-handed putting test in the laboratory and measures carried out on a nine-hole pitching course. Results indicated no significant increase in putting performance and, therefore, no successful treatment of the yips according to these selected dependent variables. Most of the golfers, however, reported improvement in their personal putting skill in interviews conducted in a follow-up (see Klämpfl, 2014; for further information).

As a follow-up to this research program, another intervention study concentrating more on client-tailored interventions was conducted (Gerland & Raab, 2015).[3] This study mainly aimed at further defining and diagnosing the yips by applying specific exercises and drills (Haney, 2006) and collecting kinematic data to analyze motor control. Three yips-affected golfers participated in this

---

3. The project was supported by the Federal Institute of Sport Science in Germany (Bundesinstitut für Sportwissenschaften, BISp) and realized in cooperation with the German Golf Association (IIA1070808/13-1).

study that lasted 10 weeks, including nine special coaching sessions. Overall, more than 30 different exercises were implemented. The basic idea shared by all of these exercises was to change the demand of the task to manipulate the focus of attention and the perception during the putting stroke. For example, the participants had to putt under water, against a fixed ball, or in slow motion. Preliminary results show a reduction of the yips within the affected strokes. A performance test for putting and a kinematic test revealed a significant improvement in putting behavior.

For future research, an action-oriented framework (Chapter 2) should be applied to investigate the yips and other types of paradoxical performance (Lobinger et al., 2014). The previously discussed origins should be addressed by applying multimethod designs. Intervention programs should (1) include a close examination of the client or patient, (2) examine the affected movement, and (3) apply a broad, interdisciplinary diagnostic approach.

## THE EDUCATION OF SOCCER COACHES

Imagine the head coach of a professional soccer club in the first division is beginning his daily tasks. At 8:00 am, he meets with his assistant coaches to plan the training session that starts at 10:00 am. They also talk about the upcoming match in two days and analyze the opponent; later that day, they will discuss tactics. Physiotherapists come along at 9:00 am and inform the coach about any injured players. Players arrive at 9:00 am, too. The team captain asks for a meeting with the head coach. A training session is held from 10:00 to 11:30 am. The head coach observes, instructs, gives feedback, and motivates. The manager critically observes the training session and wants to have a word afterward, to talk about the newly discovered talent and contracts. The press officer informs the coach of the latest news and would like to prepare a press conference the same day for 2:00 p.m. The team manager comes in to organize travel to away games, checking schedules and meals and needs for technical devices. The physician phones to talk about seriously injured players. The coaches of the Under 23 team and the Under 19 team arrive at 3:00 pm to report on players who will soon be integrated into the premier team. A second training session takes place from 5:00 to 7:00 p.m. Afterward, the coaches do video analyses and the head coach prepares his speech to the team.

How does one prepare for this job? What does a coach need to know to manage the tasks waiting to be done?

In professional soccer, the job of head coach is one of the most prominent positions and is often the most well paid, but average contract duration lies at around 1.5 years only (Schmidt & Schreyer, 2011). Irrespective of academic discussions distinguishing between performance and success, head coaches' performances are measured by success alone, and they lose their jobs when their teams run out of success (Frick, Barros, & Passos, 2009; Frick, Barros, & Prinz, 2008; Koning, 2003).

Coach education has been identified as a key vehicle for raising the standards of coaching practice (Nelson, Cushion, & Potrac, 2013) and is a growing field of interest, bringing together experts and researchers from different countries and different sports (for an overview, see Callary, Culver, Werthner, & Bales, 2014). However, the limited number of articles so far in the field of soccer coach education shows that this field is still in its infancy. Coach career development seems to be even further away from researchers' interest (Dawson & Phillips, 2013).

But at the GSU, coach education has a long tradition. Since 1947 more than 1600 coaches have been educated at the Hennes Weisweiler Academy (HWA) of the German Football Association,[4] which was located at the GSU from 1947 until 2011. At HWA coaches are able to earn the highest coaching license in Europe, the Union of European Football Associations (UEFA) pro license, a prerequisite for a job as a head coach in the first, second, and third league in Germany since 2010. Traditionally, scientists and educators of the GSU have been involved in the education program, which lasts 10 months and has 1111 h of instruction. Soccer science and coaching, movement science, and sports psychology are the main academic fields within the program.

Aiming at optimizing the education of soccer coaches at the HWA, the Institute of Psychology at GSU established a three-year research project. A task analysis of current coaches in German professional soccer leagues was conducted as the basis of requirement profiles (Kass, 2013; Lobinger, Kass, Mickler, & Raab, 2015). Profiles were established by means of a serial, multimethodological approach including observation of coaches and personal in-depth interviews with professional soccer coaches, coach educators, and managers ($N=55$). The overall aim was to evaluate coach education. Former graduates, who were by then head and assistant coaches in the first and second Bundesliga, were asked what they felt was missing from their coach education program. In addition, coaches working at youth academies as well as national association coaches were questioned to gain a deeper understanding of the needs of coaches working in other fields than Bundesliga. Interviews with managers helped to clarify the job descriptions of head coaches, and coach educators reflected on education programs and what they thought was needed for the future. On the basis of these interviews, we developed an online questionnaire on tasks, which was filled out by coaches with a UEFA pro license working in various areas ($N=87$).

Content analyses of the interviews and results of the questionnaire revealed different tasks that were integrated by triangulation into seven main required areas of expertise of a head coach in the German premier league: training (also planning and observation), coaching, leadership (also of assistant coaches and staff), tactics, media and public relations, training supervision and organization (also fitness, endurance, season planning), and scouting (players, tactics

---

4. Deutscher Fußball Bund (DFB).

opponents). The example at the beginning of this section illustrates these different tasks. The requirements were further differentiated by the affordances derived from the manager's interviews and related to the educational content of HWA pro license courses. The results of this project led to the development of best-practice models for requirement profiles of coaches. Recommendations for the education of soccer coaches completed the analyses.

In most of the coach education programs, also across different sports, academics are taught unconnectedly. Modules of the programs are domain specific and often include a first phase of basic knowledge and a second more applied phase within an individual's field of interest. On the basis of the findings, problem-based learning that integrates all academic disciplines was proposed and applied (see Table 1).

For example, candidates in pro license courses have to analyze top teams for six weeks and simulate taking over the team. They are asked to present task lists and interventions, while taking all team-related psychosocial aspects of team management, leadership style, training load, and tactical decisions for training sessions and upcoming matches into account.

In the future, the program should be evaluated, and coaches' careers should be examined in a longitudinal design. This would allow for deeper insight into career planning and aspects of individual–environment fit, which are most important for the success of coaches. Knowledge acquired in coach education courses could be linked to success, and the knowledge and leadership style of successful coaches should be examined. Coach education should be compared in different sports and for different countries to support cross-cultural research and exchange of expertise.

**TABLE 1** Examples of Problem-Based Coach Education Content in Soccer

| Topic | Soccer-Specific Lesson | Movement Science/Sports Biology | Sports Psychology |
|---|---|---|---|
| Takeover of a team | Team roster, basic tactics, match play, game philosophy | Training program, training load | Coaching philosophy, leadership style |
| Preparation of season | Team building, requirements for positions | Endurance training, rehabilitation after injury, loss of weight | Team building and designing, profiling |
| Match day | Team roster, scouting, speeches, tactics | Warming up, cool-down, preparation, nutrition | Debriefing, press conferences, speeches, coaching |

## TALENT IDENTIFICATION AND ASSESSMENT IN HIGH-PERFORMANCE SPORTS

Imagine a coach is watching an Under-10 football match at the end of a season. He will have to decide which talent to bring back for the next season and which one will be rejected and has to quit the team or club. In line with the club's profile of gifted players, the coach has to base his decision on his evaluation of the players' technical and tactical skills and psychological characteristics. What are the important psychological abilities and/or skills for becoming an expert in soccer? What factors allow one to predict later success? How should psychological characteristics be assessed?

To deal with this challenge, researchers have so far mostly focused on how expert athletes made it to the top. In characterizing talent, the differentiation between natural abilities and environmental and intrapersonal catalysts, which are relevant for the formation of general competencies, has been predominant for quite a long time (Gulbin, Oldenziel, Weissensteiner, & Gagné, 2010). Gagné's (1999) differentiated model of talent and giftedness laid the groundwork for the so-called nature versus nurture debate in this area. The focus of this debate is on general natural abilities and environmental factors that lead to general competencies, in an effort to explain (outstanding) performance.

Such models provide an overview of the factors that the athletes themselves consider most important. These factors are valuable because they may explain the route to expertise. However, models developed by information gathered retrospectively are limited because (a) they are built on (only) the information provided by the athletes who made it to the top—developed elite performance—and not, for example, by athletes who dropped out, which would allow for a specification or weighting of the parameters, and (b) retrospection provokes constructing subjective realities and causalities and, therefore, is biased.

Prospective models, on the other hand, would provide an overview of the factors that the athletes consider most important in their current sports participation. Asking coaches, for example, would allow collecting expert knowledge gathered from many years of experience for many talents. They could name relevant parameters and apply them to young athletes to forecast and predict future performance. Studies could later test the quality of the predictions to identify the most relevant factors.

Identifying talent and selecting them for financed talent development programs is a forecasting business. In contrast to employees' selection, placement, and development, where one has to predict development for periods of perhaps 2–5 years, in the field of sports, one is asked to predict, for example, whether a 10-year-old boy will be a famous football player in his mid-twenties. This is a challenging business, because not only the athletes' current performance but also their potential for development has to be taken into account. Therefore, talent development should be considered.

Whereas talent identification can be understood as the process of recognizing the potential of an athlete to excel in a particular sport, talent development, in contrast, can be understood as the process by which an athlete can realize that potential, which includes providing the most appropriate learning and training environments (Vaeyens, Lenoir, Williams, & Philippaerts, 2008). Williams and Reilly (2000) named four key stages in the talent identification and talent-development process: detection, identification, selection, and development. Vaeyens et al. (2008) later added confirmation (also see Figure 2). Detection is understood as "discovery of potential performers who are currently not involved in the sport in question" (Vaeyens et al., 2008, p. 658), and identification as "recognizing current participants with the potential to become elite players"; development means providing athletes with a learning environment; selection "involves the ongoing process of identifying players at various stages who have prerequisite levels of performance for inclusion in a given squad or team" (p. 709), and confirmation is understood as a process in which (already identified) talents "are confronted with the training requirements of the elite sports competition process" (Vaeyens et al., 2008, p. 709).

These stages should be related to different age groups and levels of expertise. Furthermore, the expected outcomes of the key stages outlined above—detecting, identifying, selecting, developing, and confirming talent and expertise—should be integrated. Acknowledging this, we propose a different way of integrating these activities (see Figure 2).

These stages can be illustrated by examples from youth soccer. Detection and identification are stages that need to be passed through to gain access to the talent-development process. Detecting potential talent means the talent developer is scouting people who are not yet involved in the sport in question (Williams & Reilly, 2000). This could be done in a primary school. Identifying (still potential) talent for professional soccer clubs is done by scouting young

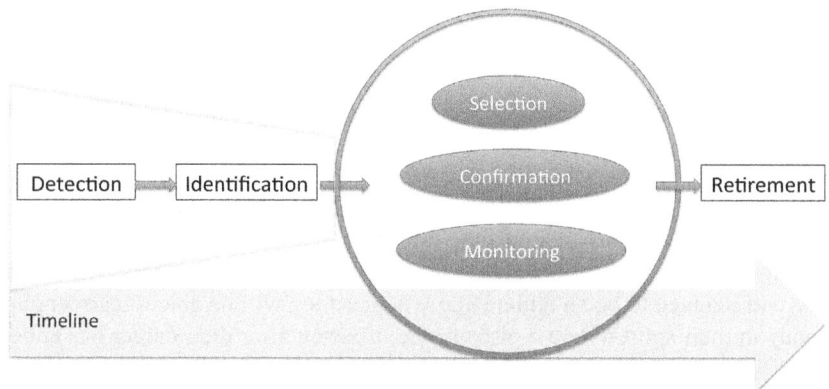

**FIGURE 2** Stages of the talent-development process.

players who already play soccer, but in amateur clubs. From this point forward, they are identified and referred to as talent and expected to show outstanding performance. Being identified as a talent most of the time is the beginning of a systematic talent-development program. This is why it is described as a sort of "entry" into the developmental process, which also takes into account that sometimes it is a question of access to be promoted. From now on the issue is not just potential but performance. The talent has to prove that he or she (really) is a talent to confirm the talent detection and identification. In soccer, for example, if the talent does not perform as expected, it is likely that he or she will be "kicked out" or rejected. This means a young athlete can be a player on the Under-11 team, but can be rejected after 1 year of practice because the coach of the Under-12 team is not convinced the athlete is a talent (outstanding, has high potential; Wolff & Lobinger, 2011). Selection and confirmation, therefore, are repeated steps in an athlete's career, which are called "transitions."

Actually, at this stage, standards are missing that will help the coach to decide when players are to be chosen, when they can be supported in their development to better meet standards, or when they must be rejected. Researchers in the field of talent development are asked to help coaches develop objective rules for when and under what circumstances of developmental progress talent should be either promoted or rejected. This decision should not be merely intuitive, meaning players simply not matching the coaches' expectations or ideas of an ideal athlete at this stage, but it should be guided by comparable values in agreed-upon categories.

This developmental process is characterized by many transitions that an athlete has to go through to be confirmed as a talent. To be selected as a talent also means the athlete has already shown to be outstanding, most of the time; as the athlete matures, specialization has already begun—the athlete joins a certain team or fills a certain position. To give a further example, the coach of the national Under-16 team might be looking for a goalkeeper. Selection will be even more professional. A similar process might happen many times in an athlete's career, which means that he has to undergo selection throughout his career—it is part of the talent-development process. The athlete's development (personal growth) should, therefore, be closely monitored.

When retiring as an expert, in most sports the player quits the system. Retirement, even if not mentioned in the models of Williams and Reilly (2000) and Vaeyens et al. (2008), is an important issue of growing interest (Wylleman, Alfermann, & Lavallee, 2004). I think it is worth taking into account this final stage of an athlete's career, because this stage also affects the ones before and would close a gap in research (see Green, 2005; Lally, 2007). For instance, we know by personal consulting experience that athletes anticipate how long they will manage to be an athlete and whether they will be able to earn enough money in their sport to live a pleasurable life even after their career has ended. These are important questions for their motivation and willingness to concentrate on sports. Furthermore, from an ethical point of view, anyone involved in talent development should feel responsible for the future of the talent. Sports systems might consider retired athletes as "senior" experts and, for example, should integrate them into the system as tutors to mentor younger athletes.

This process of selection, confirmation, and monitoring in talent development aims at increasing expertise and developing elite athletes or professional players. In his expert-performance approach, Ericsson (2007) suggested that the superior performance instead of mere behavior of experts should be examined. The expert-performance approach consists of three steps (Ericsson, 2007, p. 6f):

1. Capture the naturally observable expert performance with well-designed representative tasks that allow the superior performance to be reproduced in the laboratory.
2. Analyze the captured superior performance with standard methodologies for tracing the mediating processes, such as latencies, eye fixations, and verbal reports together with experimental procedures.
3. Once the mechanisms mediating experts' superior performance have been identified, assess if and how different types of experiments and practice activities can explain the acquisition of these mechanisms.

To fully understand the mechanisms underlying the development of these skills, such as decision-making, longitudinal studies are needed to compare different age groups. This allows for making comparisons between individuals as well as monitoring intraindividual developmental processes that need to be taken into account.

Whereas the traditional nature–nurture debate activates the analysis of natural abilities leading to competencies, recently researchers have considered the interaction of skills and situational contexts (Marasso, Laborde, Bardaglio, & Raab, 2014; de Oliveira et al., 2014; Williams & Ford, 2008). Expert athletes use rules of thumb or heuristics (Gigerenzer, Todd, & the ABC Research Group, 1999; Raab & Gigerenzer, 2005; Raab, de Oliviera, & Heinen, 2009) that are specific to the situation and can be used rapidly without much cost. The heuristics repertoire consists of psychological, morphological, physiological, and motor learning adaptations (Gigerenzer et al., 1999; Raab & Gigerenzer, 2005; Raab et al., 2009). Heuristics are acquired by deliberate practice and can be thought of as acquired personal options to solve specific problems. The most prominent ones in sport are search rules, search-stopping rules, decision rules, and execution rules (see de Oliveira et al., 2014 for further information). Search rules govern the search for and selection of relevant information cues. Search-stopping rules say when to terminate the search for information or alternatives. Decision rules define how to decide upon what to do and how to react. Execution rules finally address the execution of actions. As such, they show that cognition and action are closely related and together play a role in developing the expertise that leads to superior performance.

Future research on the cognition–action link should address heuristics, as they may help in predicting success. Longitudinal designs are needed to highlight the route to expertise and to help identify the mechanisms of acquiring expertise in a specific domain.

## CONCLUSION

In this chapter, I presented examples of research that have examined the link between action and cognition. Research on the yips has focused on the cognitive processes influencing performance and vice versa. Research on coaches' education has dealt with the importance of knowledge for performance, and research on talent development has revealed options of adding cognitive paradigms like the adaptive toolbox to talent selection and development. The following chapters of this section will provide further examples of how to bridge the gap and acknowledge the link between action and cognition in the field of performance psychology.

## REFERENCES

Adler, C. H., Crews, D., Hentz, J. G., Smith, A. M., & Caviness, J. N. (2005). Abnormal co-contraction in yips-affected but not unaffected golfers: evidence for focal dystonia. *Neurology, 64*, 1813–1814. http://dx.doi.org/10.1212/01.WNL.0000162024.05514.03.

Altenmüller, E., Ioannou, C., Raab, M., & Lobinger, B. (2014). Apollo's curse: causes and cures of motor failures in musicians: a proposal for a new classification. *Advances in Experimental Medicine and Biology, 826*, 161–178. http://dx.doi.org/10.1007/978-1-4939-1338-1_11.

Altenmüller, E., & Jabusch, H. C. (2006). Neurologische Erkrankungen bei Musikern. *Die Medizinische Welt, 57*, 569–575.

Altenmüller, E. O., & Jabusch, H. C. (2008). Focal dystonia: diagnostic, therapy, rehabilitation. In M. Grunwald (Ed.), *Human haptic perception: Basics and application* (pp. 303–311). Heidelberg, Germany: Birkhäuser.

Amelang, M., & Zielinsky, W. (2002). *Psychologische Diagnostik und Intervention.* Berlin, Germany: Springer.

Baumeister, R. F. (1984). Choking under pressure—self-consciousness and paradoxical effects of incentives on skillful performance. *Journal of Personality and Social Psychology, 46*, 610–620.

Callary, B., Culver, D., Werthner, P., & Bales, J. (2014). An overview of seven national high performance coach education programs. *International Sport Coaching Journal, 1*, 152–164. http://dx.doi.org/10.1123/iscj.2014-0094.

Chang, Y. (2014). Cognitive function. In R. C. Eklund & G. Tenenbaum (Eds.), *Encyclopedia of sport and exercise psychology* (pp. 136–139). Los Angeles, CA: SAGE.

Creswell, J. W. (2009). Mapping the field of mixed methods research. *Journal of Mixed Methods Research, 3*, 95–108. http://dx.doi.org/10.1177/1558689808330883.

Dawson, A., & Phillips, P. (2013). Coach career development: who is responsible? *Sport Management Review, 16*, 477–487.

Eklund, R. C., & Tenenbaum, G. (2014). *Encyclopedia of sport and exercise psychology.* Los Angeles, CA: SAGE.

Engel, A. K., Maye, A., Kurthen, M., & König, P. (2013). Where's the action? the pragmatic turn in cognitive science. *Trends in Cognitive Science, 17*, 202–209. http://dx.doi.org/10.1016/j.tics.2013.03.006.

Ericsson, K. A. (2007). Deliberate practice and the modifiability of body and mind: toward a science of the structure and acquisition of expert and elite performance. *International Journal of Sport Psychology, 38*, 4–34.

Ericsson, K. A., Krampe, R. T., & Tesch-Römer, C. (1993). The role of deliberate practice in the acquisition of expert performance. *Psychological Review, 100*, 363–406.

Frick, B., Barros, C. P., & Passos, J. (2009). Coaching for survival: the hazards of head coach careers in the German "Bundesliga". *Applied Economics, 41*, 3303–3311. http://dx.doi.org/10.1080/00036840701721455.

Frick, B., Barros, C. P., & Prinz, J. (2008). Analysing head coach dismissals in the German "Bundesliga" with a mixed logit approach. *European Journal of Operational Research, 200*, 151–159.

Gagné, F. (1999). My convictions about the nature of abilities, gifts, and talents. *Journal for the Education of the Gifted, 22*, 109–136.

Gerland, B., & Raab, M. *Dealing with the yips: Relearning the putting stroke in golf—A single case research*, submitted.

Gigerenzer, G., Todd, P. M., & the ABC Research Group. (1999). *Simple heuristics that make us smart*. New York: Oxford University Press.

Green, C. (2005). Building sport programs to optimize athlete recruitment, retention and transition: toward a normative theory of sport development. *Journal of Sport Management, 19*, 233–253.

Gulbin, J. P., Oldenziel, K. E., Weissensteiner, J. R., & Gagné, F. (2010). A look through the rear view mirror: developmental experiences and insights of high performance athletes. *Talent Development & Excellence, 2*, 149–164.

Hallet, M. (2011). Neurophysiology of dystonia: the role of inhibition. *Neurobiology of Disease, 42*, 177–184. http://dx.doi.org/10.1016/j.nbd.2010.08.025.

Haney, H. (2006). *Fix the yips forever: The first and only guide you need to solve the game's worst curse*. New York, NY: Penguin Group.

Hill, D. M., Hanton, S., Fleming, S., & Matthews, N. (2009). A re-examination of choking in sport. *European Journal of Sport Science, 9*, 203–212. http://dx.doi.org/10.1080/17461390902818278.

Hill, D. M., Hanton, S., Matthews, N., & Fleming, S. (2010). Choking in sport: a review. *International Review of Sport and Exercise Psychology, 3*, 24–39. http://dx.doi.org/10.1080/1750984090330119 9.

Kass, P. (2013). Die Trainertätigkeit im Profifußball—eine multimethodale Anforderungsanalyse zur Optimierung des Fußball-Lehrer-Lehrgangs (Dissertation Thesis). German Sport University Cologne. Retrieved from http://esport.dshs-koeln.de/407/2/Philipp_Kass_Dissertation_Die_Trainertaetigkeit_im_Profifussball.pdf.

Klämpfl, M.K. (2014). The yips—Diagnostics, etiology and interventions (Dissertation Thesis). German Sport University Cologne. Retrieved from http://esport.dshs-koeln.de/442/.

Klämpfl, M. K., Lobinger, B. H., & Raab, M. (2013a). How to detect the yips in golf. *Human Movement Science, 32*, 1270–1287. http://dx.doi.org/10.1016/j.humov.2013.04.004.

Klämpfl, M. K., Lobinger, B. H., & Raab, M. (2013b). Reinvestment—the cause of the yips? *PLoS One, 8*, e82470. http://dx.doi.org/10.1371/journal.pone.0082470.

Klämpfl, M. K., Philippen, P. B., & Lobinger, B. H. (2014). Self-report vs. kinematic screening test: prevalence, demographics and sports biography of yips-affected golfers. *Journal of Sports Sciences, 33*, 655–664. http://dx.doi.org/10.1080/02640414.2014.961026.

Koning, R. H. (2003). An econometric evaluation of the effect of firing a coach on team performance. *Applied Economics, 35*, 555–564.

Laborde, s., Musculus, L., Kalicinski, M., Klaempfl, M.K., Kinrade, N.P., & Lobinger, B.H. (2015). Reinvestment: Examining convergent, discriminant, and criterion validity using psychometric and behavioral measures. *Personality and Individual Differences, 78*, 77–87. http://dx.doi: 10.1016/j.paid.2015.01.020.

Lally, P. S. (2007). Identity and athletic retirement: a prospective study. *Psychology of Sport and Exercise, 8*, 85–99. http://dx.doi.org/10.1016/j.psychsport.2006.03.003.

Lobinger, B. H. (2005). *Bewegungsbezogenes Sicherheitsmanagement von Frauen im Alter— Theoretische Überlegungen und empirische Untersuchungen (Movement-Oriented Security Management of Women of Age—Theoretical Contemplations and Empirical Studies).* Hamburg. Germany: Diplomica.

Lobinger, B., Büsch, D., Werner, K., Pabst, J., Gail, S., & Sichelschmidt, P. (2014a). Erfolgsrelevante Aktionsmuster von Torhütern beim Siebenmeter-wurf im Spitzenhandball. *Zeitschrift für Sportpsychologie, 21,* 74–85. http://dx.doi.org/10.1026/1612-5010/a000116.

Lobinger, B. H., Klämpfl, M. K., & Altenmüller, E. (2014b). We are able, we intend, we act-but we do not succeed: A review on paradoxical performance in sports. *Journal of Clinical Sport Psychology.* http://dx.doi.org/10.1123/jcsp.2014-0047.

Lobinger, B. H., Kass, P., Mickler, W., & Raab, M. (2015). Welche Ausbildungsinhalte brauchen Fußball-Lehrer für den Profibereich? (Which Educational Contents are needed for Soccer Coaches in Professional Soccer?). *Leistungssport, 45,* 49–55.

Lobinger, B. H., Klämpfl, M. K., & Altenmüller, E. (2014). We are able, we intend, we act—but we do not succeed: a theoretical framework for a better understanding of paradoxical performance in sports. *Journal of Clinical Sport Psychology, 8,* 357–377. http://dx.doi.org/10.1123/jcsp.2014-0047.

Marasso, D., Laborde, S., Bardaglio, G., & Raab, M. (2014). A developmental perspective on decision making in sports. *International Review of Sport and Exercise Psychology, 7,* 251–275. http://dx.doi.org/10.1080/1750984X.2014.932424.

Marquardt, C. (2009). The vicious circle involved in the development of the yips. *International Journal of Sports Science and Coaching, 4,* 67–88. http://dx.doi.org/10.1260/174795409789577506.

Masters, R. (1992). Knowledge, knerves and know-how: the role of explicit versus implicit knowledge in the breakdown of a complex motor skill under pressure. *British Journal of Psychology, 83,* 343–358. http://dx.doi.org/10.1111/j.2044-8295.1992.tb02446.x.

Masters, R., & Maxwell, J. (2008). The theory of reinvestment. *International Review of Sport and Exercise Psychology, 1,* 160–183. http://dx.doi.org/10.1080/17509840802287218.

McDaniel, K. D., Cummings, J. L., & Shain, S. (1989). The yips: a focal dystonia of golfers. *Neurology, 39,* 192–195.

Meyers, A. W., Whelan, J. P., & Murphy, S. M. (1996). Cognitive behavioral strategies in athletic performance enhancement. *Progress in Behavior Modification, 30,* 137–164. http://dx.doi.org/10.1016/S0005-7894(05)80369-7.

Moran, A. P. (2012). Thinking in action: some insights from cognitive sport psychology. *Thinking Skills and Creativity, 7,* 85–92. http://dx.doi.org/10.1016/j.tsc.2012.03.005.

Nelson, L., Cushion, C., & Potrac, P. (2013). Enhancing the provision of coach education: the recommendations of UK coaching practitioners. *Physical Education and Sport Pedagogy, 18,* 204–218. http://dx.doi.org/10.1080/17408989.2011.649725.

de Oliviera, R. F., Lobinger, B. H., & Raab, M. (2014). An adaptive toolbox approach to the route to expertise in sport. *Frontiers in Psychology, 5,* 709. http://dx.doi.org/10.3389/fpsyg.2014.00709.

Philippen, P. B., & Lobinger, B. H. (2012). Understanding yips in golf: thoughts, feelings, and focus of attention in yips-affected golfers. *The Sport Psychologist, 26,* 325–340.

Pont-Sunyer, C., Martí, M. J., & Tolosa, E. (2010). Focal limb dystonia. *European Journal of Neurology, 17,* 22–27. http://dx.doi.org/10.1111/j.1468-1331.2010.03046.x.

Protevi, J. (Ed.). (2006). *A dictionary of continental philosophy.* New Haven, CT: Yale University Press.

Raab, M. (2007). Think SMART, not hard—a review of teaching decision making in sport from an ecological rationality perspective. *Physical Education and Sport Pedagogy, 12,* 1–22.

Raab, M., & Gigerenzer, G. (2005). Intelligence as smart heuristics. In R. J. Sternberg & J. E. Pretz (Eds.), *Cognition and intelligence. Identifying the mechanisms of the mind* (pp. 188–207). Cambridge, England: Cambridge University Press.

Raab, M., de Oliviera, R. F., & Heinen, T. (2009). How do people perceive and generate options? *Progress in Brain Research, 174*, 49–59. http://dx.doi.org/10.1016/S0079-6123(09)01305-3.

Sachdev, P. (1992). Golfers' cramp: clinical characteristics and evidence against it being an anxiety disorder. *Movement Disorders, 7*, 326–332. http://dx.doi.org/10.1002/mds.870070405.

Schmidt, S. L., & Schreyer, D. (2011). *In the Line of Fire: Verweildauer von Bundesligatrainern und CEOs in Deutschland (In the Line of Fire: Turnaround Time of Bundesliga Coaches and CEOs in Germany)*. Oestrich Winkel: Institute for Sports, Business & Society.

Smith, A. M., Adler, C. H., Crews, D., Wharen, R. E., Laskowski, E. R., Barnes, K., et al. (2003). The "yips" in golf: a continuum between a focal dystonia and choking. *Sports Medicine, 33*, 13–31.

Smith, A. M., Malo, S. A., Laskowski, E. R., Sabick, M., Cooney, W. P., III, Finnie, S. B., et al. (2000). A multidisciplinary study of the "yips" phenomenon in golf: an exploratory analysis. *Sports Medicine, 30*, 423–437.

Stinear, C. M., Fleming, M. K., & Byblow, B. D. (2006). Lateralization of unimanual and bimanual motor imagery. *Brain Research, 1095*, 139–147.

Tenenbaum, G., & Filho, F. (2014). Cognitive styles. In R. C. Eklund & G. Tenenbaum (Eds.), *Encyclopedia of Sport and exercise psychology* (pp. 141–143). Los Angeles, CA: SAGE.

Thomas, O., Maynard, I., & Hanton, S. (2007). Anxiety responses and psychological skill use during the time preceding competition: theory to practice I. *Journal of Applied Sport Psychology, 19*, 379–397.

Vaeyens, R., Lenoir, M., Williams, A. M., & Philippaerts, R. M. (2008). Talent identification and development programmes in sport—current models and future directions. *Sports Medicine, 38*, 703–714.

Ward, P. (2014). Cognitive task analysis. In R. C. Eklund & G. Tenenbaum (Eds.), *Encyclopedia of sport and exercise psychology* (pp. 143–145). Los Angeles, CA: SAGE.

Williams, A. M., & Ford, P. R. (2008). Expertise and expert performance in sport. *International Review of Sport and Exercise Psychology, 1*, 4–18. http://dx.doi.org/10.1080/17509840701836867.

Williams, A. M., & Reilly, T. (2000). Talent identification and development in soccer. *Journal of Sport Sciences, 18*, 657–667.

Windholz, G. (1996). Pavlov's conceptualization of the dynamic stereotype in the theory of higher nervous activity. *American Journal of Psychology, 109*, 287–295.

Wolff, F., & Lobinger, B. H. (2011). Nicht-übernahme jugendlicher Leitungsfußballer in Bundesliganachwuchsmannschaften (Non-Take-Over of Adolescent Competitive Soccer Players in Bundesliga Junior Teams). In J. Ohlert, & J. Kleinert (Eds.), *Sport Köln* (p. 154). Hamburg, Germany: Czwalina.

Wulf, G., & Su, J. (2007). An external focus of attention enhances golf shot accuracy in beginners and experts. *Research Quarterly for Exercise and Sport, 78*, 384–389. http://dx.doi.org/10.1080/02701367.2007.10599436.

Wylleman, P., Alfermann, D., & Lavallee, D. (2004). Career transitions in sport: European perspectives. *Psychology of Sport and Exercise, 5*, 7–20.

## Chapter 6

# Improving Performance by Means of Action–Cognition Coupling in Athletes and Coaches

**Gloria B. Solomon**
*Department of Kinesiology, Texas Christian University, Fort Worth, TX, USA*

---

*A new coach, Eddie, has been hired to lead a college basketball team. Coach Eddie played basketball in college and served as a voluntary assistant before earning the head coach position. Coach Eddie has a lot of decisions to make and many of these decisions will be made in accordance with his past playing and coaching experiences. What type of leadership style will Coach Eddie employ? Will he simply replicate what he has experienced, or will he seek new methods for developing high-skilled performers? How will his athletes judge Coach Eddie's coaching ability? Will Coach Eddie be prone to offer equitable feedback to his athletes or treat his higher skilled players better? What type of additional coach training might Coach Eddie seek in order to enhance his abilities as a leader?*

Throughout the world, the coaching profession appears to be unique among careers. Regardless of the setting, whether small-scale youth sport or highly visible professional sport, coaches tend to be public figures who perform their work in public venues. This is quite unlike most professionals who conduct their business without spectators engaging in the event and officials enforcing rules. Furthermore, coaches have the capacity to significantly impact the athletic and personal development of their athletes. There is ample evidence in the scholarship on coaching, suggesting that coaches are highly influential mentors and role models (Solomon & Buscombe, 2013). Subsequently, the athletes in their care are vulnerable to the coaches' influence (Horn, Lox, & Labrador, 2001). The first purpose of this chapter is to explore the development of coaching expertise. The second purpose is to examine the impact of coaches on athlete development and performance. While the duties and responsibilities of coaches are common regardless of nation or nationality, the process of training and developing coaches varies considerably across country and region. Because of this, it is deemed to be wise to focus this chapter on the current coaching conditions in the United States.

## THE PROFESSION OF COACHING

There are over 100,000 coaches in the United States today. Ranging from the volunteer youth sport coach to the highly paid coach of professional sport, coaches serve millions of youth, adolescents, and adults each year. Among youth sport coaches, the most common role is that of a volunteer coach. The majority of youth sport coaches in the most popular team sports (baseball, basketball, football, soccer) serve in a voluntary capacity. Minimal qualifications typically include first aid training and clearance of a criminal background check. While most of these coaches serve for virtuous reasons (the love of children, opportunity to coach their own child, etc.), they lack systematic training in such basic skills as sport skill instruction and child development. Some private sport organizations (i.e., PeeWee Football, Little League Baseball) mandate coach-training workshops, which do provide basic knowledge on the coaching process. There are also organizations (i.e., Club Volleyball) which hire part-time paid coaches. In sum, the vast majority of people serving as coaches of the youth are unpaid volunteers with minimal training in coaching or child psychology.

Interscholastic, or high school, coaches typically take over from the youth sport experience as children progress from youth to adolescence. Unique among many countries, the high school programs in the United States offer three levels of interscholastic athletic competition: Freshmen, Junior Varsity, and Varsity. While there are multiple levels of competition (commonly based on school size), these programs can be highly organized, competitive, and expensive. However, high school coaches are most commonly teachers who choose to coach along with their instructional duties. The vast majority of these coaches, again, lack systematic training in the profession of coaching. In fact, few of these coaches earn more than a few thousand dollars as a coaching stipend. It is the rare coach (i.e., high school coaches in some football programs in Texas) who earns enough income from coaching to meet the economics of living expenditures.

College coaches typically hold university degrees, but rarely in majors related to coaching. In fact, most job descriptions only require a bachelor's degree but not in any specific field. Many of these coaches played the sport themselves and rely on this experience to guide their coaching behaviors. Head coaches at all levels of competition (i.e., NCAA Division I, II, III, etc.) usually serve as an assistant coach for several years prior to promotion to head coach.

The coach of professional sport is enigmatic. While they share in common many of the roles of other coaches, they are unique in one simple way. Professional coaches are not tasked with developing the athlete as a person; their sole focus is the development of athletic talent, which translates to performance. Coaches from the youth level through college are tasked with both developing the person (i.e., character) and the player (i.e., goal blocking). Since the goal of this chapter is to examine the impact of coaches on athlete development and performance, the professional coach will be excluded hereon.

Another key distinction among coaches is their role within the team. Typically a coaching staff is represented by a head coach and an assortment of assistant coaches. While the roles are distinct, they vary considerably across teams. Most commonly, the head coach is the final arbiter of most major decisions within the team; assistant coaches serve more specific roles, such as scout, position coach, or mental coach. Numerically, there are considerably more assistant coaches than head coaches at most levels of competition. However, the coaching science favors research on the head coach. When relevant, the level (youth, interscholastic, intercollegiate) and role (head, assistant) of coaches are distinguished in the coaching literature.

## PURPOSE

Despite the vast numbers of coaches in the United States, at this time, there are no national-level training requirements. Each sporting entity creates criteria for their coaches; these range from weekend workshops to college degrees in coaching. Much of the preparation and training for coaches involve varied experiences such as coaching in a volunteer capacity to playing competitively in that sport. Having good mentors and establishing a network of contacts appears more important to coach training than a formal educational process. Thus, this chapter will not represent a specific critique of standardized coach-training programs, as they do not exist. The purpose of this chapter is twofold. One, an exploration of the development of coaching expertise is conducted. Two, an examination of the impact of coaches on athlete development and performance is performed.

## DEVELOPMENT OF COACHING EXPERTISE

There is a significant body of literature on the development of coaching expertise. While much of this work is atheoretical, there is a recent emphasis among coaching scholars to apply and develop theories and models to guide their inquiry. An exploration of the leadership literature will explore the historical evolution of leadership theory and provide empirical evidence on coach behavior.

### History of Leadership in Sport

Like many theoretical and technological advances, progress in the study of leader behaviors emerged in response to the events of World War II (Guetzkow, 1951). The majority of these early theoretical outcomes focused on military, political, or industrial leadership. The concept of leadership gained broad acceptance as "the behavioral process of influencing individuals or groups toward set goals" (Barrow, 1977, p. 232). It was also in the 1970s when scholars began to apply leader behavior concepts to the sport leader (Chelladurai & Saleh, 1978). Coaches are universally recognized as the leaders in sport. Early research in the sport setting indicated that coaches demonstrate identifiable behaviors when leading their teams (Chelladurai & Saleh, 1980). Unfortunately, there was a lack of importance assigned to athletic leadership and efforts to further understand this unique role. Two major categories of theories were employed as the early sport leadership literature was born: behavioral and situational.

## Behavioral Theories

The position taken by those condoning the behavioral approaches to leadership directs the focus on behaviors exhibited by effective leaders. It was suggested that successful leaders demonstrate similar behaviors such as communication, organization, and initiation (Hemphill & Coons, 1957). This lent an optimistic perspective; if these behaviors could be identified, it follows that they could then be taught. Thus, leaders could be trained to become effective by learning and exhibiting specific behaviors (Cox, 1998).

The majority of this classic research was conducted at The Ohio State University (OSU) and the University of Michigan (UM) during the 1950s and 1960s. The goal of both of these ambitious projects was to determine universal behaviors of successful leaders. The longitudinal work performed at OSU focused on identifying, describing, and evaluating leader behaviors in varied settings (military, business, industrial, education); the researchers at UM aimed to identify leader behaviors specifically associated with production in industrial environments (Kahn & Katz, 1951). Both groups of researchers concluded their work by identifying two key leadership factors. For the OSU team, those factors included Initiating Structure and Consideration; for their counterparts at UM the two factors included Production Emphasis and Employee Orientation. These two major outcomes paralleled one another, and the terms have been used interchangeably in research (Cox, 1998).

## Situational Theories

While behavioral theorists were busy identifying effective leader behaviors, the situational approach was being developed to accommodate contextual demands. In short, leadership demands change in certain environments, so situational variables (i.e., experience, goals, environment) must be taken into account. Several theories emerged; there are four, in particular, that served as the framework for contemporary theoretical development in coach leadership in the sport setting. These theories are the following: the Contingency Theory of Leader Effectiveness (Fiedler, 1967), Path-Goal Theory (Evans, 1970), Adaptive-Reactive Theory (Osborn & Hunt, 1975), and Discrepancy Theory (Yukl, 1971). Using these theories as the basis for studying leadership in sport, researchers integrated concepts gleaned from these situational approaches to create theories relevant to the sport environment (Chelladurai, 1978; Chelladurai & Carron, 1978).

## Multidimensional Model of Leadership

In the late 1970s, researchers began applying leader behavior concepts to sports. However, there is a considerable gap between the importance assigned to athletic leadership and efforts to systematically understand the leader role in this unique competitive setting. One factor contributing to this gap is that early sport researchers interested in coaching behavior tested their hypotheses using theories developed for the corporate environment. However, there are differences between these two settings, and these differences may contribute to the inconsistent results on coach behaviors in early research. One difference is the disproportionate amount of time spent training and being assessed. In sports, the majority of time is spent practicing, while in the work setting, employees may have one training session a year (Chelladurai, 1978). Furthermore, in sports, there is only one winner, and the life of intact teams is relatively short compared to employment stability in work environments. Therefore, a sport-specific model was needed to study coach behaviors. The integration of the four situational theories delineated above served as the basis for Chelladurai's Multidimensional Model of Leadership (MML; Chelladurai, 1978; Chelladurai & Carron, 1978).

The MML posits that group performance and member satisfaction are functions of the congruence among three leader behaviors: required, actual, and preferred (Chelladurai, 1978). Situational, leader, and member characteristics are treated as antecedents to the three leader behaviors, while performance and satisfaction are consequences. See Figure 1 for a depiction of the MML (Figure 2).

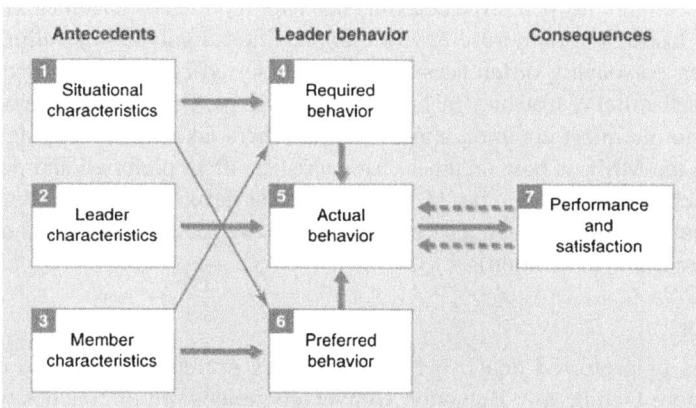

**FIGURE 1** The multidimensional model of leadership (Chelladurai, 1978; Chelladurai & Carron, 1978).

As demonstrated by the model, situational characteristics (i.e., goals of the team, organizational structure, social norms, cultural values, governmental regulations) directly affect the coach's required behavior and the team's preferred behavior. The individual characteristics of group members directly affect the coach's required behavior and the group members' preferred behavior. Furthermore, the coach's actual behavior is influenced by the coach's personal characteristics, requirements demanded by the situation, and the preferred behaviors of the team members. Consequently, the degree of congruence among the required, actual, and preferred behaviors directly affects the levels of performance and satisfaction of the group members. The embedded feedback loop projects how the actual coach behavior may be a consequence of performance and satisfaction of the group. In order to test the efficacy of this model in sport, Chelladurai and Saleh (1978, 1980) created the Leadership Scale for Sport (LSS). This questionnaire is utilized to measure athlete perceptions of coach behavior (required, actual, preferred) via five dimensions. The dimension of Training and Instruction measures coaches' teaching behavior; both Democratic Behavior and Autocratic Behavior measure coaches' decision-making styles; and the two dimensions of Social Support and Positive Feedback measure coaches' motivational tendencies.

The majority of research using the MML as a framework and the LSS as a tool addresses three distinct facets. The most widely studied facet is the predominantly descriptive work influence of member characteristics on preferred and perceived coach behaviors. The second, and more complex, approach is to determine the congruence between preferred and perceived coach behaviors in relation to athlete satisfaction. Finally, researchers attempt to relate coach behaviors to athlete performance.

## Member Characteristics and Coach Behavior

The most commonly perceived coaching behavior reported by all athletes is Training and Instruction (Amorose & Horn, 2000; Chelladurai, 1984; Wallin, 2003). However, personality differences impacting preferred coaching are also identified (Chelladurai & Carron, 1981, pp. 87–101; Serpa, Unpublished manuscript). Therefore, the most common approach researchers take when studying leadership via the MML is how member characteristics affect preferred and perceived coach behaviors. The member characteristics that have been subjected to scientific inquiry include gender, age, experience, and various psychological qualities (confidence and motivation).

### Gender

In terms of preferred coach behavior, females prefer coaches who demonstrate more Democratic Behavior. Conversely, males prefer coaches who use an Autocratic Behavioral style while providing Social Support (Amorose & Horn, 2000; Chelladurai & Saleh, 1980; Gardner, Shields, Bredemeier, &

Bostrom, 1996). When exploring perceived coach behavior, there are both commonalities and disparities by gender. While both males and females reported perceiving high levels of Training and Instruction, males perceived more Autocratic Behavior and females perceived more Democratic Behaviors (Gardner et al., 1996). Males and Females may hold differing perceptions of coach behavior due to a vast array of factors including previous sporting experiences.

## Age and Experience

Empirical evidence demonstrates that athlete's age and athletic experience influence preferences for coach behavior. Younger athletes preferred greater amounts of Social Support and Democratic Behavior (Serpa, Pataco, & Santos, 1991); older, more experienced, athletes preferred less Training and Instruction and more Social Support (Chelladurai & Carron, 1983), Positive Feedback (Erle, 1981), and Autocratic Behavior (Chelladurai & Carron, 1981, pp. 87–101). It appears that older and more experienced athletes prefer less instruction from the coach. Conversely, younger, less-experienced athletes prefer more instruction.

## Psychological Qualities

The psychological variables of motivation and confidence have been explored as predictors of behavioral preferences. Motivation serves as a moderator of athlete's preferences regarding coach behavior (Erle, 1981). Specifically, athletes high in intrinsic motivation preferred more Training and Instruction and Positive Feedback. Athletes reporting high levels of extrinsic motivation preferred more Social Support. Player perceptions of coach behaviors were explored, and the researchers discovered that athletes high in confidence perceived greater amounts of Democratic Behavior, Social Support, and Positive Feedback from their coaches (Garland & Barry, 1988). Athletes reporting low levels of sport confidence perceived coaches as displaying high rates of Autocratic Behavior.

# Coach Behavior and Satisfaction

As predicted by the MML, the greater the congruency between preferred and actual coach behavior the greater the levels of satisfaction (Chelladurai, 1978; Chelladurai & Carron, 1978). Supportive results reveal that low levels of discrepancy between coach behavior and athletes' preferences are strongly associated with satisfaction in leadership (Riemer & Chelladurai, 1995; Madeson, 2005; Weiss & Friedrichs, 1986). Furthermore, high congruency between preferred and actual Training and Instruction, Democratic Behavior, and Social Support increases satisfaction (Schliesman, 1987; Weiss & Friedrichs, 1986). These results offer evidence that the premise of the MML is reliable. When coach behaviors are consistent, satisfaction is enhanced.

## Coach Behavior and Performance

This third line of inquiry attempts to determine congruencies between specific coach behaviors and performance outcomes. The results of much of this research are inconsistent. One reason may be the diverse methods for measuring performance (win-loss percentage, point differential, and individual statistics). Furthermore, performance outcomes may be influenced by factors such as weather, opponent's great performance, and poor officiating. Thus, the relationship between coach behavior and performance posed in the MML has not been confirmed (Chelladurai, 1978; Garland & Barry, 1988; Wallin, 2003). In fact, one study reported that high levels of Social Support were related to low levels of team performance (Weiss & Friedrichs, 1986). Summers (1983) found that coaches who use Training and Instruction through the season actually decreased in team performance. Clearly, further exploration on the impact of coach behaviors (preferred, perceived) on performance is warranted.

## Summary of Leadership in Sport

There is concrete empirical evidence, which supports the utility of the MML in sport. There are recurring themes in the coach behavior literature. Training and Instruction is consistently identified as the primary perceived leader behavior. Furthermore, the effects of member characteristics (i.e., gender, age, experience) on preferred leader behavior are evident. In support of the model, congruency between preferred coach behavior and actual coach behavior is found to correlate with athletes' satisfaction. However, minimal evidence demonstrates the relationship between coach behaviors and performance. Further exploration into athlete perceptions of coaching has ensued and will be examined here.

## ATHLETE PERCEPTIONS OF SUCCESSFUL COACHING

The sport literature, to date, has failed to directly address the qualities that make for successful coaching. Becker (2009) sought to rectify this omission in her Model of Great Coaching. In short, Becker (2009) critiques the previous work based on how successful coaching behavior is defined. Much of the work directed at successful coaching behaviors defines this by win-loss record and/or the public attention received by coaches. However, this work has also been limited by the focus on high visibility sports (baseball, basketball, and football) at high levels of competition (collegiate and professional).

The idea of coaching greatness has not been directly addressed in the sport psychology literature. Much of the coaching research focuses on factors associated with effective coaching. There is general consensus that effective coaching is defined as follows:

> ...that which results in successful performance outcomes (measured in terms of either win-loss percentages or degree of self-perceived performance abilities) or

*positive psychological responses on the part of the athletes (e.g., high perceived ability, high esteem, an intrinsic motivational orientation, high level of sport enjoyment).*

Horn (2002, p. 309)

Furthermore, many studies examine coach effectiveness as it may relate to coaching behaviors (e.g., Bloom, Crumpton, & Anderson, 1999; Lacy & Goldston, 1990; Tharp & Gallimore, 1976). The majority of this research explores observable phenomena (coaching behaviors) in relation to measurable outcomes (athletes' self-perceptions and performance). Absent from the coaching literature was the study of coaching from an experiential perspective or athletes' experiences of great, or effective, coaching.

With this goal in mind, Becker (2009) queried 18 former elite-level athletes and identified six dimensions that characterized the lived experience of great coaching. These six dimensions include Coach Attributes, The Environment, Relationships, The System, Coaching Actions, and Influences. See Figure 2 for the representation of the model.

The most significant finding according to Becker (2009) is the interaction between the dimensions. Specifically, Coaching Actions and Influences take center stage, but the other dimensions continue to impact athletes' experiences. The ability to maintain a stable interaction between these dimensions elevates coaches to the status of greatness. Athletes described their coaches' Attributes, The Environment, The System, and Relationships as consistent throughout their experiences. Great coaches were consistent in every aspect of coaching; managing personnel, running their system, preparing for games, and communicating with their athletes. As a result, these athletes knew exactly what to expect from their coaches. The dimensions comprising the background of athletes' experiences only became figural when there was a lack of consistency of stability. This signified a breakdown in the coaching process as athletes became preoccupied with factors that detracted from learning and performance.

Ultimately, coaching greatness was not about what coaches do (gleaned from the research on observable coach behaviors) but rather how they do it.

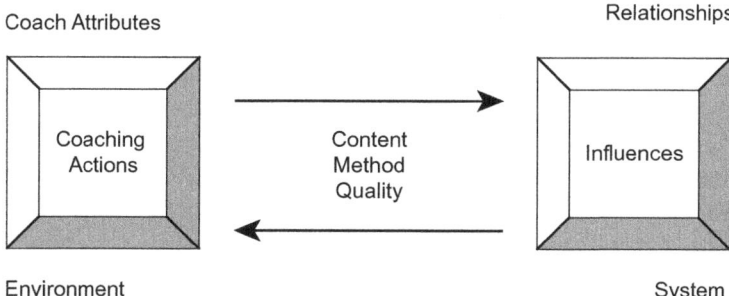

**FIGURE 2** Model of great coaching (Becker, 2009).

According to the Model of Great Coaching (Becker, 2009), it is the content, method, and quality of coach actions and interactions that set great coaches apart. Further investigation using this model as a framework will help elucidate the influence of coaching style on athlete development and performance. The impact of coaching behaviors on athletes is explored in numerous ways, including the lens of expectancy theory.

## EXPECTANCY EFFECTS IN COMPETITIVE SPORT

From the early roots in sociology, based on the work of Robert Merton (1948), expectancy theory has served as a lens to access the influence of leaders on their subject's behavior. Termed the self-fulfilling prophecy, Merton (1948) defines this process as "a situation whereby an expectation serves as the stimulus to behavior which causes the initial expectation to come true" (p. 193). As applied to the coach–athlete relationship, a four-step model was developed to explain and explore this process (Martinek, 1981; Horn et al., 2001; Solomon, 2001). In step one, coaches form expectations of their athlete's ability based on qualities, termed impression cues, housed in three major categories: personal (i.e., age, gender, and height), performance (i.e., past performance, speed, and agility), and psychological (i.e., confidence, anxiety, and motivation). During step 2, coaches instruct athletes (via quantity and quality of feedback) based on the impressions gleaned in step 1. In step 3, athletes respond to the coaches' instruction and gain information about their competence, which subsequently impacts performance. Finally in step 4, athletes respond based on the coaches' initial assessments, thus confirming to the coach that they are effective judges of athletic ability.

### Step 1—Coach Develops Expectations for Athlete Performance

Directing attention to step 1, Solomon explored the sources of information coaches rely on to evaluate athletic ability (Solomon, 2008a). A series of studies querying hundreds of coaches identified four distinct factors coaches employ in developing athlete expectations: Coachability, Team Player, Physical Ability, and Maturity (Solomon, 2008a; Solomon & Rhea, 2008). The Solomon Expectancy Sources Scale (SESS; Solomon, 2008a) was created to explore the information most essential for assisting coaches in developing athletes. The findings both reinforce the three impression cues identified in the expectancy cycle and increase the breadth of understanding the explicit psychological factors coaches adopt to evaluate athlete ability.

When comparing more and less successful coaches on their utilization of the four SESS sources, the results showed that regardless of win percentage, coaches prioritized similar qualities in their evaluations (Becker & Solomon, 2005). However, athletes playing for more successful coaches (win > 60%) were aware of the qualities coaches employed to evaluate them; athletes playing for less successful coaches (win < 50%) were not able to identify the criteria upon which they were being evaluated. This confirms evidence from the Model of Great Coaching that it is not what coaches do (in this case, the sources of evaluation used to evaluate athletes) but how they do it (which will be explored in step 2).

## Step 2—Expectations Influence Coaching Behaviors

Empirical inquiry into step 2 of the cycle is primarily addressed by comparing the quantity and quality of feedback coaches issue to their athletes. Case studies of highly regarded and successful coaches from various sports and competitive levels have indicated that these coaches utilize training and instruction more frequently than any other coaching behavior (Bloom et al., 1999; Lacy & Darst, 1985; Segrave & Ciancio, 1990; Tharp & Gallimore, 1976). Expectancy theorists hypothesize that high-expectancy athletes will be afforded better and better quality instruction than their low-expectancy counterparts (Solomon, 2010). These patterns of differential feedback are evidenced in high school (Solomon, 2008b; Solomon, DiMarco, Ohlson, & Reece, 1998; Solomon, Golden, Ciapponi, & Martin, 1998), college (Solomon, 2008b; Solomon & Kosmitzki, 1996; Solomon, Striegel, et al., 1996), and elite levels of competition (Sinclair & Vealey, 1989).

Factors such as coaching experience (Solomon, DiMarco, et al., 1998) and player ethnicity (Solomon, Wiegardt, et al., 1996) have not been found to influence quantity nor quality of feedback to high- and low-expectancy athletes. However, coach status (head versus assistant coach) had a significant impact on feedback patterns. Researchers found that though head coaches followed the hypothesized pattern of issuing differential feedback to high- and low-expectancy athletes, assistant coaches were more equitable in their feedback (Solomon, Striegel, et al., 1996).

There also appears to be a discrepancy between coaches' and athletes' perceptions of coach behaviors (Salminen, Liukkonen, & Telama, 1992; Serpa et al., 1991). Furthermore, athletes are more accurate in their recall of coach feedback (Solomon, Striegel, et al., 1996), whereas coaches are fairly inaccurate in their perceptions of their own feedback (De Marco, Mancini, & Wuest, 1997). As team size increases, coaches' accuracy decreases as they have more personnel under their tutelage (Hansen & Gould, 1988).

## Step 3—Perceptions of Coach Behavior Affects Athletes

Although a discrepancy exists between coaches' and athletes' impressions of coach treatment, it is important to explore how perceptions of coach behavior affect athletes. Both consciously and unconsciously, players become aware of differences in coach behavior to high- and low-expectancy athletes. Coach behavior, via feedback, provides information to the athletes regarding their level of perceived competence and satisfaction (Allen & Howe, 1998). Coach feedback becomes competency information informing athletes about their ability, effort, and future expectations (Amorose & Smith, 2003). Increased praise and instruction will lead to increased feelings of competence; increased criticism leads to decreased competence (Black & Weiss, 1992). Furthermore, positive coach behaviors (such as training and instruction, praise, and corrective information) increase player satisfaction (Allen & Howe, 1998; Dwyer & Fischer, 1990; Riemer & Chelladurai, 1995). Overall, the relevance of feedback depends on the skill being learned, the effects of different types of feedback, and individual characteristics (such as skill level) (Magill, 1994). Thus, it remains important for coaches to understand how their feedback influences athlete perceptions and, ultimately, performance.

### Step 4—Athlete Performance Conforms to Coach Expectations

With much of the expectancy literature focused on differential feedback patterns, few studies have explored the actual impact of coach expectations on athlete performance. In terms of the individual athlete, coaches have emphasized the importance of confidence in achieving performance success (Gould, Guinan, Greenleaf, & Chung, 2002). In addition, Solomon (2001) examined the relationship between coach perceptions and athlete performance. Results demonstrated that coach evaluation of athlete confidence was the only significant predictor of actual athletic ability. In a follow-up study, Solomon (2002) replicated the study and obtained the same results.

## SUMMARY AND CONCLUSION

There were two purposes to this chapter. The goal was to begin with an exploration of the development of coaching expertise. Then, an examination of the impact of coach behavior on athlete development and performance was conducted. The ultimate goal was to elucidate the influence of coaches, informed by their training, on athlete development and performance. The development of coaching expertise was explored through a perusal of the leadership literature and athletes' perceptions of great coaching. The lens of expectancy theory was evoked to explore the impact of coach behavior on athlete development and performance. Though numerous factors contribute to the development of competent athletes, ultimately coaching behaviors are a reflection of coach philosophies and leadership styles.

## REFERENCES

Allen, J. B., & Howe, B. (1998). Player ability, coach feedback, and female adolescent athletes' perceived competence and satisfaction. *Journal of Sport & Exercise Psychology, 20,* 280–299.

Amorose, A. J., & Horn, T. S. (2000). Intrinsic motivation: relationships with collegiate athletes' gender, scholarship status, and perceptions of their coaches' behavior. *Journal of Sport & Exercise Psychology, 22,* 63–84.

Amorose, A. J., & Smith, P. J. K. (2003). Feedback as a source of physical competence information: effects of age, experience, and type of feedback. *Journal of Sport & Exercise Psychology, 25,* 341–359.

Barrow, J. C. (1977). The variables of leadership: a review and conceptual framework. *Academy of Management Review, 2,* 231–251.

Becker, A. J. (2009). It's not what they do, it's how they do it: athlete experiences of great coaching. *International Journal of Sports Science & Coaching, 4,* 93–119.

Becker, A. J., & Solomon, G. B. (2005). Expectancy information and coach effectiveness in intercollegiate basketball. *The Sport Psychologist, 19,* 251–266.

Black, J. S., & Weiss, M. R. (1992). The relationship among perceived coaching behaviors, perceptions of ability, and motivation in competitive age-group swimmers. *Journal of Sport & Exercise Psychology, 14,* 309–325.

Bloom, G. A., Crumpton, R., & Anderson, J. E. (1999). A systematic observation of the teaching behaviors of an expert basketball coach. *The Sport Psychologist, 19,* 251–266.

Chelladurai, P. (1978). *A contingency model of leadership in athletics* (Unpublished doctoral dissertation). Ontario, Canada: University of Waterloo.

Chelladurai, P. (1984). Discrepancies between preferences and perceptions of leader behavior and satisfaction of athletes in varying sports. *Journal of Sport Psychology, 6,* 27–41.

Chelladurai, P., & Carron, A. V. (1978). *Leadership [Monograph].* Ottawa, Canada: Canadian Association for Health, Physical Education, and Recreation.

Chelladurai, P., & Carron, A. V. (1981). *Task characteristics and individual differences and their relationship to preferred leadership in sports. Psychology of motor behavior and sport—1982.* College Park, MD: North American Society for the Psychology of Sport and Physical Activity.

Chelladurai, P., & Carron, A. V. (1983). Athletic maturity and preferred leadership. *Journal of Sport Psychology, 5,* 371–380.

Chelladurai, P., & Saleh, S. D. (1978). Preferred leadership in sport. *Canadian Journal of Applied Sport Sciences, 3,* 85–92.

Chelladurai, P., & Saleh, S. D. (1980). Dimension of leader behavior in sports: development of a leadership scale. *Journal of Sport Psychology, 2,* 34–45.

Cox, R. H. (1998). *Sport psychology: Concepts and applications.* Champaign, IL: Human Kinetics.

De Marco, G. M. P., Mancini, V. H., & Wuest, D. A. (1997). Reflections on change: a qualitative and quantitative analysis of a baseball coach's behavior. *Journal of Sport Behavior, 20,* 135–163.

Dwyer, J. M., & Fischer, D. G. (1990). Wrestlers' perceptions of coaches' leadership as predictors of satisfaction with leadership. *Perceptual and Motor Skills, 71,* 511–517.

Erle, F. J. (1981). *Leadership in competitive and recreational sport* (Unpublished master's thesis). London, Canada: University of Western Ontario.

Evans, M. G. (1970). Leadership and motivation: a core concept. *Academy of Management Journal, 13,* 91–102.

Fiedler, F. E. (1967). *A theory of leadership effectiveness.* New York: McGraw-Hill.

Gardner, D. E., Shields, D. L. L., Bredemeier, B. J. L., & Bostrom, A. (1996). The relationship between perceived coaching behaviors and team cohesion among baseball and softball players. *The Sport Psychologist, 10,* 367–381.

Garland, D. J., & Barry, J. R. (1988). The effects of personality and perceived leader behavior on performance in collegiate football. *The Psychological Record, 38,* 237–247.

Gould, D., Guinan, D., Greenleaf, C., & Chung, Y. (2002). A survey of US Olympic coaches: variables perceived to have influenced athlete performances and coach effectiveness. *The Sport Psychologist, 16,* 229–250.

Guetzkow, H. (1951). *Groups, leadership, and men: Research in human relations*. Pittsburg, PA: Carnegie Press.

Hansen, T. W., & Gould, D. (1988). Factors affecting the ability of coaches to estimate their athletes' trait and state anxiety levels. *The Sport Psychologist, 2*, 298–313.

Hemphill, J. K., & Coons, A. E. (1957). A factorial study of the Leader Behavior Description Questionnaire. In R. M. Stogdill, & A. E. Coons (Eds.), *Leader behavior: Its description and measurement* (pp. 6–37). Columbus, OH: The Ohio State University Press.

Horn, T. S. (2002). *Advances in sport psychology*. Champaign, IL: Human Kinetics.

Horn, T. S., Lox, C., & Labrador, F. (2001). The self-fulfilling prophecy theory: when coaches' expectations become reality. In J. Williams (Ed.), *Applied sport psychology: Personal growth to peak performance* (pp. 68–81). Mountain View, CA: Mayfield.

Kahn, R. L., & Katz, D. (1951). Human organization and worker motivation. In L. R. Tripp (Ed.), *Industrial productivity* (pp. 146–171). Madison, WI: Industrial Relations Research Association.

Lacy, A. C., & Darst, P. W. (1985). Evolution of a systematic observation system: the ASUOI observation instrument. *Journal of Teaching in Physical Education, 3*, 59–66.

Lacy, A. C., & Goldston, P. O. (1990). Behavioral analysis of male and female coaches in high school girls' basketball. *Journal of Sport Behavior, 13*, 29–39.

Madeson, M. N. (2005). *The influence of coach-athlete compatibility on satisfaction and performance* (Unpublished master's thesis). Sacramento, CA: California State University.

Magill, R. A. (1994). The influence of augmented feedback on skill learning depends on characteristics of the skill and the learner. *Quest, 46*, 314–327.

Martinek, T. J. (1981). Pygmalion in the gym: a model for the communication of teacher expectations in physical education. *Research Quarterly for Exercise and Sport, 52*, 58–67.

Merton, R. K. (1948). The self-fulfilling prophecy. *The Antioch Review, 8*, 193–210.

Osborn, R. N., & Hunt, J. G. (1975). An adaptive-reactive theory of leadership: the role of macro variables in leadership research. In J. G. Hunt, & L. L. Larson (Eds.), *Leadership frontiers* (pp. 27–44). Kent, OH: Kent State University.

Riemer, H. A., & Chelladurai, P. (1995). Leadership and satisfaction in athletics. *Journal of Sport & Exercise Psychology, 17*, 276–293.

Salminen, S., Liukkonen, J., & Telama, R. (1992). The differences between coaches' and athletes' perception of leader behavior of Finnish coaches. In T. Williams, L. Almond, & A. Sparkes (Eds.), *Sport and physical activity: Moving towards excellence* (pp. 517–522). London: E. & F.N. Spoon.

Schliesman, E. S. (1987). Relationship between the congruence of preferred and actual leader behavior and subordinate satisfaction with leadership. *Journal of Sport Behavior, 10*, 157–166.

Segrave, J. O., & Ciancio, C. A. (1990). An observational study of a successful Pop Warner football coach. *Journal of Teaching in Physical Education, 9*, 294–306.

Serpa, S. *Research work on sport leadership in Portugal*. Lisbon, Portugal: Lisbon Technical University, Unpublished manuscript.

Serpa, S., Pataco, V., & Santos, F. (1991). Leadership in handball international competition. *International Journal of Sport Psychology, 22*, 78–89.

Sinclair, D. A., & Vealey, R. S. (1989). Effects of coaches' expectations and feedback on the self-perceptions of athletes. *Journal of Sport Behavior, 12*, 77–91.

Solomon, G. B. (2001). Performance and personality impression cues as predictors of athletic performance: an extension of expectancy theory. *International Journal of Sport Psychology, 32*, 88–100.

Solomon, G. B. (2002). Confidence as a source of expectancy information: a follow-up investigation. *International Sports Journal, 6*, 119–127.

Solomon, G. B. (2008a). The assessment of athletic ability in intercollegiate sport: instrument construction and validation. *International Journal of Sports Science and Coaching, 2*, 161–179.

Solomon, G. B. (2008b). Expectations and perceptions as predictors of coaches' feedback in three competitive contexts. *Journal for the Study of Sports and Athletics in Education, 2*, 161–179.

Solomon, G. B. (2010). The influence of coach expectations on athlete development. *Journal of Sport Psychology in Action, 1*, 76–85.

Solomon, G. B., & Buscombe, R. M. (2013). Expectancy effects in sports coaching. In P. Potrac, J. Denison, & W. Gilbert (Eds.), *Handbook of sports coaching* (pp. 247–258). London: Routledge.

Solomon, G. B., DiMarco, A. M., Ohlson, C. J., & Reece, S. D. (1998). Expectations and coaching experience: is more better? *Journal of Sport Behavior, 21*, 444–455.

Solomon, G. B., Golden, A. J., Ciapponi, T. M., & Martin, A. D. (1998). Coach expectations and differential feedback: perceptual flexibility revisited. *Journal of Sport Behavior, 19*, 163–177.

Solomon, G. B., & Kosmitzki, C. (1996). Perceptual flexibility among intercollegiate basketball coaches. *Journal of Sport Behavior, 19*, 163–177.

Solomon, G. B., & Rhea, D. J. (2008). Sources of expectancy information among college coaches: a qualitative test of expectancy theory. *International Journal of Sports Science and Coaching, 3*, 251–268.

Solomon, G. B., Striegel, D. A., Eliot, J. F., Heon, S. N., Maas, J. L., & Wayda, V. K. (1996). The self-fulfilling prophecy in college basketball: implications for effective coaching. *Journal of Applied Sport Psychology, 8*, 44–59.

Solomon, G. B., Wiegardt, P. A., Yusuf, F. R., Kosmitzki, C., Williams, J., Stevens, C. E., et al. (1996). Expectations and ethnicity: the self-fulfilling prophecy in college basketball. *Journal of Sport & Exercise Psychology, 18*, 83–88.

Summers, R. J. (1983). *A study of leadership in a sport setting* (Unpublished master's thesis). Ontario, Canada: University of Waterloo.

Tharp, R. G., & Gallimore, R. (1976). What a coach can teach a teacher. *Psychology Today, 9*, 75–78.

Wallin, J. (2003). *The influence of team cohesion and coach behavior on individual performance* (Unpublished master's thesis). Sacramento, California: California State University.

Weiss, M. R., & Friedrichs, W. D. (1986). The influence of leader behaviors, coach attributes, and institutional variables on performance and satisfaction of collegiate basketball teams. *Journal of Sport Psychology, 8*, 332–346.

Yukl, G. A. (1971). Toward a behavioral theory of leadership. *Organizational Behavior ad Human Performance, 6*, 414–440.

Chapter 7

# Music Performance: Expectations, Failures, and Prevention

Eckart Altenmüller, Christos I. Ioannou
*Institute of Music Physiology and Musicians' Medicine, Hanover University of Music, Drama and Media, Hanover, Germany*

---

*Imagine a concert pianist sitting in the green room and getting ready to go on stage. More than a thousand people are eager to listen to his playing. He is well prepared, he has practiced the pieces a thousand times, and yet, he still feels insecure about the performance. His thoughts are focused on his anxieties; he becomes aware how his forearm muscles become stiff and how his heart is beating. He wants to reassure himself by playing a difficult passage on the small upright piano in the green room, but he fails in finding the right fingerings and loses track, he repeats the same passage hectically over and over, always hitting the same wrong notes—he is panicking. Will he find means to reduce overactivity and to perform positive self-talk? Will he be prepared to guide his focus of attention to the beautiful sounds he is producing? Will he regain control over the highly refined motor programs?*

There can be no doubt that performing music at a professional level is one of the most demanding and fascinating human experiences. It involves the precise execution of very fast and, in many instances, extremely complex physical movements that must be structured and coordinated with continuous auditory, visual, and somatosensory feedback. Furthermore, it requires retrieval of musical, motor, and multisensory information from long-term memory and relies on continuous planning of an ongoing performance in the working memory system. The consequences of motor actions have to be anticipated and monitored, with adjustments made accordingly (Brown, Penhune, & Zatorre, 2015). At the same time, music should be "expressive," that is, the performance has to be enriched with a complex set of innate and culture-dependent emotional gestures.

Extensive practice is required to develop the necessary auditory, sensorimotor, and social skills for performing at a professional level. Ericsson, Krampe, and Tesch-Romer (1993) proposed the concept of "deliberate practice" as a means of studying goal-oriented, structured, and effortful facets of practice. In this framework, the amount and the quality of practice undertaken are determined by motivation, resources, and focus of attention. The authors argued that a major distinction between professional and amateur musicians is the number of years of deliberate practice required to develop a high level of instrumental proficiency (Ericsson & Lehmann, 1996). Consequently, musicians' skills are usually acquired by prolonged practice, involving numerous repetitions under intrinsic and extrinsic (augmented) feedback. As holds for all complex skills, these are not represented in isolated brain areas, but rather depend on the multiple connections and interactions established during training within and between different regions of the brain. The general ability of our central nervous system (CNS) to adapt to changing environmental conditions and newly imposed tasks is referred to as plasticity. In music, planning and learning through experience and training are accompanied by development and changes, which take place not only in the brain's neuronal networks as a result of strengthening of connections but also in its overall gross structure (Münte, Altenmüller, & Jäncke, 2002). It is still not completely understood how practice habits and sensorimotor maturation interact and how genetic predisposition and environmental influence shape musicians' skills (Mosing, Madison, Pedersen, Kuja-Halkola, & Ullén, 2014). In recent years, it has become clear that not only the amount of practice but also the period of life during which intense practice begins are important factors. For instance, early practice (before age seven) leads to optimized and more stable motor programs (Furuya & Altenmüller, 2013) and specific auditory skills, such as the development of "absolute" pitch (naming pitches without an external reference), as opposed to practice that starts after puberty. It is, therefore, probable that sensitive periods exist during development and maturation of the CNS for the acquisition of auditory and sensorimotor skills (Steele, Bailey, Zatorre, & Penhune, 2013).

## COMMUNALITIES AND DIFFERENCES BETWEEN SPORTS AND MUSIC PERFORMANCE

Clearly, music and sports performance share many features: both activities need prolonged training, discipline, physical strength and endurance, social skills, strategic decisions, and emotional expressivity; although in sports, this last element is frequently subsumed under "elegance" and "charisma." There are, however, important differences between music performance and most sport disciplines, which render the former particularly challenging and fragile in terms of motor control:

1. Musical training usually starts very early, sometimes before age six when the adaptability of the CNS is highest. This feature is not unique to music, as some sport activities, for example, skiing, benefit from an early start in a similar way.
2. Making music is linked to sound production. As in speech, the auditory system provides very precise feedback of the movement effects, with a temporal

resolution superior to kinaesthetic and visual feedback. Furthermore, in the frame of classical music, which is notated and available as sheet music, the target parameters are precisely predefined: namely, temporal accuracy (correct tempo, rhythm, swing, beat, pulse, etc.) and spatial accuracy (correct key or finger position on the fingerboard, sound quality).
3. Most musicians work at the upper limit of their sensorimotor capabilities and strive to push their limits in order to be faster, louder, and more expressive. Given the complexity of music, the demands of composers, especially in the last 100 years, and the role of outstanding peers as models, such as the Chinese pianist Lang Lang, or the "speed record breaking" violinist David Garrett, musicians are aware that the only limit of movement accuracy and speed is the temporal and spatial resolution of the auditory system. As musicians say, "There is always a colleague, who plays this piece faster, louder, and more expressively." In part, this is also true for sports, but in music, fine motor skills predominate. Therefore, musicians are frequently colloquially referred to as "small muscle athletes."
4. Music performance takes place in an extremely competitive environment under high social pressures and a strong reward–punishment mechanism. The ubiquitous availability of music recorded by outstanding peers, both online and in more traditional formats, has increased the demands of the audience with respect to the expected quality of performances. This process, in turn, augments anxiety, tension, and competition among musicians, making their lives more stressful. Even outstanding soloists frequently have to cope with severe performance anxiety (Wilson, 1997).
5. Nevertheless, making music is frequently linked to highly positive emotions, to feelings of joy, satisfaction, and to strong physiological reactions, known as "chill responses" (Altenmüller, Kopiez, & Grewe, 2013). These qualities are known to enhance plastic adaptations of the brain and can even lead to a sort of addictive behavior, causing younger musicians to over-practice and ignore warning signals from the body, such as fatigue and musculoskeletal pain.

## PERFORMANCE FAILURES IN MUSICIANS

In the following, we will discuss the different forms of failures in musicians, their phenomenology, and the underlying pathophysiological mechanisms.

### Music Performance Anxiety

Music performance anxiety (MPA) describes a particular state of arousal, which occurs when musicians present themselves before an audience in performance situations (Spahn, 2015). MPA is a common symptom and regularly reported by performers. In a survey of 2536 orchestra musicians in Germany, about 90% indicated that they suffer from MPA (Gembris & Heye, 2012). Estimates of severe and debilitating MPA in orchestral musicians range from 15 to 25% (Fishbein,

Middlestadt, Ottati, Strauss, & Ellis, 1988; James, 1997; van Kemenade, van Son, & van Heesch, 1995). Approximately one-third of performers with severe MPA have other comorbid disorders such as a generalized anxiety disorder or social anxiety (Sanderson, DiNardo, Rapee, & Barlow, 1990).

MPA extends along a continuum of severity, whereby performance may be better, worse, or even impossible, and the affected musician may suffer in various ways. If MPA is debilitating in professional contexts, it is considered to be pathological and requires therapy (Spahn, 2012). MPA, like any emotional state, is expressed on the physical level as well as through motor and cognitive behaviors. Physical symptoms mainly arise from sympathetic activation of the autonomic nervous system (ANS) and manifest as accelerated heartbeat, increased blood pressure, tremors, rapid and shallow breathing, dry mouth, bladder dysfunction, etc. Symptoms may also include narrowing of thought, heightened alertness, and concentration on fear of failure, as well as behavior like rituals, general unrest, and avoidance (Endo, Juhlberg, Bradbury, & Wing, 2014; Studer, Danuser, Wild, Hildebrandt, & Gomez, 2014). Interestingly, the emotional component of MPA does not necessarily co-vary with the autonomic symptoms, that is, performers with highly elevated levels of blood pressure and heart rate can report low anxiety, whereas performers with normal physical parameters can experience intense anxiety (Spahn, Echternach, Zander, Voltmer, & Richter, 2010; Steptoe & Fidler, 1987). With respect to the underlying mechanisms, behavioral models of MPA emphasize general learning theories, for example, of classical and operant conditioning as relating to musical performance in public. Why persons under comparable learning conditions develop different forms of MPA may then be explained by a genetic contribution, for example, linked to personality factors such as trait anxiety (Kenny, 2011, p. 120).

Cognitive theories focus on the importance of cognitions on behavior and physical symptoms. Here, Beck and Clark (1988) identified attention binding and catastrophizing as primary contributors to MPA. Physiological stress models can best explain the aforementioned physical symptoms of MPA. MPA accordingly represents an evolutionarily ancient program developed for situations of particular threat. These physical reactions are mediated by activation of the sympathetic ANS. This series of physiological reactions may enable a state of increased performance for fight or flight within a very short time as a response to the emotional signal of fear. The evolutionarily programmed reactions toward a threat to life are today largely replaced by the fear of negative social consequences, such as exposure, failure, losing face, and disappointment of important peers.

## Choking under Pressure

With respect to fine motor control and movement planning, high-pressure situations can also lead to motor failures during performances. In sports psychology this condition is frequently termed choking under pressure (CuP, Hill, Hanton,

Matthews, & Fleming, 2010). It is a condition characterized by acute motor disturbances, an increased level of anxiety, and a decrease in self-confidence, and thus, it may be conceptualized as a motor symptom of MPA (Hill, Hanton, Fleming, & Matthews, 2009). CuP is a well-known condition in performing musicians, and it may afflict 5–60% of performing musicians according to epidemiological studies (Brugués, 2011a; Steptoe & Fidler, 1987). Here, it is characterized by loss of agility, heightened muscular stiffness accompanied by increased co-contraction of antagonist muscles (Yoshie, Kudo, Murakoshi, & Ohtsuki, 2009), leading to a reduction of temporo-spatial precision of movements and sound quality. Furthermore, especially in brass players, it frequently manifests as "tongue-stopper," an inability to release the tongue and allow air passage prior to tone attack.

Several accounts have been given for the choke mechanism. They can be mainly divided into drive and attentional theories (Hill et al., 2010). Drive theories are based on the assumption that increased arousal, resulting from striving to excel under pressure, will have a detrimental effect on performance (Spence & Spence, 1966). They are based on the classical inverted U model of performance anxiety. Increased arousal will first lead to improved performance, whereas further increase could deteriorate performance quality (Yerkes & Dodson, 1908). However, drive theories cannot explain why some musicians thrive but others fail under pressure.

Attentional theories try to account for individual differences. They can be separated into distraction (Carver & Scheier, 1981) and self-focus (Baumeister, 1984) theories. Distraction theories assume that pressure-induced anxiety will occupy the performer's working memory, restricting the processing of task-relevant information (Eysenck & Calvo, 1992). Self-focus theories propose that performance anxiety causes the performer to shift the focus of attention inward or to consciously monitor the skill, which detrimentally affects the well-learned automated skill (Baumeister, 1984). Crucial for the occurrence of choking is the perceived control of a situation, which is influenced by both situational and personality factors (Otten, 2009). Personality factors that increase the likelihood of choking include low self-confidence (Baumeister & Showers, 1986) and high trait anxiety (Wilson, 2008).

## Dynamic Stereotype

When the lack of motor control persists for longer than four weeks and performance failures occur more regularly, a more serious alteration of the sensorimotor networks may be suspected, leading to a deterioration of motor programs in the CNS. We have called this condition dynamic stereotype (DS), a term borrowed from the eminent Russian physiologist Ivan Pavlov (Pavlov, 1951; for a critical review, see Windholz, 1996). Originally, this condition was understood as a reflection of fatal compensation strategies, which became automated. In the words of the exponents of Russian behaviourism, DS is defined as a type of integral activity performed by the cerebrum of higher animals and man, and manifested by a fixed, or stereotyped, succession of conditioned reflexes. DS is influenced by external factors that are repeated in a certain order. Accordingly, DS is the most vivid manifestation of the extremely subtle analysis and synthesis activity of the cerebral cortex and is the product of very complex interactions between cortical areas. It can be at least partly conceptualized as a consequence of long-term CuP when these dysfunctional movements are stored in procedural memory traces, maybe as a consequence of conditioned reactions to previous choking experiences and procedural memory formation under stress (Klämpfl, Lobinger, & Raab, 2013; Lobinger, Klämpfl, & Altenmüller, 2014).

The phenomenology of DS resembles in many aspects of focal, task-specific dystonia (see below). However, in contrast to the latter, it seems to be more modifiable and more fluctuating, especially during stressful performances (Ioannou, Furuya, & Altenmüller, submitted for publication). Sometimes, "islands" of well-being and complete motor control occur, although only for hours or a few days. It responds occasionally to sensory trick maneuvers, such as alterations of tactile input from the body parts affected by dystonia or of auditory input, for example, by delay of the produced sound. We have demonstrated that improvement of motor control when playing with a latex glove is related to better outcome of retraining and behavioral therapies (Paulig, Jabusch, Großbach, Boullet, & Altenmüller, 2014). It should be mentioned, however, that responses to sensory tricks and objective improvement are rare and highly variable (Cheng, Großbach, & Altenmüller, 2013). We, therefore, prefer to consider this phenomenon a "soft-sign" when classifying motor problems in musicians.

## Musician's Dystonia

The most severe movement failure is task-specific musician's dystonia (MD) (Altenmüller, 2003), also known as musician's cramp. Commonly, two types of this movement disorder are distinguished: focal hand dystonia (FHD) and embouchure dystonia (ED). MD is characterized by persistent muscular incoordination or loss of voluntary motor control during highly trained, task-specific movements, such as playing a musical instrument (Altenmüller, 2003; Jankovic & Ashoori, 2008). In most cases, pain does not accompany the disorder; occasionally, some muscular strain can occur when patients attempt to compensate for the dystonic movement by overactivating the antagonist muscles; however, lack of pain distinguishes it from overuse injury. It is important to make this distinction while bearing in mind that prolonged pain syndromes may lead to symptomatic dystonia. MD frequently terminates professional careers and is highly disabling amongst musicians (Altenmüller, 2003; Altenmüller & Jabusch, 2010; Brandfonbrener & Robson, 2004; Lederman, 1991).

Various symptoms can mark the beginning of the disorder, for example, subtle loss of control in fast passages, irregularity of trills, and general lack of temporo-spatial precision of the fingers. Its phenomenology is characterized by

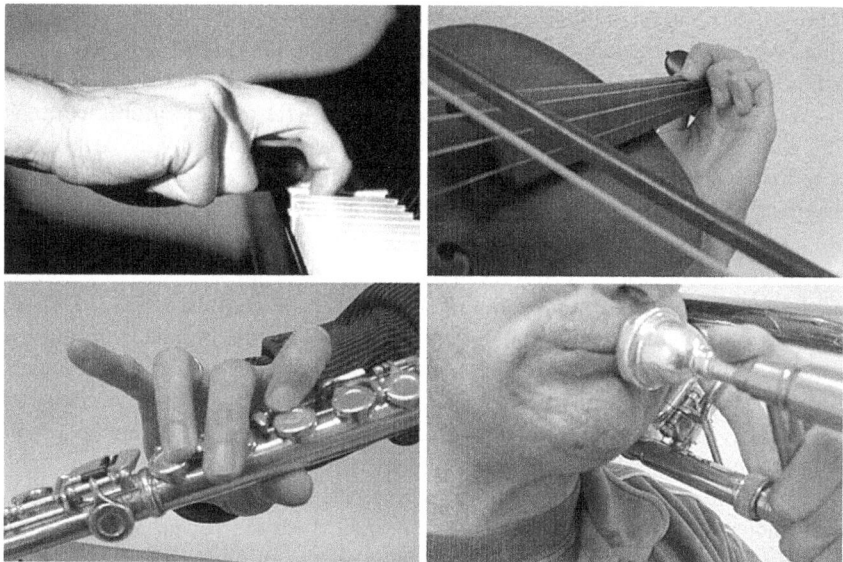

**FIGURE 1** Typical patterns of dystonic posture in a pianist, a violinist, a flautist, and a trombone player.

an involuntary finger flexion or extension, loss of control of the embouchure in certain registers for wind players, or even flexion of the bowing thumb and uncontrolled bowing tremors for string players (Figure 1). At this stage, most musicians believe that the reduced precision of their movements is due to a technical problem or lack of practice. As a consequence, they intensify their efforts, but this reaction often exacerbates the problem. The loss of muscular coordination is frequently accompanied by a co-contraction of antagonist muscle groups.

According to recent estimates, 1% of all professional musicians are affected by MD (Altenmüller & Jabusch, 2010). In comparison to other activities and their corresponding dystonic movements, such as writing (writer's cramp), playing golf (the yips), or playing darts (dartism), playing classical music has the highest risk of producing focal dystonia (Frucht, 2009). Demographic data demonstrate a preponderance of male musicians with a male/female ratio of about 4:1 (Lim & Altenmüller, 2003). The majority of patients have solo positions and often have an anxiety-prone, perfectionist, control-type personality (Ioannou & Altenmüller, 2014). Hereditary factors play a role in the etiology of MD, as a positive family history of dystonia exists in up to 36% of affected musicians (Schmidt et al., 2009). According to epidemiological data, the probability of developing MD depends on the instrument played: guitar players, pianists, and brass instrument players have the highest risk of developing dystonia (Altenmüller & Jabusch, 2010). Repetitive use, controllability of motor actions, temporo-spatial demands, and extra-instrumental fine-motor burdens, such as writing, are triggering factors (Altenmüller, Baur, Hofmann, Lim, &

Jabusch, 2012; Baur, Jabusch, & Altenmüller, 2011). Furthermore, those musicians who start practicing after age 10 are at a much higher risk of developing MD (Schmidt, Jabusch, Altenmüller, Kasten, & Klein, 2013).

While there is probably an overlap between MD and the aforementioned DS, each of these disorders is distinct from the other. Generally, focal dystonia is more severe, and the dysfunctional movements are more obvious and more resistant to any attempt to correct them voluntarily. Furthermore, movements are less responsive to the sensory tricks described for DS. Sudden spontaneous improvements are rare exceptions, and psychological stressors do not influence the loss of motor control to a major degree. Finally, in contrast to DS, MD has a tendency to spread from the specific movement when playing the instrument to general, daily-life movements. For example, pianists with MD may first experience problems only in the ring and little fingers when playing scales, as depicted in Figure 1, but the disorder may extend to typing on a computer keyboard and buttoning up a shirt, finally leading to permanent cramping of the hand.

The etiology of FHD remains incompletely understood, but it is probably multifactorial. In brief, most studies of focal dystonia reveal abnormalities in three main areas: (1) reduced inhibition in the motor system at cortical, subcortical, and spinal levels; (2) deficits in sensory perception and integration; and (3) impaired sensorimotor integration. The latter changes are mainly believed to originate from dysfunctional brain plasticity. There is growing evidence for an abnormal cortical processing of sensory information as well as degraded representation of motor functions in patients with focal dystonia. Furthermore, additional factors such as gender and genetic predisposition appear to play an important role in the development of focal dystonia (Schmidt et al., 2009).

## IMPROVING PERFORMANCE IN MUSICIANS

In the following section, we will briefly discuss means of improving performance in musicians and treatment options to overcome different types of performance failures. Disabling MPA can be treated with a psychodynamic approach to make both the conscious and unconscious meaning and origin of the symptoms accessible to the artist; however, cognitive behavioral therapy (CBT) is more frequently applied and has been positively evaluated. Because CBT assumes that emotions and behaviors are influenced by beliefs or ideas about oneself and others, the therapeutic approach is designed to change thinking patterns through cognitive restructuring and shifting the attentional focus. In recent years, a multimodal treatment model of MPA has become popular (Spahn, 2011). It follows a personal and problem-oriented approach, in which various psychotherapeutic methods are applied, based on the individual needs of each patient. Furthermore, it combines the psychodynamic understanding of the person with respect to their individual history with elements of CBT and exercises related to body-oriented methods and mental techniques. According to a large interventional study at the Freiburg Music University, this approach is effective in the reduction of MPA (Spahn, 2011).

Other treatment approaches include the prescription of beta-adrenergic receptor blockers and tranquilizers. Whereas the former may reduce the disabling symptoms arising from hyperarousal and ANS activation in severe cases of MPA, tranquilizers such as benzodiazepines are mostly not recommended due to their potential addictive risk (Spahn, 2015). Alternative treatment strategies from the field of meditation, such as yoga (Khalsa, Butzer, Shorter, Reinhardt, & Cope, 2013), or interventions using guided imagery music (Martin, 2007) have been applied. The results provide evidence for a reduction of MPA, but overall beneficial effects are relatively small (Kenny, 2011, p. 195). Promising therapies come from performance research. In sports psychology, the model of individual zones of optimal functioning (IZOF) was developed by Harmison (2006). To achieve a high performance output, an individual profile is created, as performance anxiety at high output is pronounced and accessed in accordance with the competitive situation. In sports, the relationship between goal setting and performance output has also been investigated. The best performance output was achieved when the goal was specific, challenging, realistic, acceptable, and measurable (SCARP, Mahoney, 1992).

Extreme levels of MPA or CuP in musicians are likely to be best prevented by mental training and specific cognitive strategies, which are already being applied in sports psychology (e.g., hemispheric priming, Beckmann, Gröpel, & Ehrlenspiel, 2013). Efficient treatment probably again includes CBT, mindfulness training, and various breathing and relaxation techniques (for a review, see Brugués, 2011b). In these cases, medications such as beta-adrenergic receptor blockers (such as propanolol) and benzodiazepines are potentially helpful. Both drugs prevent dysfunctional motor memory formation (Soeter & Kindt, 2013).

As far as the treatment of DS is concerned, we predict that psychological techniques, such as the prevention of dysfunctional reinvestment and cognitive interference, will be helpful. The latter could be influenced by guiding attention from an internal (body-related) to an external (sound-related) focus. These techniques have been shown to be efficient in complex motor studies with normal and patient groups (see Wulf, 2007, for an overview). Furthermore, psychotherapeutic techniques reducing anxiety and perfectionism will probably improve the condition. One candidate intervention is learning-based sensorimotor training based on redefining spatial and temporal processing capacities in the sensory and motor cortices to restore task-specific skills (Byl, Negajaran, & McKenzie, 2003). Prolonged pedagogical retraining has been successfully applied to pianists suffering from loss of motor control (van Vugt, Boullet, Jabusch, & Altenmüller, 2014). Useful medications include selective serotonin-reuptake inhibitors (e.g., escitalopram) to overcome reinvestment, over focusing, and depression; anticholinergic drugs to reduce dysfunctional motor memories; and finally, local injections of botulinum toxin into the cramping muscles as a symptomatic treatment (Schuele, Jabusch, Lederman, & Altenmüller, 2005). These medications are also applied in MD.

Finally, treatment for MD is mostly symptomatic and depends on the type of dystonia. Psychological interventions seem to be less effective for the treatment of MD than for DS. Various oral medications have been used, and the anticholinergic drug trihexyphenidyl has proven to be the most effective agent (Jabusch, Zschucke, Schmidt, Schuele, & Altenmüller, 2005). Chemical denervation using botulinum toxin has found considerable success against many forms of MD. Botulinum toxin blocks the transmission of nerve impulses to the muscle and weakens the overactive muscles involved. In our series, injections of botulinum toxin were applied to 71 musicians suffering from hand dystonia. Fifty-seven percent of patients reported long-term improvement (Schuele et al., 2005). A promising new therapy is bihemispheric transcranial direct current stimulation during the execution of functional movements at the instrument. In a double blind randomized prospective trial, we demonstrated that stimulation with inhibition of the dysfunctional network of the "dystonic" motor cortex and the activation of the "healthy" network of the contralateral motor cortex improved performance significantly (Furuya, Nitsche, Paulus, & Altenmüller, 2014).

## A HEURISTIC MODEL OF PERFORMANCE FAILURES IN MUSICIANS

Summarizing the previous sections, we can classify the four types of failures in performance according to psychological symptoms, persistence of motor failures, accompanying symptoms, the underlying neurophysiological mechanisms, and the response to treatment. In Figure 2, we depict a heuristic model, which goes from a continuous worsening of motor control from temporary failure due to MPA to increasingly unstable motor control and, finally, fully developed focal dystonia. We have added triggering factors, identified in our previous epidemiological studies (Altenmüller, 2003; Altenmüller & Jabusch, 2010; Altenmüller et al., 2013).

At one end of the scale, MPA may produce a temporary degradation of motor skills, or in highly skilled experts, additional/alternative recruitment of muscles contributing to co-activation and dysfunctional movements. This mechanism has been convincingly demonstrated in skilled table tennis players (Aune, Ingvaldsen, & Ettema, 2008) and most likely also applies to instrumental musicians. For example, in skilled piano players, MPA produces co-activation of forearm flexor muscles (Yoshie et al., 2009). These changes are putatively accompanied by central nervous adaptations due to short-term plasticity, and they result in short-term alteration of the topography of motor and premotor cortex activations (Dirnberger, Duregger, Trettler, Lindinger, & Lang, 2004).

Under conditions with heightened anxiety and other stressors, such as high professional workload, these dysfunctional motor patterns may stabilize in procedural memory. Here, psychological stress might bring about the cascade of emotionally induced memory consolidation, which has been previously described for different forms of memory and mainly relies on noradrenergic activation of the

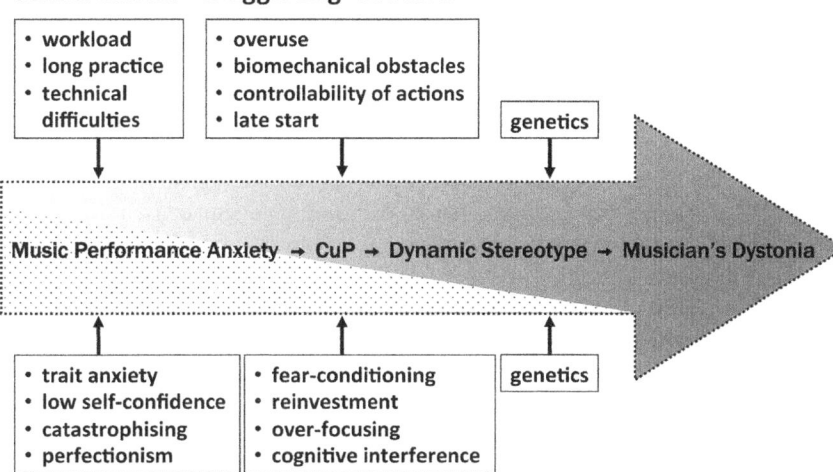

FIGURE 2 Heuristic model that assumes a stepwise progression of performance failures from music performance anxiety (MPA) to choking under pressure (CuP), dysfunctional motor memories in dynamic stereotype, and finally to musician's dystonia with structural memory consolidation and maladaptive brain adaptations. Potential triggering factors provoking the transition from one step to the next are categorized according to motor demands (above the arrow) and psychological contributors (below the arrow). The increasing gray shade in the arrow symbolizes the increasing degree of loss of motor control.

basolateral amygdala (BLA) (McGaugh, 2000; Packard, Cahill, & McGaugh, 1994). The primary motor cortex, which is essentially involved in the storage of motor memories (Karni, Meyer, & Rey-Hipolito, 1998), receives a basolateral amygdala projection (Sripanidkulchai, Sripanidkulchai, & Wyss, 1984). Thus, it may be assumed that consolidation of these dyscoordinated movements as dysfunctional motor programs is a BLA-mediated process in the primary motor cortex (Jabusch, Müller, & Altenmüller, 2004). This may be also the link to conditions, which we discuss in the following paragraphs, namely CuP and DS.

CuP is common in sports and has been investigated in golfers suffering from the yips. This condition is defined as involuntary movements during the execution of putting strokes, resulting in a serious decrease in the success rate in putting (Lobinger et al., 2014). In music, the term CuP has been introduced only recently by Altenmüller, Ioannou, Raab, and Lobinger (2014), describing phenomena like the aforementioned tongue-stopper felt by brass players, or the short-action bowing tremor of violinists playing soft notes with the tip of the bows. Here, it is most likely that anxiety-induced "reinvestment" leads to cognitive interference, resulting in dysfunctional movements due to the attempt to prevent or even correct feared errors. In an EEG study, we demonstrated that the brains of highly trained professional pianists anticipate errors of motor execution, and that these "pre-error" related brain waves arise about 50 ms prior to the wrong keystroke (Herrojo-Ruiz, Jabusch, & Altenmüller, 2009). Such a rapid error-anticipating mechanism cannot be cognitively controlled; rather, it is highly susceptible to disturbances via cognitive control. Therefore, it is plausible that reinvestment and CuP may lead to a deterioration of motor control.

DS is characterized by a more permanent reduction of motor control. As outlined above, it differs from focal dystonia only in part, but it is more likely linked to psychological triggering factors than to underlying genetic causes (Ioannou & Altenmüller, 2014). In terms of underlying neurophysiological mechanisms, we speculate that these musicians have a deficit in the so-called limbic loops of the basal ganglia, linking movements and motor control to emotions. However, this question remains to be addressed in future investigations in a patient population, which is correctly classified a priori (Ioannou & Altenmüller, 2014).

MD can be distinguished from DS by more pronounced worsening of motor control, the lack of sudden, but short-term, improvement, and the tendency to progress to dystonic cramps. The pathophysiological basis of MD can only briefly be summarized here. For more details, we refer to a review by Altenmüller and Jabusch (2010). Numerous studies have revealed abnormalities in three main areas: (1) reduced inhibition in the sensorimotor system, (2) altered sensory perception, and (3) impaired sensorimotor integration. In recent years, an increasing number of brain imaging studies on focal dystonias and musicians more generally have demonstrated that these alterations are probably not task specific. Functional connectivity (Moore, Gallea, Horovitz, & Hallett, 2012), cortical activation patterns (Haslinger, Altenmüller, Castrop, Zimmer, & Dresel, 2010), and basal ganglia anatomy (Walter et al., 2012) have been proven to be abnormal, although behavioral tests of musicians in other motor skill domains indicate a degree of task specificity (Rosset-Llobet, Candia, Fàbregas, Ray, & Pascual-Leone, 2007). We argue that musician's focal dystonia is the product of a hereditary susceptibility, probably related to a general lack of central nervous inhibition and the aforementioned triggering factors. This leads not only to task-specific functional alterations of CNS networks—as we propose also occurs in patients suffering from DS—but also to a range of structural alterations.

In previous publications, we have emphasized psychological conditions as underlying triggering factors. In several questionnaire studies, we found elevated anxiety and extreme perfectionism in MD patients (Altenmüller & Jabusch, 2009; Enders et al., 2011). According to Jabusch et al. (2004), these psychological characteristics are typically already present in patients before the onset of dystonia. We now argue that musicians with a major psychological burden can most likely be subsumed under the classification of DS.

## CONCLUSION: SOME IMPLICATIONS FOR PREVENTION

Prevention of failures in music performance should begin during musical training in childhood and adolescence. Instrumental and singing teachers have a responsibility to provide their students with knowledge and skills to overcome dysfunctional MPA and to employ successful self-management strategies, and avoid dysfunctional thoughts, behaviors, and negative conditioning. We hope that the implementation of teaching programs dedicated to music physiology and performance psychology in music conservatories will continue and provide sufficient strategies to cope with MPA, CuP, and DS.

Because successful treatment of focal dystonia is still a challenge, preventing MD is enormously important. On the basis of the heuristic model outlined above, we now have the theoretical means at hand to intervene at an early stage. From the first lesson onwards, music educators should strive to create a friendly, supportive atmosphere, to introduce reasonable practice schedules, and to teach energy-efficient techniques. Furthermore, they should prevent overuse and pain by including mental practice and variations of movement patterns. By avoiding mechanical repetitions and, hence, frustration, teachers can help maintain students' motivation. Students should be taught to adopt healthy living habits, which include warm-up and cool-down exercises, regular physical exercise, sufficient breaks, and sleep as the cornerstones of healthy musical practice.

We would like to conclude our chapter with a general remark: as outlined at the outset, music performance on a professional level is one of the most complex of all human activities. Importantly, it is not restricted to reliable functioning of sensorimotor brain circuits but also involves emotion, memory, and imagination. The best trained musicians with the best working sensorimotor networks will not move their listeners if imagination, color, fantasy, and emotion are not a part of their artistic expression. These qualities are rarely trained solely within a practice studio, but depend on and are linked to experiences from daily life, human relationships, a rich artistic environment, empathy, and emotional depth. Personal, emotional growth is, thus, an important prerequisite of a touching music performance.

## ACKNOWLEDGMENT

Both authors are generously supported by a grant of the German Research Foundation (Al 278/7-1).

## REFERENCES

Altenmüller, E. (2003). Focal dystonia: advances in brain imaging and understanding of fine motor control in musicians. *Hand Clinics*, *19*(3), 523–538.

Altenmüller, E., Baur, V., Hofmann, A., Lim, V. K., & Jabusch, H.-C. (2012). Musician's cramp as manifestation of maladaptive brain plasticity: arguments from instrumental differences. *Annals of the New York Academy of Sciences*, *1252*, 259–265.

Altenmüller, E., Ioannou, C. I., Raab, M., & Lobinger, B. (2014). Apollo's curse: causes and cures of motor failures in musicians: a proposal for a new classification. *Advances in Experimental Medicine and Biology*, *826*, 161–178.

Altenmüller, E., & Jabusch, H.-C. (2009). Focal hand dystonia in musicians: phenomenology, aetiology, and psychological trigger factors. *Journal of Hand Therapy*, *22*(2), 144–155.

Altenmüller, E., & Jabusch, H.-C. (2010). Focal dystonia in musicians: phenomenology, pathophysiology, triggering factors, and treatment. *Medical Problems of Performing Artists, 25*(1), 3–9.

Altenmüller, E., Kopiez, R., & Grewe, O. (2013). A contribution to the evolutionary basis of music: lessons from the chill response. In E. Altenmüller, S. Schmidt, & E. Zimmermann (Eds.), *Evolution of emotional communication from sounds in nonhuman mammals to speech and music in man Series in affective sciences* (pp. 13–335). Oxford: Oxford University Press.

Aune, T. K., Ingvaldsen, R. P., & Ettema, G. J. (2008). Effect of physical fatigue on motor control at different skill levels. *Perceptual and Motor Skills, 106*(2), 371–386.

Baumeister, R. F. (1984). Choking under pressure: self-consciousness and paradoxical effects of incentives on skillful performance. *Journal of Personality and Social Psychology, 46*(3), 610–620.

Baumeister, R. F., & Showers, C. J. (1986). A review of paradoxical performance effects: choking under pressure in sports and mental tests. *European Journal of Social Psychology, 16*(4), 361–383.

Baur, V., Jabusch, H.-C., & Altenmüller, E. (2011). Behavioral factors influence the phenotype of musician's dystonia. *Movement Disorders, 26*(9), 1780–1781.

Beck, A. T., & Clark, D. A. (1988). Anxiety and depression: an information processing perspective. *Anxiety Research, 1*(1), 23–36.

Beckmann, J., Gröpel, P., & Ehrlenspiel, F. (2013). Preventing motor skill failure through hemisphere-specific priming: cases from choking under pressure. *Journal of Experimental Psychology: General, 142*(3), 679–691.

Brandfonbrener, A. G., & Robson, C. (2004). Review of 113 musicians with focal dystonia seen between 1985 and 2002 at a clinic for performing artists. *Advances in Neurology, 94*, 255–256.

Brown, R. M., Penhune, V. B., & Zatorre, R. (2015). Expert music performance: cognitive, neural, and developmental bases. *Progress in Brain Research, 217*, 57–86.

Brugués, A. O. (2011a). Music performance anxiety-part 1. A review of its epidemiology. *Medical Problems of Performing Artists, 26*(2), 102–105.

Brugués, A. O. (2011b). Music performance anxiety-part 2. A review of treatment options. *Medical Problems of Performing Artists, 26*(3), 164–171.

Byl, N. N., Negajaran, S., & McKenzie, A. L. (2003). The effect of sensory discrimination training on structure and function in patients with focal hand dystonia: a case series. *Archives of Physical Medicine and Rehabilitation, 84*(10), 1505–1514.

Carver, C. S., & Scheier, M. F. (1981). *Attention and self-regulation: A control theory approach to human behavior*. New York: Springer.

Cheng, F. P., Großbach, M., & Altenmüller, E. (2013). Altered sensory feedbacks in pianist's dystonia: the altered auditory feedback paradigm and the glove effect. *Frontiers in Human Neuroscience, 17*(7), 868.

Dirnberger, G., Duregger, C., Trettler, E., Lindinger, G., & Lang, W. (2004). Fatigue in a simple repetitive motor task: a combined electrophysiological and neuropsychological study. *Brain Research, 1028*(1), 26–30.

Enders, L., Spector, J. T., Altenmüller, E., Schmidt, A., Klein, C., & Jabusch, H.-C. (2011). Musician's dystonia and comorbid anxiety: two sides of one coin? *Movement Disorders, 26*(3), 539–542.

Endo, S., Juhlberg, K., Bradbury, A., & Wing, A. M. (2014). Interaction between physiological and subjective states predicts the effect of a judging panel on the postures of cellists in performance. *Frontiers in Psychology, 7*(5), 773.

Ericsson, K. A., Krampe, R. T., & Tesch-Romer, C. (1993). The role of deliberate practice in the acquisition of expert performance. *Psychological Review, 100*(3), 363–406.

Ericsson, K. A., & Lehmann, A. C. (1996). Expert and exceptional performance: evidence of maximal adaptation to task. *Annual Review of Psychology, 47*, 273–305.

Eysenck, M. W., & Calvo, M. G. (1992). Anxiety and performance: the processing efficiency theory. *Cognition and Emotion, 6*(6), 409–434.

Fishbein, M., Middlestadt, S. E., Ottati, V., Strauss, S., & Ellis, A. (1988). Medical problems among ICSOM musicians: overview of a national survey. *Medical Problems of Performing Artists, 3*(1), 1–8.

Frucht, S. J. (2009). Focal task-specific dystonia of the musicians' hand-a practical approach for the clinician. *Journal of Hand Therapy, 22*(2), 136–143.

Furuya, S., & Altenmüller, E. (2013). Flexibility of movement organization in piano performance. *Frontiers in Human Neuroscience, 7*, 173.

Furuya, S., Nitsche, M. A., Paulus, W., & Altenmüller, E. (2014). Surmounting retraining limits in musicians' dystonia by transcranial stimulation. *Annals of Neurology, 75*(5), 700–707.

Gembris, H., & Heye, A. (2012). Älter werden im Orchester: Eine empirische Untersuchung. *Schriften des Instituts für Begabungsforschung in der Musik (IBFM)* (Bd. 5).

Harmison, R. J. (2006). Peak performance in sport: identifying ideal performance states and developing athletes' psychological skills. *Professional Psychology: Research and Practice, 37*(3), 233–243.

Haslinger, B., Altenmüller, E., Castrop, F., Zimmer, C., & Dresel, C. (2010). Sensorimotor overactivity as a pathophysiologic trait of embouchure dystonia. *Neurology, 74*(22), 1790–1797.

Herrojo-Ruiz, M., Jabusch, H.-C., & Altenmüller, E. (2009). Detecting wrong notes in advance: neuronal correlates of error monitoring in pianists. *Cerebral Cortex, 19*(11), 2625–2639.

Hill, D. M., Hanton, S., Fleming, S., & Matthews, N. (2009). A re-examination of choking in sport. *European Journal of Sport Science, 9*(4), 203–212.

Hill, D. M., Hanton, S., Matthews, N., & Fleming, S. (2010). Choking in sport: a review. *International Review of Sport and Exercise Psychology, 3*(1), 24–39.

Ioannou, C. I., & Altenmüller, E. (2014). Psychological characteristics in musician's dystonia: a new diagnostic classification. *Neuropsychologia, 61*, 80–88.

Ioannou, C. I., Furuya, S., & Altenmüller, E. Consequences of stress on skilled motor performance in musicians suffering from focal dystonia: a pilot study, submitted for publication.

Jabusch, H.-C., Müller, S. V., & Altenmüller, E. (2004). Anxiety in musicians with focal dystonia and those with chronic pain. *Movement Disorders, 19*(10), 1169–1175.

Jabusch, H. C., Zschucke, D., Schmidt, A., Schuele, S., & Altenmüller, E. (2005). Focal dystonia in musicians: treatment strategies and long-term outcome in 144 patients. *Movement Disorders, 20*(12), 1623–1626.

James, I. (1997). *Federation Internationale des Musiciens 1997 Survey of 56 orchestras worldwide*. London: British Association for Performing Arts Medicine.

Jankovic, J., & Ashoori, A. (2008). Movement disorders in musicians. *Movement Disorders, 23*(14), 1957–1965.

Karni, A., Meyer, G., & Rey-Hipolito, C. (1998). The acquisition of skilled motor performance: fast and slow experience-driven changes in primary motor cortex (M1). *Proceedings of the National Academy of Sciences of the United States of America, 95*, 861–868.

van Kemenade, J. F., van Son, M. J., & van Heesch, N. C. (1995). Performance anxiety among professional musicians in symphonic orchestras: a self-report study. *Psychological Reports, 77*(2), 555–562.

Kenny, D. T. (2011). *The psychology of music performance anxiety*. UK: Oxford University Press.

Khalsa, S. B., Butzer, B., Shorter, S. M., Reinhardt, K. M., & Cope, S. (2013). Yoga reduces performance anxiety in adolescent musicians. *Alternative Therapies, Health and Medicine, 19*(2), 34–45.

Klämpfl, M. K., Lobinger, B. H., & Raab, M. (2013). How to detect the yips in golf. *Human Movement Science, 32*(6), 1270–1287.
Lederman, R. J. (1991). Focal dystonia in instrumentalists: clinical features. *Medical Problems of Performing Artists, 6*(4), 132–136.
Lim, V., & Altenmüller, E. (2003). Musicians cramp: instrumental and gender differences. *Medical Problems of Performing Artists, 18*(1), 21–27.
Lobinger, B., Klämpfl, M., & Altenmüller, E. (2014). We are able, we intend, we act - but we do not succeed: a theoretical framework for a better understanding of paradoxical performance in sports. *JCSP, 8*(4), 357–377.
Mahoney, M. J. (1992). Performing under pressure. *American Psychological Association, 37*(4), 312.
Martin, R. The effect of a series of guided music imaging sessions on music performance anxiety (Unpublished Master of Music thesis). University of Melbourne.
McGaugh, J. L. (2000). Memory - a century of consolidation. *Science, 287*(5451), 248–251.
Moore, R. D., Gallea, C., Horovitz, S. G., & Hallett, M. (2012). Individuated finger control in focal hand dystonia: an fMRI study. *Neuroimage, 61*(4), 823–831.
Mosing, M. A., Madison, G., Pedersen, N. L., Kuja-Halkola, R., & Ullén, F. (2014). Practice does not make perfect: no causal effect of music practice on music ability. *Psychological Science, 25*(9), 1795–1803.
Münte, T. F., Altenmüller, E., & Jäncke, L. (2002). The musician's brain as a model of neuroplasticity. *Nature Reviews Neuroscience, 3*(6), 473–478.
Otten, M. (2009). Choking vs. clutch performance: a study of sport performance under pressure. *Journal of Sport & Exercise Psychology, 31*(5), 583–601.
Packard, M. G., Cahill, L., & McGaugh, J. L. (1994). Amygdala modulation of hippocampal-dependent and caudate nucleus-dependent memory processes. *Proceedings of the National Academy of Sciences of the United States of America, 91*(18), 8477–8481.
Paulig, J., Jabusch, H.-C., Großbach, M., Boullet, L., & Altenmüller, E. (2014). Sensory trick phenomenon improves motor control in pianists with dystonia: prognostic value of glove-effect. *Frontiers in Psychology, 23*(5), 1012.
Pavlov, I. P. (1951). Dinamicheskaya stereotipiya vysshego otdela golovnogo mozga (The dynamic stereotype of the higher section of the brain). In I. P. Pavlov (Ed.), *polnoe sobraniie sochineniy* (2nd ed.) *Enlarged: Vol. 3(Pt. 2).* (pp. 240–244). Moscow: Izdatel'stvo Akademii Nauk SSSR.
Rosset-Llobet, J., Candia, V., Fàbregas, S., Ray, W., & Pascual-Leone, A. (2007). Secondary motor disturbances in 101 patients with musician's dystonia. *Journal of Neurology, Neurosurgery, and Psychiatry, 78*(9), 949–953.
Sanderson, W. C., DiNardo, P. A., Rapee, R. M., & Barlow, D. H. (1990). Symptom comorbidity in patients diagnosed with DSM-III-R anxiety disorders. *Journal of Abnormal Psychology, 99*(3), 308–312.
Schmidt, A., Jabusch, H.-C., Altenmüller, E., Hagenah, J., Brüggemann, N., Lohmann, K., et al. (2009). Etiology of musicians' dystonia: familial or environmental? *Neurology, 72*(14), 1248–1254.
Schmidt, A., Jabusch, H.-C., Altenmüller, E., Kasten, M., & Klein, C. (2013). Challenges of making music: what causes musician's dystonia? *JAMA Neurology, 70*(11), 1456–1459.
Schuele, S. U., Jabusch, H.-C., Lederman, R. J., & Altenmüller, E. (2005). Botulinum toxin injections of musician's dystonia. *Neurology, 64*(2), 341–343.
Soeter, M., & Kindt, M. (November 18, 2013). High trait anxiety: a challenge for disrupting fear memory reconsolidation. *PLoS One, 8*(11), e75239.
Spahn, C. (2011). Psychosomatische Medizin und Psychotherapie. In C. Spahn, B. Richter, & E. Altenmüller (Eds.), *MusikerMedizin. Diagnostik, Therapie und Prävention von musikerspezifischen Erkrankungen.* Stuttgart: Schattauer.

Spahn, C. (2012). *Lampenfieber: Handbuch für den erfolgreichen Auftritt. Grundlagen, Analyse, Maßnahmen*. Leipzig: Henschel Verlag.
Spahn, C. (2015). Treatment and prevention of music performance anxiety. *Progress in Brain Research, 217*, 129–140.
Spahn, C., Echternach, M., Zander, M. F., Voltmer, E., & Richter, B. (2010). Music performance anxiety in opera singers. *Logopedics Phoniatrics Vocology, 35*(4), 175–182.
Spence, J. T., & Spence, K. W. (1966). The motivational component of manifest anxiety: drive and drive stimuli. In C. Spielberger (Ed.), *Anxiety and behavior* (pp. 291–326). New York: Academic Press.
Sripanidkulchai, K., Sripanidkulchai, B., & Wyss, J. M. (1984). The cortical projection of the basolateral amygdaloid nucleus in the rat: a retrograde fluorescent dye study. *Journal of Comparative Neurology, 229*(3), 419–431.
Steele, C. J., Bailey, J. A., Zatorre, R. J., & Penhune, V. B. (2013). Early musical training and white-matter plasticity in the corpus callosum: evidence for a sensitive period. *Journal of Neuroscience, 33*(3), 1282–1290.
Steptoe, A., & Fidler, H. (1987). Stage fright in orchestral musicians: a study of cognitive and behavioural strategies in performance anxiety. *British Journal of Psychology, 78*(Pt 2), 241–249.
Studer, R. K., Danuser, B., Wild, P., Hildebrandt, H., & Gomez, P. (2014). Psychophysiological activation during preparation, performance, and recovery in high- and low-anxious music students. *Applied Psychophysiology and Biofeedback, 39*(1), 45–57.
van Vugt, F. T., Boullet, L., Jabusch, H.-C., & Altenmüller, E. (2014). Musician's dystonia in pianists: long-term evaluation of retraining and other therapies. *Parkinsonism & Related Disorders, 20*(1), 8–12.
Walter, U., Buttkus, F., Benecke, R., Grossmann, A., Dressler, D., & Altenmüller, E. (2012). Sonographic alteration of lenticular nucleus in focal task-specific dystonia of musicians. *Neurodegenerative Diseases, 9*(2), 99–103.
Wilson, G. D. (1997). Performance anxiety. In D. J. Hargreaves, & A. C. North (Eds.), *The social psychology of music* (pp. 229–248). Oxford: Oxford University Press.
Wilson, M. (2008). From processing efficiency to attentional control: a mechanistic account of the anxiety-performance-relationship. *International Review of Sport and Exercise Psychology, 1*(2), 184–201.
Windholz, H. (1996). Pavlov's conceptualization of the dynamic stereotype in the theory of higher nervous activity. *American Journal of Psychology, 109*(2), 287–295.
Wulf, G. (2007). *Attention and motor skill learning*. Champaign, IL: Human Kinetics.
Yerkes, R. M., & Dodson, J. D. (1908). The relation of strength of stimulus to rapidity of habit-formation. *Journal of Comparative Neurology, 18*(5), 459–482.
Yoshie, M., Kudo, K., Murakoshi, T., & Ohtsuki, T. (2009). Music performance anxiety in skilled pianists: effects of social-evaluative performance situation on subjective, autonomic, and electromyographic reactions. *Experimental Brain Research, 199*(2), 117–126.

Chapter 8

# Motor Imagery and Mental Training in Older Adults

Michael Kalicinski[1], Monika Thomas[1], Babett H. Lobinger[2]
[1]Institute of Physiology and Anatomy, German Sport University Cologne, Cologne, Germany;
[2]Institute of Psychology, German Sport University Cologne, Cologne, Germany

> *Imagine having a walk through the park on a sunny afternoon remembering how fit you felt after your last long walk. Perhaps it is spring and the ground is soft. After a while, you become a bit tired, but you keep going. Suddenly, you stumble and begin to fall. You immediately imagine a step to the side that helps you break your fall and a counter rotation movement that restores your balance. You cope with this risky situation and continue your walk.*

Falls due to loss of balance are a major source of loss of mobility and frailty in older people (between 65 and 80 years old). Hip fractures and other fall-related injuries can necessitate admission to hospitals and nursing homes and result in large health care expenditures. Deficits in postural control are largely behind the increase in risk of falling among older adults (Horak, 2006; Segev-Jacubovski et al., 2011). Given that the population is aging in many countries, it is crucial to develop beneficial and cost-effective approaches for encouraging postural control among older adults.

Strength and balance training is a well-established method for improving postural control. Here, we focus on a less common training among older adults: mental training. Athletes have long included mental training in their regimens, and several studies have documented its beneficial effects on motor performance and mental skills (Driskell, Copper, & Moran, 1994; Feltz & Landers, 1983; Weinberg, 2008). Combining mental training with physical practice increases this effect and has a greater impact than physical training alone (Allami, Paulignan, Brovelli, & Boussaoud, 2007). For healthy older adults, mental training can provide the opportunity to learn and practice strategies for coping with risky situations without the risk of physical harm. In the domain of postural control, mental training can be used to acquire, practice, and correct balance skills and to encourage self-efficacy beliefs. Mental training can be performed even when a person is physically fatigued. Yet to date, there is little knowledge of the feasibility and benefits of mental training for older adults.

In this chapter, we provide a brief overview of research on the mechanisms underlying motor imagery (MI), an important prerequisite of mental training. We discuss the main findings regarding whether older adults are able to perform MI and fulfil this prerequisite. Finally, we present the procedures and results of a recent pilot intervention study aimed at improving postural control in older adults with mental training.

## MOTOR IMAGERY AS A PREREQUISITE FOR MENTAL TRAINING

MI is a dynamic covert state during which an internal representation of a motor action is mentally reproduced in working memory (Jeannerod, 2001). Mental training is the systematic and repeated application of MI, with the intention to improve motor execution (ME) (Malouin & Richards, 2010).

In his simulation theory, Jeannerod (2001) suggested that MI is a simulation of physical activity that relies on similar action representations. MI and ME are represented in shared neural networks (Munzert, Lorey, & Zentgraf, 2009), have similar autonomic nervous system responses (Decety, Jeannerod, Durozard, & Baverel, 1993; Decety, Jeannerod, Germain, & Pastene, 1991), and are temporally similar (Decety, Jeannerod, & Prablanc, 1989; Guillot & Collet, 2005).

The concept of internal forward models explains these aspects of MI and ME (Jeannerod, 2006): The intention to execute a movement produces a motor command and an efference copy. The motor command is sent to the motor system for overt execution, and the efference copy is sent to the forward model for covert representation. The forward model thus functions to predict the sensory feedback and the desired effect of an action (Wolpert & Miall, 1996). An emulator (Grush, 2004) that uses short-term storage of action information and predictions of changes of the motor system and the external world produces the desired effect. This is usually a non-conscious process in the pre-execution stage of movements, but it can also be accessed consciously by MI. As the motor representation contains all elements of the movement, a well-developed motor representation leads to accurate MI and, thus, increases the amount of functional equivalence between MI and ME. In the following section, we provide a brief overview of the similarities of MI and ME.

Several investigations (Guillot & Collet, 2005; Jeannerod, 2001; Munzert et al., 2009) have found support for the hypothesized behavioral and neurophysiological similarities between MI and ME. Neuroimaging studies using functional magnetic resonance imaging, magnetoencephalography, positron emission tomography, and transcranial magnetic stimulation (Lotze & Halsband, 2006) have shown an overlap of active brain regions during MI and ME. In a study of the autonomic nervous system, Decety et al. (1991) found that respiration and heart rate significantly increased during imagined walking. On the behavioral side, researchers have measured the temporal congruence of MI and ME (the similarity of real and imagined times) with the mental chronometry paradigm. Researchers found that the ability to precisely estimate the duration of a movement depends on individual capabilities, movement focus, environmental constraints, and task difficulty (Guillot & Collet, 2005).

These similarities between MI and ME lead to the opportunity to systematically improve ME by using MI techniques. To do so, the actual performance and environment should be simulated as accurately as possible. Holmes and Collins (2001) developed the PETTLEP model for improving MI, which takes into account physical, environmental, timing, task, learning, emotion, and perspective factors. It remains unclear whether older adults are able to consider these factors adequately in their MI processes. In the following section, we discuss whether MI ability is preserved in older adults.

## MOTOR IMAGERY AND MENTAL TRAINING IN OLDER ADULTS

Several studies have investigated age-related changes in MI abilities (for a review, see Saimpont, Malouin, Tousignant, & Jackson, 2013). The results have shown that older and younger adults perform similarly on MI tasks when the movements and tasks are simple. Age-related MI differences (e.g., in mental chronometry, controllability, and vividness; Schott, 2012) emerge, however, when task difficulty increases (Saimpont et al., 2013).

For example, healthy younger adults showed higher temporal congruence of MI and ME compared to older adults, especially when imagining more complex tasks (Saimpont et al., 2013). In a walking task, older adults' performance declined as the width of a track decreased (Personnier, Kubicki, Laroche, & Papaxanthis, 2010). Specifically, older adults systematically overestimated actual walking durations along three paths with different widths (50, 25, and 15 cm), and the narrower the track, the more pronounced the effect. Temporal congruence also decreased when the length of a path increased (Schott, 2012; Schott & Munzert, 2007). In contrast, Kalicinski, Kempe, and Bock (2015) found no age-related differences in walking tasks with short distances, but Kalicinski and Raab (2014) found an interaction of age and task requirements, suggesting that some task requirements lead to an increase of age difference, while others lead to a decrease. Moreover, body posture seems to have a large impact on MI performance in older adults (Saimpont, Malouin, Tousignant, & Jackson, 2012). For example, when body posture was congruent with an intended motor action (e.g., standing while imagining a walking task), older adults showed better MI performance than when they adopted an incongruent body posture.

The controllability of MI has been investigated with the Controllability of Motor Imagery (CMI) test (Schott, 2004, 2013). Participants have to imagine five body-part positions in response to verbal instructions and integrate them into a whole-body position. As in the mental chronometry studies, younger adults outperformed older adults (Schott, 2012).

The vividness of MI has been investigated using standardized questionnaires that ask participants to rate the content of their MI and sensations associated with it.

Interestingly, age-related differences in vividness were observed only when the MI was performed from a first-person and not from a third-person perspective (Mulder, Hochstenbach, Heuvelen, & den Otter, 2007). In another study on the effects of aging on MI vividness, visual MI scores were found to be higher than kinesthetic scores for young and middle-aged adults but older adults showed a loss of this visual dominance (Malouin, Richards, & Durand, 2010). However, when they performed MI from the third-person perspective, older adults reached MI scores comparable to those of the younger groups. In a recent study, older adults rated the vividness of imagined walking significantly higher than younger adults (Kalicinski & Raab, 2014). This was not the case for the ratings on the CMI tasks and is further evidence that task requirements play a role in MI of older adults.

In older adults, there are several age-related alterations for MI. Unfortunately, most of the published research does not take into account that the reported decline in MI performance among older adults should not be interpreted as an unavoidable consequence of aging. Researchers have suggested that working memory capacity (Saimpont, Pozzo, & Papaxanthis, 2009; Schott, 2012), attention (Decety, 1996), and action planning are important moderating factors, leading to the conclusion that age should not be seen as a disqualifier for mental training (Schott, 2012). In addition, the findings on the influence of task requirements lead us to the conclusion that the prerequisite is fulfilled at least for simple and short movement sequences. The broader question remains: Is mental training beneficial for older adults?

The few published reports on mental training for this target group suggest that older adults could benefit from mental training in terms of postural oscillation and MI vividness (Hamel & Lajoie, 2005), retention processes of motor learning (Jarus & Ratzon, 2000), and sequence learning (Stoter, Scherder, Kamsma, & Mulder, 2008). However, due to the limited number of intervention studies, clear evidence to support the usefulness of mental training is still needed (Kalicinski & Lobinger, 2013). To gain insight into the feasibility and benefits of mental training for older adults, we designed a pilot intervention study addressing postural control, a promising field of application as illustrated at the beginning of this chapter.

## MENTAL BALANCE TRAINING FOR POSTURAL CONTROL

Various intervention studies have shown that balance exercises have a beneficial effect on postural control and mobility in older adults (see Granacher, Mühlbauer, Zahner, Gollhofer, & Kressig, 2011). Three types of balance training have been established: (1) traditional balance training based on the principle of performing static and dynamic exercises under various surface, stance, and vision conditions; (2) perturbation-based balance training, which addresses the recovery of balance in a perturbed or destabilized situation by using specific movement strategies (e.g., a step strategy to bring the support base back into alignment under the center of pressure); and (3) multitask balance training, consisting of multitask balance exercises (e.g., walking while talking). All three approaches have been shown to be effective in encouraging postural control in older adults (Granacher et al., 2011). We implemented these three approaches in a mental balance training program and conducted a pilot intervention study to test the efficiency for older adults. In this section, we present the methods, procedure and results of the pilot study.

## Method

### Participants

Twenty-five healthy older adults (15 men, 10 women) with a mean age of 67.8 years (SD=4.2) took part in the study. Participants had to meet the following criteria: (1) 60–80 years old; (2) physically active at least once a week; (3) no acute orthopedic disorders; (4) no cardiovascular disorders; and (5) no neurological disorders. We assigned participants to either the experimental group, which received mental training for balance tasks, or to a wait-list control group. To assess the benefits of mental training, we asked all participants to engage in tests of postural control and complete questionnaires before (pre) and after a 6-week training phase (post).

### Postural Control

To measure postural control, we asked participants to assume two test postures on a moving platform (Posturomed, Haider-Bioswing, Pullenreuth, Germany): the tandem stand (one foot in front of the other) and the one-leg stand on the preferred leg. Participants executed both postures on the moving platform in two conditions (with and without perturbation). For the condition with perturbation, a standardized excursion of 1 cm in the mediolateral direction was triggered after 1 s. Participants were asked to hold each posture in each condition two times for as long as possible up to a maximum of 30 s. As a parameter, we determined the average balance time of each posture in both conditions.

### Self-Efficacy

Self-efficacy beliefs have the potential to influence imagery outcomes and can show if an intervention has had an effect. We measured general perceived self-efficacy with the German version of the Generalized Self-Efficacy Scale (Schwarzer & Jerusalem, 1995), estimated balance-related self-efficacy in simple and complex tasks with the German version of the Activities-Specific Balance Confidence Scale (Schott, 2008), and measured gait-related self-efficacy with a translated version of the modified Gait Efficacy Scale (Newell, VanSwearingen, Hile, & Brach, 2012).

### Interviews

As a manipulation check, we asked participants to comment on the content of the intervention and its relation to their daily life. After the first and last session, we asked participants to fill out a six-item questionnaire asking for their opinion on the helpfulness of the training for their balance. To get an idea of the effects of the intervention on everyday life, we asked participants about the impact the training had on their daily life after the last session. We asked participants (1) what kind of people they would advise to take part in such an intervention and (2) if something had changed in their daily life since they had started the intervention or if they had experienced situations in which they profited from techniques they had learned in the intervention. As we also wanted to test participant satisfaction with the newly developed program, we asked participants to indicate after each training session their satisfaction with the content.

## Intervention Procedures

The intervention was scheduled two times a week (Tuesday and Friday mornings) for 1 h, resulting in a total of 12 sessions. We divided the experimental group into three smaller groups ($n=5$) to enable individualized support during the sessions. The aim of the intervention was to increase postural control in a typical balance task by using the mental training approach. Each training session consisted of five parts: relaxation, MI ability, balance experience, mental training for balance tasks, and reflection.

### Relaxation

After the first session and at the beginning of each subsequent session we invited questions and comments on the previous session. After clarifying any issues, the instructor led the participants in a relaxation exercise, as relaxation is a precondition for mental training. Relaxation techniques in the six sessions included progressive muscle relaxation, guided imagery, and breathing exercises.

### MI Ability

Following the relaxation exercise, participants performed various exercises that demanded imagery abilities. For instance, we used the CMI test (Schott, 2013) to foster the production and manipulation of mental images. In this test, participants had to create a mental picture of themselves and afterward assume the body position that was reached at the end of the MI. Participants stood with their eyes closed and received six consecutive instructions for moving a body part into a specific position. We asked them to remain physically still and only to imagine moving their body parts as requested. After the sixth instruction, participants opened their eyes and arrange themselves in the assumed final body configuration. We increased the number and complexity of instructions during the training phase according to the participants' abilities. Other examples of MI exercises include walking a triangle with eyes closed or reaching a high temporal accuracy of imagined and actual walking through an obstacle course.

## Balance Experience

In this part of the training session, participants experienced various static and dynamic balance situations. Static balance tasks included standing on different surfaces (e.g., soft, uneven, confined space), in different postural conditions (e.g., tandem stand, one-leg stand), or while simultaneously performing an additional task (e.g., catching a ball, reacting to an acoustic signal). Dynamic balance tasks were walking/balancing on different surfaces with decreasing track width, with sudden changes of direction, or while simultaneously performing an additional task. We progressively increased the difficulty of the tasks through the 6 weeks. One very complex dynamic balance task was crossing a slackline, a flat line made of webbing that is strung above ground between two anchors. Various aspects of crossing a slackline challenge the ability to balance (e.g., the line is narrow, it oscillates, and the tension of the line varies). Experiencing balance is an important prerequisite for the mental training for balance tasks.

## Mental Training for Balance Tasks

The main part of the training consisted of MI of balance situations. We asked participants to imagine the experienced balance situations, visualizing their successful performance. Participants stood mostly with eyes closed and imagined their movements from the first-person perspective. Every experienced balance situation was accompanied by several MI trials. After three trials, the instructor asked if the participants were able to simulate the situation and the appropriate movement. The instructor then gave hints on how to solve the problems and also asked participants to focus on specific aspects during the next trials ("Can you feel how your center of gravity shifts when you lift the other leg?"). Besides creating mental images of the experienced balance situations, the participants also mentally simulated movement sequences, such as a sequence of several one-leg stands ("lift your right leg→direct it forward and put it on the ground→shift your weight to the right leg→lift your left leg→direct it in a semicircle forward to the left→shift your weight to the left leg→lift your right leg→direct it in a semicircle backward to the left→shift your weight to the right leg→lift your left leg→set it beside your right leg→afterward, do the same sequence starting with the left leg so that you end on your starting position"). At the beginning, the instructor guided the MI of this sequence and also manipulated movement speed when necessary. For instance, when the instructor took a long pause after the instruction "lift your right leg," the instructor asked them to mentally hold this position until the next instruction. At the end of each session, participants mentally crossed a slackline. Because of the complexity of this task, participants followed a self-made imagery script. Besides describing the situation and the movements, the script also imparted positive messages, such as "you are starting to stumble, but you successfully bring yourself back to a standstill." After this, the instructor asked participants to imagine crossing the slackline without the script.

**TABLE 1** Sample Training Session

| Part | Duration | Content | Aim |
|---|---|---|---|
| Relaxation | 5 min | Guided imagery | Relaxation and concentration |
| Motor imagery (MI) ability | 15 min | Controllability of MI<br>• 5 trials | Learn to produce and manipulate MI |
| Balance experience | 10 min | Walking/balancing:<br>• Soft ground<br>• Uneven ground<br>• Narrow line<br>• Dual task<br>Slackline experience | Learn the differences in environmental requirements |
| Mental training | 25 min | Walking/balancing:<br>• Soft ground<br>• Uneven ground<br>• Narrow line<br>• Dual task<br>Guided slackline | Integrate the differences in the mental simulation of walking |
| Reflection | 5 min | • Feedback<br>• Questions | Learn to verbalize imagery experience |

*Reflection*

After each exercise and at the end of each session, the instructor asked participants to reflect on and share their MI experiences. A sample training session is illustrated in Table 1.

## Results and Discussion

The mental training group increased their balance time for the tandem-stand task in comparison to the control group (Figure 1). These results are in keeping with the results of Hamel and Lajoie (2005) and suggest that the mental training of balance tasks can foster postural control in older adults. However, we did not observe a significant effect on postural control for the one-leg-stand task (Figure 2). A possible explanation is that the training program contained more imagination of the tandem stand. In particular, to imagine walking a slackline, participants had to focus on balancing in a tandem-stand position. Another explanation is that the one-leg-stand position is more difficult to hold than the tandem stand and would require more frequent training sessions.

The self-efficacy measures revealed no significant effect of mental training, in line with Hamel and Lajoie's earlier results (2005). Hamel and Lajoie suggested that effects on confidence may take longer to appear. Additionally,

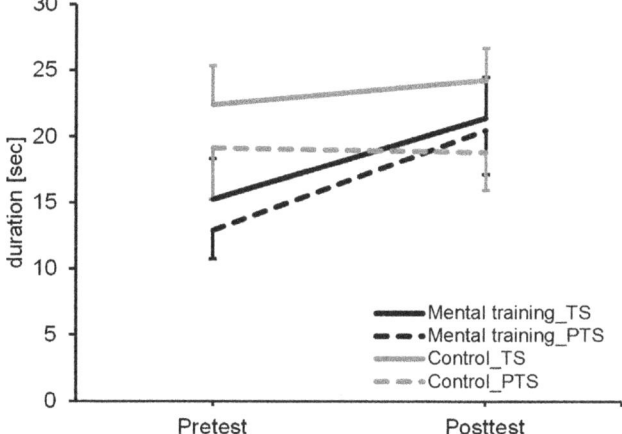

**FIGURE 1** Mean duration remaining in the tandem stand position without (TS) and with (PTS) perturbation, separated by group (mental training, control) before and after the 6-week training. Standard errors are represented by the error bars.

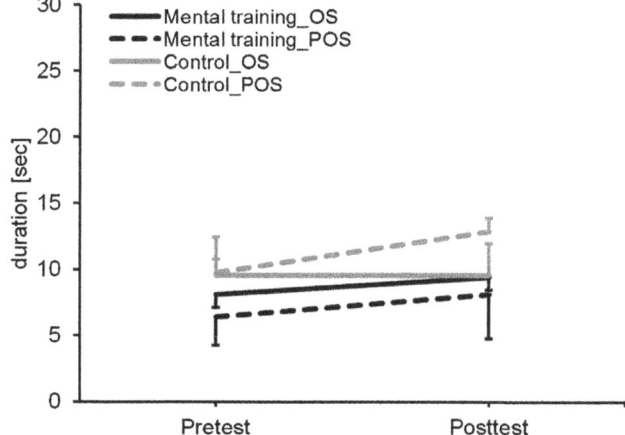

**FIGURE 2** Mean duration remaining in the one-leg-stand position without (OS) and with (POS) perturbation, separated by group (mental training, control) before and after the 6-week training. Standard errors are represented by the error bars.

the self-efficacy measures in the pre-test showed no relation to the changes in performance of the mental training group, which contradicts Munroe-Chandler, Gammage, and Estabrooks's (2005) proposed theory of self-efficacy.

Although mental training was not effective for improving performance in all measures, the participants nonetheless perceived the training program as effective for postural control, concentration, self-confidence, and awareness of movements. It is conceivable that some benefits of the mental training program were not detected by the performance measures because transfer of the broad training to the specific tasks would have required a longer training period. Furthermore, the participants rated the mental training techniques as especially helpful for older adults ($n=5$), people with low physical activity ($n=3$), people with degeneration due to illness ($n=2$), in general ($n=2$), and for the purpose of better concentration ($n=1$). This shows that participants differentiated the potential benefits for age, skill level, and purpose (motivation, concentration). When asked if they were able to apply the mental training techniques in their daily lives, 9 of 13 participants answered "yes." Four participants experienced improvements in physical activities and also in sports. Three talked about better mental preparation and more concentration in other activities of daily living. What is striking in the participants' personal reports is that they were overall satisfied with the training program and each individual session, suggesting that older adults take kindly to mental training.

## CONCLUSION AND FUTURE RESEARCH

To the best of our knowledge, the current study is one of the first to use the three types of balance training classified by Granacher et al. (2011) in a mental training program for healthy older adults. Although, overall, there is little evidence that mental training is beneficial for older adults, the significant improvements in postural control in our participants and their positive ratings of the intervention lead us to conclude that the mental training program we developed is useful for encouraging postural control and thereby reducing the risk of falling. As mental training is most effective when combined with physical practice, future studies should use a combined training procedure. Physical balance training has been shown to be highly effective for improving postural control. A combination of mental and physical approaches could take advantage of both strategies.

## REFERENCES

Allami, N., Paulignan, Y., Brovelli, A., & Boussaoud, D. (2007). Visuo-motor learning with combination of different rates of motor imagery and physical practice. *Experimental Brain Research*, *184*, 105–113. http://dx.doi.org/10.1007/s00221-007-1086-x.

Decety, J. (1996). The neurophysiological basis of motor imagery. *Behavioural Brain Research*, *77*, 45–52.

Decety, J., Jeannerod, M., Durozard, D., & Baverel, G. (1993). Central activation of autonomic effectors during mental simulation of motor actions in man. *Journal of Physiology*, *461*, 549–563.

Decety, J., Jeannerod, M., Germain, M., & Pastene, J. (1991). Vegetative response during imagined movement is proportional to mental effort. *Behavioural Brain Research*, *42*, 1–5.

Decety, J., Jeannerod, M., & Prablanc, C. (1989). The timing of mentally represented actions. *Behavioural Brain Research*, *34*, 35–42.

Driskell, J. E., Copper, C., & Moran, A. (1994). Does mental practice enhance performance? *Journal of Applied Psychology*, *79*, 481–492. http://dx.doi.org/10.1037/0021-9010.79.4.481.

Feltz, D. L., & Landers, D. M. (1983). The effects of mental practice on motor skill learning and performance—a meta-analysis. *Journal of Sport Psychology*, *5*, 25–57.

Granacher, U., Mühlbauer, T., Zahner, L., Gollhofer, A., & Kressig, R. W. (2011). Comparison of traditional and recent approaches in the promotion of balance and strength in older adults. *Sports Medicine*, *41*, 377–400.

Grush, R. (2004). The emulation theory of representation: motor control, imagery, and perception. *Behavioral and Brain Sciences*, *27*, 377–396. http://dx.doi.org/10.1017/S0140525X04000093.

Guillot, A., & Collet, C. (2005). Duration of mentally simulated movement: a review. *Journal of Motor Behavior*, *37*, 10–20.

Hamel, M. F., & Lajoie, Y. (2005). Mental imagery. Effects on static balance and attentional demands of the elderly. *Aging Clinical and Experimental Research*, *17*, 223–228.

Holmes, P. S., & Collins, D. J. (2001). The PETTLEP approach to motor imagery: a functional equivalence model for sport psychologists. *Journal of Applied Sport Psychology*, *13*, 60–83. http://dx.doi.org/10.1080/10413200109339004.

Horak, F. B. (2006). Postural orientation and equilibrium: what do we need to know about neural control of balance to prevent falls? *Age and Ageing*, *35*(Suppl. 2), ii7–ii11. http://dx.doi.org/10.1093/ageing/afl077.

Jarus, T., & Ratzon, N. Z. (2000). Can you imagine? The effect of mental practice on the acquisition and retention of a motor skill as a function of age. *Occupational Therapy Journal of Research*, *20*, 163–178.

Jeannerod, M. (2001). Neural simulation of action: a unifying mechanism for motor cognition. *NeuroImage*, *14*, S103. http://dx.doi.org/10.1006/nimg.2001.0832.

Jeannerod, M. (2006). *Motor cognition: What actions tell the self*. Oxford, England: Oxford University Press.

Kalicinski, M., Kempe, M., & Bock, O. (2015). Motor imagery: effects of age, task complexity, and task setting. *Experimental Aging Research*, *41*, 25–38. http://dx.doi.org/10.1080/0361073X.2015.978202.

Kalicinski, M., & Lobinger, B. H. (2013). Benefits of motor and exercise imagery for older adults. *Journal of Imagery Research in Sport and Physical Activity*, *8*, 1–15. http://dx.doi.org/10.1515/jirspa-2012-0003.

Kalicinski, M., & Raab, M. (2014). Task requirements and their effects on imagined walking in elderly. *Aging Clinical Experimental Research*, *26*, 387–393. http://dx.doi.org/10.1007/s40520-013-0184-9.

Lotze, M., & Halsband, U. (2006). Motor imagery. *Journal of Physiology-Paris*, *99*, 386–395. http://dx.doi.org/10.1016/j.jphysparis.2006.03.012.

Malouin, F., & Richards, C. L. (2010). Mental practice for relearning locomotor skills. *Physical Therapy*, *90*, 240–251. http://dx.doi.org/10.2522/ptj.20090029.

Malouin, F., Richards, C. L., & Durand, A. (2010). Normal aging and motor imagery vividness: implications for mental practice training in rehabilitation. *Archives of Physical Medicine and Rehabilitation*, *91*, 1122–1127. http://dx.doi.org/10.1016/j.apmr.2010.03.007.

Mulder, T., Hochstenbach, J. B. H., Heuvelen, M. J. G., & den Otter, A. R. (2007). Motor imagery: the relation between age and imagery capacity. *Human Movement Science*, *26*, 203–211. http://dx.doi.org/10.1016/j.humov.2007.01.001.

Munroe-Chandler, K. J., Gammage, K. L., & Estabrooks, P. A. (2005). A new vision of exercise imagery. *Journal of Sport & Exercise Psychology*, *27*, S11–S12.

Munzert, J., Lorey, B., & Zentgraf, K. (2009). Cognitive motor processes: the role of motor imagery in the study of motor representations. *Brain Research Reviews*, *60*, 306–326. http://dx.doi.org/10.1016/j.brainresrev.2008.12.024.

Newell, A. M., VanSwearingen, J. M., Hile, E., & Brach, J. S. (2012). The modified Gait Efficacy Scale: establishing the psychometric properties in older adults. *Physical Therapy*, *92*, 318–328. http://dx.doi.org/10.2522/ptj.20110053.

Personnier, P., Kubicki, A., Laroche, D., & Papaxanthis, C. (2010). Temporal features of imagined locomotion in normal aging. *Neuroscience Letters, 476*, 146–149. http://dx.doi.org/10.1016/j.neulet.2010.04.017.

Saimpont, A., Malouin, F., Tousignant, B., & Jackson, P. L. (2012). The influence of body configuration on motor imagery of walking in younger and older adults. *Neuroscience, 222*, 49–57. http://dx.doi.org/10.1016/j.neuroscience.2012.06.066.

Saimpont, A., Malouin, F., Tousignant, B., & Jackson, P. L. (2013). Motor imagery and aging. *Journal of Motor Behavior, 45*, 21–28. http://dx.doi.org/10.1080/00222895.2012.740098.

Saimpont, A., Pozzo, T., & Papaxanthis, C. (2009). Aging affects the mental rotation of left and right hands. *PLoS One, 4*, e6714. http://dx.doi.org/10.1371/journal.pone.0006714.

Schott, N. (2004). Controllability of visual motor imagery in older adults. *Journal of Aging and Physical Activity, 12*, 352–353.

Schott, N. (2008). German adaptation of the "Activities-Specific Balance Confidence (ABC) scale" for the assessment of falls-related self-efficacy. *Zeitschrift für Gerontologie und Geriatrie, 41*, 475–485. http://dx.doi.org/10.1007/s00391-007-0504-9.

Schott, N. (2012). Age-related differences in motor imagery: working memory as a mediator. *Experimental Aging Research, 38*, 559–583. http://dx.doi.org/10.1080/0361073X.2012.726045.

Schott, N. (2013). Test zur Kontrollierbarkeit der Bewegungsvorstellungsfähigkeit (TKBV) bei älteren Erwachsenen. *Zeitschrift für Gerontologie und Geriatrie, 46*, 663–672. http://dx.doi.org/10.1007/s00391-013-0520-x.

Schott, N., & Munzert, J. (2007). Temporal accuracy of motor imagery in older women. *International Journal of Sport Psychology, 38*, 304–320.

Schwarzer, R., & Jerusalem, M. (1995). Generalized self-efficacy scale. In J. Weinman & S. J. M. Wright (Eds.), *Measures in health psychology: A user's portfolio. Causal and control beliefs* (pp. 35–37). Windsor, England: NFER-NELSON.

Segev-Jacubovski, O., Herman, T., Yogev-Seligmann, G., Mirelman, A., Giladi, N., & Hausdorff, J. M. (2011). The interplay between gait, falls and cognition: can cognitive therapy reduce fall risk? *Expert Review of Neurotherapeutics, 11*, 1057–1075. http://dx.doi.org/10.1586/ern.11.69.

Stoter, A. J. R., Scherder, E. J. A., Kamsma, Y. P. T., & Mulder, T. (2008). Rehearsal strategies during motor-sequence learning in old age: execution vs motor imagery. *Perceptual and Motor Skills, 106*, 967–978.

Weinberg, R. (2008). Does imagery work? Effects on performance and mental skills. *Journal of Imagery Research in Sport and Physical Activity, 3*. http://dx.doi.org/10.2202/1932-0191.1025.

Wolpert, D. M., & Miall, R. C. (1996). Forward models for physiological motor control. *Neural Networks: The Official Journal of the International Neural Network Society, 9*, 1265–1279.

Section C

# Dysfunctional Learning, Errors, and Other Performance Phenomena of Perception–Cognition Interactions

9. Bridging the Gap between Perception and Cognition: An Overview (Sven Hoffmann)   135
10. Performance and Error Monitoring: Causes and Consequences (Sven Hoffmann)   151
11. Committing Errors as a Consequence of an Adverse Focus of Attention (Daniel Schneider)   169
12. Lifestyle and Interventions for Improving Cognitive Performance in Older Adults (Patrick D. Gajewski and Michael Falkenstein)   189

## Overview

We evaluate our own actions permanently; though this evaluation does not always reach consciousness, such evaluations are crucial for goal-directed behavior. Basically, this type of evaluation marks the endpoint of the solving of one of the main problems every human being is confronted with constantly: the choice between behavioral alternatives or between competing information processing systems seeking access to restricted resources simultaneously. Such action selection is obviously error prone, since we do not always select the adequate action in a given situation. However, we use these errors to adapt our behavior to changes not only in environmental but also in internal demands. This short-term behavioral adaption closely related to learning requires constant monitoring, evaluation, and adaptation of one's own actions with respect to these internal and environmental demands.

Obviously, response monitoring is a key function if we think of behavioral performance, especially if timing is an issue, for example, in speeded tasks where performance is reflected in a trade-off of speed and accuracy. However, a key pre-processing stage of action selection and performance monitoring is an adequate perceptual and cognitive integration of relevant information (Chapter 9), followed by an overview of key neurophysiological correlates and theories about performance monitoring (Chapter 10). Schneider will provide an overview of how misallocated attention might lead to errors (Chapter 11). Finally, Chapter 12 will conclude with a more differential perspective on the individual differences with respect to performance monitoring.

Chapter 9

# Bridging the Gap between Perception and Cognition: An Overview

**Sven Hoffmann**
*Department of Performance Psychology, Institute of Psychology, German Sport University Cologne, Cologne, Germany*

---

*Maybe you know this situation: Imagine you are playing tennis, and you are observing your opponent serve. You watch the ball cross the court with unexpected speed, and you are preparing your return of this hard-to-see ball when, in the moment your racket hits the ball, what comes to mind is "this will not end well." Indeed, the ball ricochets off your racket and flies into the sky. However, you are now prepared to adapt your play to your opponent's hard service; you modify your strategy and thus your behavior.*

As this example shows, to respond to something or to build up expectations about what is yet to come, a person first has to see it, or to put it more generally and scientifically, there must be a *percept* of a stimulus being relevant for the task at hand. If this *perception* is deficient, it is indeed hard to adapt. However, even if adaptation is not possible, that is, behavior is maladaptive, building up expectations or focusing on the relevant information via allocation of attention and/or increase cognitive control of their own action helps people to improve their perception and control of their own actions. Obviously, people differ not only with respect to perceptual capacities but also in how efficiently they build up cognitive control to adapt to a particular situation. Before we dive into this, however, this theme has to be defined in more detail. So: what is meant by such maladaptive behavior? In the context of this section, maladaptive behavior refers to errors. An error is defined as a deficient response with respect to some choice alternatives. This error type is often referred to as "slips." Obviously, there are many more errors types, for example, "mistakes." This term refers to an erroneous choice in a situation where the decision-maker actually does not have a clue about the correct choice. Typically, when committing a slip, the decision-maker knows the correct choice but selects an inappropriate alternative. The focus of this section is on slips, as they are central not only to sports with high time constraints, but also to everyday life.

Hopefully, people commit more slips that can be relatively easily remedied than serious mistakes. However, slips can be very dangerous. Just consider traffic situations: A slip here could be deadly.

This section also provides an overview of the core functions underpinning sensorimotor and cognitive processing, which are the prerequisites of error processing and, thus, performance on a neurophysiological level (Chapter 10), followed by an overview of how misallocated attention might lead to errors (Chapter 11).

Furthermore, besides the relevance of errors in everyday life, it is important to know that performance with respect to perception and cognition varies considerably across the life span within but also between individuals. The most interesting research questions in this regard have to do with (1) what induces developmental differences between individuals, and (2) how cognitive performance can be maintained as people age. Gajewski and Falkenstein (Chapter 12) round out this section with an overview and empirical findings pertaining to this topic.

## NEUROPHYSIOLOGICAL IMPLEMENTATION OF COGNITIVE MECHANISMS GUIDING PERCEPTION AND ACTION

I start by considering the neurophysiological mechanisms that are involved in the integration of perception and action selection (and thus cognition), because these are crucial with respect to the question of why errors occur in situations requiring immediate action. As already stated in the Introduction, the sources of errors in situations where a person has to react immediately, or at least fast, are manifold. For example, the target[1] might not be sufficiently detected, or the available time was too short, or an inappropriate response was chosen due to what is termed *response conflict* (I will come back to this later).

Obviously, this is not the whole story. Unfortunately, humans are not "perfect" task-processing units in a technical sense: For example, attention, motivation, and concentration fluctuate with time-on-task, and thus, the probability of an error varies with time-on-task as well. From a neuroscientific perspective, the likelihood of committing errors is likely related to the stability of task representations in prefrontal networks, attentional selection mechanisms, and mechanisms of action selection in basal ganglia circuits. As already stated, the likelihood of committing an error is not stable over time but rather changes depending on the interplay of different functional neuroanatomical and neurobiological systems. This interplay is reflected, for example, by fluctuations of concentration or motivation during an experiment.

With respect to the neural systems involved in allocation of attention and action control, several theories are important. The most influential are the

---

1. In a typical experiment from cognitive psychology, a *target* refers to the stimulus to which the participant of such an experiment has to respond in a predefined way.

reinforcement learning theory (Holroyd & Coles, 2002) and conceptions proposing that the basal ganglia play an important role in action selection (Bar-Gad, Morris, & Bergman, 2003; Gurney, Humphries, Wood, Prescott, & Redgrave, 2004). Moreover, the dual-state theory of dopamine function (Durstewitz & Seamans, 2008; Seamans & Yang, 2004) and the guided activation theory (Miller, 2000; Miller & Cohen, 2001) seem to be important when trying to elucidate mechanisms that lead to an error.

These theories are important because they describe how task-relevant representations stored in prefrontal cortex (PFC) networks are kept "online" and used to support the selection of actions. Thus, these are important to consider when attempting to describe the neurophysiological processes that are involved in response selection or the occurrence of errors. In particular, in tasks that require continuous and rapid processing of stimuli, errors might emerge because of attentional lapses or even conflicts in stimulus processing. Thus, aspects of attentional selection processes have to be considered, too. In this regard, the biased competition model of attention (Desimone & Duncan, 1995), as Schneider discusses in Chapter 11, is also of particular relevance.

However, when describing the neural mechanisms of information processing and action selection (or decisions), well-established cognitive approaches should be described as well, because these provide a solid framework for exploring how the cognitive system comes to a decision, be it efficient or inefficient. One of the most popular models is the drift diffusion model (DDM; Ratcliff, 1978, 1979, 1980, 2006; Vandekerckhove & Tuerlinckx, 2007), which is one of the mathematical models of decisions with the strongest empirical support.

In the next section, I start with the neurobiological mechanisms involved in committing errors. The perspective presented herein has been partly derived from an integrative review by Hoffmann and Beste (2015). That publication provides a much more detailed introduction.

## Prefrontal Cortex and Attentional Selection

When a person conducts an everyday task and tries to fulfil a goal, the relevant information and actions have to be kept in working memory, which has to be checked continually to verify if the action that was selected with respect to some intention is productive. The central anatomical structure involved in this issue is the PFC. The PFC is, thus, central to the question of why errors that result from deficient stimulus processing (slips) occur: To process information effectively, there has to be a stable representation of such information in the PFC.

Consider typical daily situations where people have a goal that they try to fulfil or a decision they need to make given some information. Often, this fulfilment or decision relies on the establishment of adequate stimulus–response mappings with respect to the task at hand, for example, "go if the traffic light is green, and stop if the traffic light is red." Such stimulus–response representations are stored in working memory within PFC networks (Jonides et al., 2008). These are able to keep and manipulate information (i.e., task-goal representations) via activity states that can persist over time (Seamans & Yang, 2004). Within the PFC, the neurotransmitter dopamine plays a crucial role, because it is involved in maintaining such activity states in the PFC networks (Miller, 2000). More importantly, it seems that dopamine is involved in strengthening current representations against interference by irrelevant distracting information (Miller, 2000). Obviously, this is of interest in the context of emerging errors. In order not to be deleted by information that is competing with the relevant information for the task at hand, the relevant information somehow has to be biased. For example, consider sitting in a car at a red light. You are dreamily awaiting the green light when suddenly your child in the back seat exclaims, "Now it's green!" meaning his or her popsicle is showing green after the red surface has been licked off. Obviously, it is quite dangerous if you immediately follow this acoustic cue. In this situation, it would be helpful if the information cascade of the distracting information is somehow suppressed or inhibited and attention is switched quickly to the relevant information.

How is this actually implemented on a neurobiological level? In the cortex, several neurotransmitter systems are involved in information processing. In the context of this chapter, the dopamine system plays an important role, because it is of great relevance to the PFC or executive functions in general. Several theories exist regarding dopamine function in the brain, but the dual-state theory is of high relevance: according to this theory, the state of the dopamine system is responsible for letting information access working memory (State 1) or stabilizing current representations within PFC networks (State 2; Durstewitz & Seamans, 2008; Seamans & Yang, 2004). Within these networks, basically, four receptor types exist for dopamine (D1–D4), of which types D1 and D2 are relevant for the current discussion. In the first state (State 1), D2-receptor-mediated neural transmission dominates and allows multiple inputs to access working memory networks. In contrast, the second state (State 2) is dominated by D1-receptor-related neural transmission, and the information flow in the working memory networks are relatively closed, but information held within these networks is more stable and controls the output of prefrontal networks (Miller, 2000; Seamans & Yang, 2004). It is likely that these different states serve different brain functions in relation to performance monitoring (Hoffmann & Beste, 2015).

Recent evidence has shown that D2-receptor-mediated neural transmission is associated with exploitative learning that adjusts response times as a function of positive and negative outcomes (Frank, Doll, Oas-Terpstra, & Moreno, 2009). This, in turn, implies that this type of transmission is relevant for error commission and monitoring, as well. However, the D1 and D2 states fluctuate spontaneously and work in an antagonistic way, meaning that during phases in which the D1 state dominates, D2-receptor-related neural transmission is decreased, and vice versa. It has been suggested that increases in D1-receptor activation initially increase the robustness of representations within the PFC (Seamans & Yang, 2004), but a further increase above an optimal level shifts the system away from robustness. Several theories exist with respect to dynamic gating of representation in the PFC (for a selection, compare, e.g., Hazy, Frank, & O'Reilly, 2010; O'Reilly, Herd, & Pauli, 2010). They commonly conclude that the biophysical properties of the PFC determine the stability of information as well as the stability of task goals in the PFC. This, in turn, results in a fluctuating varying error likelihood over time.

The question remains as to how the PFC activity related to task-goal representations contributes to information processing in other cortical areas, for example, to focusing or allocating attention. One theory that provides an answer to this question is the guided activation theory (Miller, 2000; Miller & Cohen, 2001). This theory proposes that representations of task-specific rules stored in PFC networks have an impact on attentional templates and goals to enable goal-directed behavior (Miller, 2000). In other words, the PFC is able to control processing in other brain systems and direct that processing toward task-relevant information (Miller, 2000). This control is particularly important in situations where stimuli are ambiguous, that is, when stimuli activate more than one input representation (Miller & Cohen, 2001). The function of the PFC is, in this regard, to bias neural implementation of rules or goals depending on the target. If these "bias signals" affect sensory modalities, this might well have an effect on mechanisms of attentional control and selection, both factors playing key roles in committing errors.

Indeed, frontal regions, including the anterior cingulate (Lawrence, Ross, Hoffmann, Garavan, & Stein, 2003), the right frontal cortex, and bilateral parietal regions, are relevant with respect to the control of attention over time (Coull, 1998; Posner & Petersen, 1990). Furthermore, they have connections to areas involved in sensory processing and can likely bias information processing in these areas (Knudsen, 2007). For example, distracting information can be inhibited or representations that are relevant for performing a task can be biased. However, because task-goal representation in PFC networks might fluctuate over time (e.g., due to boredom or fatigue), this fluctuation likely affects processing in these sensory areas as well (Desimone & Duncan, 1995; Knudsen, 2007). Obviously, this has an effect on the relative saliency of stimuli that are not linked to correct performance (i.e., distracting stimuli) and thus might determine error likelihood. The biased competition theory of attention (Desimone & Duncan, 1995; Knudsen, 2007) assumes that the outcome of perceptual competition on a neural level is determined by the saliency of the stimuli and intentional biases (top-down influences; e.g., Beste & Dinse, 2013; Schneider, Beste, & Wascher, 2012). When the result of this competition gives the erroneous Stimulus B stronger weight, Stimulus A loses the competition, is not detected, and is not linked to some appropriate action (Desimone & Duncan, 1995). More simply, an error occurs and is not detected.

To sum up, fluctuations on the neural level in the stability of the task-goal representation in the PFC affect attentional selection processes and, therefore, are a major property that determines the likelihood of committing an error in tasks dealing with simple stimulus–response mappings. Thus, errors do not occur simply because of deficient attentional selection processes (e.g., "I have not seen it."), but also because of fluctuations in the task-goal representation (e.g., "Huh, is it over yet?") and the interplay of both.

## "The Winner Takes It All"

However, errors also occur because of an inappropriate action selection ("Oops"). Consider a situation where a stimulus is linked to several response alternatives, and in different tasks, the same stimulus might occur, but a different response might be appropriate. A simple example for this is a Stroop task (Stroop, 1935). Here, for example, the (written) *word* "green" might be associated with the right hand, and the *word color* green might be associated with the left hand. Now, attention might be misguided, but also if time pressure is high, the incorrect response becomes simultaneously activated and, unfortunately, sometimes selected. Thus, the error might occur because of inappropriate action selection.

The key brain regions with respect to action selection are the basal ganglia and the striatum. Indeed, the basal ganglia have a functional connection to the PFC: They are related to processing modes of task-goal representations in PFC networks (O'Reilly et al., 2010). However, to achieve this, the basal ganglia must receive information from the PFC and from areas involved in processing sensory information. It is well established that the PFC is closely connected to the basal ganglia (Chudasama & Robbins, 2006), and more important, they are the key regions with respect to response selection mechanisms (Gurney et al., 2004; Gurney, Prescott, & Redgrave, 2001; Redgrave, Prescott, & Gurney, 1999). Comparable to the biased competition account described above, it has been suggested that the selection of actions (motor commands) occurring at the basal ganglia level depends on the relative saliency of competing actions and that the most salient competitor wins this selection process (Redgrave et al., 1999). More specifically, it has been suggested that the selection of a correct action may be inhibited or overridden by a more salient response alternative than the desired one (Redgrave et al., 1999). This mechanism was termed "the winner take all" (WTA) mechanism (Plenz, 2003; Redgrave et al., 1999). This means that if the saliency of a relevant or irrelevant action falls below that of a competing action on a neural level, the competing action will be executed (Redgrave et al., 1999). Indeed, it was shown that this WTA mechanism can be implemented in neural networks (Bar-Gad et al., 2003; Gurney et al., 2004). The action-selection process takes place within the striatum, and the chosen action is submitted to the basal ganglia and subsequently transmitted to the cortex where the selected response is executed (Bar-Gad et al., 2003). In other words: *the activation with the strongest representation is selected by the basal ganglia WTA network* (Hoffmann & Beste, 2015). In turn, the basal ganglia WTA mechanism most likely selects another action simply because the relative weights, that is, the "presetting" of the network, favor that action (Hoffmann & Beste, 2015). This reweighting of task representations, in turn, may be due to a strengthening of the correct task representation in the PFC or to the sensitivity of the neural networks that process the error-favoring information being reduced. In this context, it is important to know that the aforementioned adjustment of task representations in the PFC is made possible by the neurotransmitter dopamine (Miller, 2000).

In addition, not only goals, or top-down information via prefrontal representations, play a role. It has been shown that even basic visual inputs enter the basal ganglia networks, as well (Hikosaka, 1989; Hikosaka & Wurtz, 1989). This is supported by a growing body of evidence indicating that a subcortical structure, the superior colliculus (a key station in visual processing), has a functional connection to dopaminergic neurons (McHaffie et al., 2006; Redgrave & Gurney, 2006; Silkis, 2000). In sum, several brain areas forward convergent information to the basal ganglia, and this information is central to action selection mechanisms there (Bar-Gad et al., 2003) because it consists basically of a neural representation of stimuli relevant for the task at hand, a task set that has been established in the PFC, and finally a corresponding efference copy.

Moreover, recent models (Humphries, Stewart, & Gurney, 2006) assume that different dopaminergic receptor systems are central to response-selection processes in the basal ganglia. This assumption has also been incorporated in models of the basal ganglia implementing response-inhibition mechanisms (and thus response selection)[2] via go–no-go neuronal networks (Wiecki & Frank, 2013). In sum, the described WTA mechanism appears to be central to the occurrence of errors but also generally to response selection and, thus, performance in tasks requiring processing a number of stimuli in quick succession. Basically, the basal ganglia can be conceptualized as a kind of "central response selection unit" that receives different types of information and has an influence on action selection and, hence, the mechanisms of error commissioning and performance.

As already indicated, the neuromodulator dopamine plays a key role here; thus, I will go into a bit more detail about why this is the case. It can be assumed that the basal ganglia and the action-selection processes mediated via this transmitter system play an important role when considering factors that are possibly important for performance.

---

2. Response selection can be interpreted in terms of inhibition because to select a response, the other one(s) are being inhibited. For example, during execution of a left-hand response in a two-alternative forced choice task, the contralateral motor cortex is activated, but the ipsilateral motor cortex is inhibited.

## The Role of Dopamine

Another aspect of efficient response monitoring, as already noted, is the stability of task-goal representations in PFC structures. As described, this stability is linked to dopamine signals that bind to D1 and D2 receptors. One problem in neurocognitive control models is that the dopamine system is considered to be too slow to support the rapid processes of action selection that lead to erroneous responses; thus, the term neuromodulator is more appropriate for dopamine (Hoffmann & Beste, 2015). However, Jocham and Ullsperger (2009) earlier proposed that error monitoring and the behavioral adaptation processes that follow an error, such as those indicated by the error-related negativity (Falkenstein, Hohnsbein, Hoormann, & Blanke, 1991; Gehring, Goss, Coles, Meyer, & Donchin, 1993),[3] are modulated via slower dopamine responses *after* the error monitoring system has been activated through other (faster) neurotransmitter systems.

Besides this, it seems that the stability of task-goal representations depends on those properties of the dopamine system that seem to provide long latency responses in prefrontal structures that can last several seconds (Seamans & Yang, 2004). However, the core neurotransmitter in corticostriatal synapses is glutamate (Bolam, Hanley, Booth, & Bevan, 2000), and indeed, states in the striatal system can be changed by these glutamatergic inputs and influence action selection processes (Tomkins, Vasilaki, Beste, Gurney, & Humphries, 2013). It is likely that changes in the stabilities of task-goal representations in PFC networks affect the striatal mechanisms of action control via glutamatergic projections. However, sensory input via the superior colliculi modulates activity in striatal dopamine neurons as well (e.g., May et al., 2009; Redgrave, Vautrelle, & Reynolds, 2011). These findings show that dopamine-sensitive neurons are already active prior to behavioral responses or decisions. In sum, it appears as if the dopamine system is involved not only in error processing or behavioral adaptation, but also already prior to action selection, that is, during information accumulation with respect to the stimulus at hand. At this point, it would be useful to consider a cognitive model that can be utilized to quantify different processing stages in the information-processing stream and can be linked conceptually to the neurobiology of error commissioning, or in other terms, fast decisions with respect to some stimulus–response mapping.

## COGNITIVE MODELS: DRIFT DIFFUSION

Having outlined the neurobiological mechanisms that may be important to consider when interested in the neural mechanisms underlying the commission of errors, the question arises: How can the involved constructs, such as stimulus

---

3. The error-related negativity or error negativity is a negativity in an electroencephalogram (EEG) that can be observed immediately after committing an error at fronto-central electrode positions.

processing, action selection, decisions, and erroneous and correct information processing, be quantified so they can be related to these neurophysiological mechanisms? Note, however, that initially it has to be stipulated that in the context of this chapter, information processing and action selection are interpreted in a wider sense as processes that are involved in decision-making (also see Werner, Chapter 20).

There are well-established models related to this topic. In the field of cognitive modeling of decision-making, one perspective assumes that evidence related to stimulus processing is accumulated via a stochastic, noisy process until a decision criterion is reached (for a recent overview, cf. Ratcliff, 2013). The DDM (Ratcliff, 1978) decomposes reaction time distributions and error rates (or more generally decisions) into several parameters. Basically, it is assumed that information is being accumulated during the performance of a binary choice until a certain threshold is reached and a decision is made. This information accumulation is modeled as a stochastic process that drifts into different decisional outcomes, for example, errors (compare Figure 1).

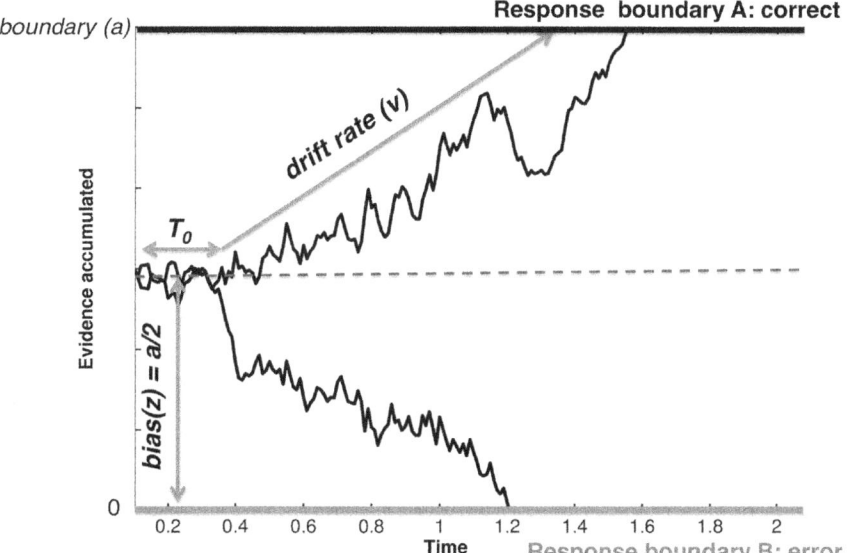

**FIGURE 1** **Illustration of the diffusion process of two decisions (correct, error) between two response alternatives A and B.** The curved line represents the accumulation of information over time (drift rate, $v$) until boundary separation $a$ is reached, which is the time point of the decision for the corresponding response. The reaction time is, therefore, a function of the boundary separation, the speed of information processing (as reflected by the steepness of $v$), the non-decisional time $T_0$ (which reflects basic stimulus processing and/or motor processes), and the starting point $z$ (which refers to how conservative or liberal the subject is with respect to one of the response alternatives). Note that the scales of the x-axis (Evidence accumulated) and y-axis (Time) are arbitrary because these data had been simulated for illustration purposes via Fast-dm (Voss & Voss, 2007).

The four key parameters of the DDM (Figure 1) are the drift rate ($v$), the boundary or threshold separation ($a$), the starting point ($z$), and the duration of the non-decisional process ($T_0$). In terms of cognitive psychology, the drift rate reflects the speed of evidence accumulation or the speed of information processing. Thus, the drift rate is a performance measure. It reflects, for example, the difficulty of the task. The smaller the drift rate, the more difficult the task. Vice versa; if the drift rate is high, the decision threshold is reached faster, which is reflected in short response times. The next parameter, the boundary separation ($a$), describes how much information is needed until a decision can be made. Large values indicate rather conservative decision strategies, and small values indicate rather liberal decision strategies. Indeed, the boundary separation is sensitive to speed and accuracy instructions (Voss, Rothermund, & Voss, 2004), that is, $a$ increases if participants are instructed to respond as precisely as possible and decreases if they are instructed to respond as fast as possible. The starting point $z$ describes the a priori bias associated with one of the choices, for example, if one of the responses is associated with a larger reward than the other (Voss et al., 2004). The fourth parameter, that is, the duration of non-decisional processes, quantifies processes such as basic encoding and/or response execution (more specifically, motor activity). Obviously, all these parameters are related to the question of how liberally or conservatively participants process errors. For example, obsessive–compulsive disorder patients tend to increase their response monitoring, as indicated by the error-related negativity (Endrass, Klawohn, Schuster, & Kathmann, 2008). In summary, these parameters partly answer the question of how the time course and cognitive processes involved in error commissioning can be quantified and predicted.

At this point, it is very important that the DDM can also be linked to biological models that link frontal cortex functions with the basal ganglia (Ratcliff & Frank, 2012). A link (at least a correlative) between the DDM and the functions of the medial PFC and basal ganglia seems to be obvious: it has been suggested that following stimulus presentation, the medial PFC generates possible actions with different probabilities of execution that depend on the specific stimulus representation (Cavanagh et al., 2011). This is supported by the finding that an EEG correlate of response monitoring, that is, the correct response negativity (Vidal, Hasbroucq, Grapperon, & Bonnet, 2000), varies as a function of the response given a certain stimulus probability (Meckler et al., 2011). Furthermore, response monitoring appears to be attenuated if the probability of a certain response is inhibited. This can happen if prior to response selection, a defined visual spatial stimulus position is inhibited (Hoffmann & Wascher, 2012).

Let us return to the link between the DDM and neural mechanisms of response monitoring. Cavanagh et al. (2011) suggested that in the presence of response conflict, the medial PFC–STN (prefrontal cortex–subthalamic nucleus) network should increase the decision threshold (i.e., *the boundary separation*) to enable the corticostriatal network to process the reward values of the response alternatives (Cavanagh et al., 2011). This hypothesis was investigated via integration of the core ideas of Frank's (2006) basal ganglia model and the DDM. More specifically, Cavanagh et al. (2011) investigated whether the STN has an inhibitory influence on corticostrial functions during decision conflict. They found that single-trial[4] medial PFC activity, as measured based on EEG theta power,[5] correlated with the threshold for evidence accumulation, and this relationship was modulated by conflict. Interestingly, besides these correlative findings, results show that deep-brain stimulation of the STN in Parkinson's patients strongly modulated this relationship, indicating a causal link between STN and PFC functions.

Finally, the DDM provides opportunities to test the key assumption of the integrative view presented by Hoffmann and Beste (2015), that is, the modulation of error likelihood, on a behavioral level. The DDM allows quantification not only of the four previously mentioned parameters but also of the parameters that reflect inter-trial fluctuations in decision processes. These fluctuations are captured by the variability in the drift rate (e.g., variability due to fluctuations in attention), starting point (e.g., due to online reward or punishment), and the nondecisional parameter. Apparently, there is a close link between the DDM and the error-commissioning mechanisms described in the previous section: Fluctuations in error likelihood are determined by the strengths of the representations in prefrontal networks. These are conceptually reflected in the boundary separation and starting point variability in the DDM. Moreover, the mechanisms associated with attentional selection are conceptually similar to the drift rate parameter proposed in the DDM. Aspects of vigilance are closely related to the perceptual processing of stimuli, which is modeled by the non-decisional parameter and its variability. Thus, if these processes (via experimental manipulation) or structures (via, e.g., manipulation via transcranial magnetic stimulation deep brain stimulation) are varied, one would expect effects in the corresponding DDM parameters, and in line with this, changes in the fluctuations in error likelihood. However, there are many approaches to the cognitive processes involved in stimulus processing and response selection. For example, the dual-stage model (Hübner, Steinhauser, & Lehle, 2010) assumes two phases (an early and a late phase) of response selection: Initially (Phase 1) stimulus information has an effect on the categorical selection and response-selection processes. During the second phase, response selection is driven by categorical selection. However, an open debate about these competing models exists (White, Ratcliff, & Starns, 2011) because

---

4. In EEG or more general cognitive neuroscience experiments, a "single-trial" parameter refers to a parameter that is derived in each trial of the experiment. For an excellent paper about integration of behavioral and brain dynamics, consider Makeig, Debener, Onton, and Delorme (2004).
5. The EEG signal of a channel at given time points can be decomposed into different frequency bands via wavelet transformation. The most relevant frequency in this context is theta (which ranges from 4 to 7 Hz) at frontal electrode positions. Power refers, simply, to an increase or decrease of the amplitude or "loudness" of the frequency at given time points relative to pre-stimulus baseline.

the dual-stage model has been exemplified with a flanker task and, therefore, might not be generalizable to other types of tasks. This is not the case for the DDM, which not only has been applied to a wide range of cognitive tasks (e.g., memory) but also has the advantage of being able to quantify a wide range of different parameters.

## CONCLUSION AND OUTLOOK

In this introductory chapter, a perspective has been laid down with respect to the neurobiological mechanisms underpinning errors in situations where stimuli have to be processed rapidly. However, the involved mechanisms and brain functions also play a role in higher order cognition, for example, in error monitoring (see also Hoffmann, Chapter 10). There, the involved mechanisms and brain functions overlap with those described in this chapter, since adaptation with respect to an error might require behavioral adaptation as well as adaptation on a perceptual and cognitive level. In other words, if you commit an error, you better have a closer look at what and how you are doing it, and what you are actually seeing.

However, the allocation of attention and response monitoring are key processes with respect to adequate performance. Schneider (Chapter 11) goes into more detail with respect to attention mechanisms (e.g., biased competition) as a prerequisite for performance. Additionally, it was suggested that neurobiological models could be linked to behavioral or cognitive models as exemplified by the DDM. However, this requires sophisticated methods, and it will be a challenge to develop and extend models with respect to this topic. A key step in this regard has been the "renaissance" of the EEG in the last 10 years. The high temporal resolution of the EEG (Debener, Ullsperger, Siegel, & Engel, 2006; Makeig et al., 2004) has provided a huge number of novel methods that now make it possible to investigate and integrate brain dynamics and behavioral dynamics. My aim in presenting the DDM model was to exemplify that these neurophysiologic mechanisms can be related to cognitive modeling. This might be an important aspect if considering integrating neuroscience methods, that is, EEG methods and the investigation of cognition, perception, and action in the context of performance science. In this context, the challenge will be to select the appropriate parameter that reflects the activity of the neural network by which it is assumed to be linked to the cognitive model of interest.

On a theoretical level, the focus in this chapter was on a certain process that leads to an error on the neural level. The focus in Hoffmann (Chapter 10), in contrast, is on the response level, that is, on *error processing*. Obviously, there are many more error types that have not been included in this short introduction. However, even errors emerging from those situations relevant in this chapter have a close connection to the applied context: the neural mechanisms described herein are core mechanisms of action selection and behavioral adaptation in this regard. Thus, they are also relevant for the final chapter in this section (Falkenstein & Gajewski, Chapter 12): The dopamine system is at the core of learning and executive functions. These executive functions or fluid intelligence are of key interest with respect to maintaining cognitive performance across the life span.

## REFERENCES

Bar-Gad, I., Morris, G., & Bergman, H. (2003). Information processing, dimensionality reduction and reinforcement learning in the basal ganglia. *Progress in Neurobiology*, *71*, 439–473. http://dx.doi.org/10.1016/j.pneurobio.2003.12.001.

Beste, C., & Dinse, H. R. (2013). Learning without training. *Current Biology*, *23*, R489–R499. http://dx.doi.org/10.1016/j.cub.2013.04.044.

Bolam, J. P., Hanley, J. J., Booth, P. A., & Bevan, M. D. (2000). Synaptic organisation of the basal ganglia. *Journal of Anatomy*, *196*, 527–542.

Cavanagh, J. F., Wiecki, T. V., Cohen, M. X., Figueroa, C. M., Samanta, J., Sherman, S. J., et al. (2011). Subthalamic nucleus stimulation reverses mediofrontal influence over decision threshold. *Nature Neuroscience*, *14*, 1462–1467. http://dx.doi.org/10.1038/nn.2925.

Chudasama, Y., & Robbins, T. W. (2006). Functions of frontostriatal systems in cognition: comparative neuropsychopharmacological studies in rats, monkeys and humans. *Biological Psychology*, *73*, 19–38. http://dx.doi.org/10.1016/j.biopsycho.2006.01.005.

Coull, J. T. (1998). Neural correlates of attention and arousal: insights from electrophysiology, functional neuroimaging and psychopharmacology. *Progress in Neurobiology*, *55*, 343–361.

Debener, S., Ullsperger, M., Siegel, M., & Engel, A. K. (2006). Single-trial EEG-fMRI reveals the dynamics of cognitive function. *Trends in Cognitive Sciences*, *10*, 558–563.

Desimone, R., & Duncan, J. (1995). Neural mechanisms of selective visual attention. *Annual Review of Neuroscience*, *18*, 193–222.

Durstewitz, D., & Seamans, J. K. (2008). The dual-state theory of prefrontal cortex dopamine function with relevance to catechol-o-methyltransferase genotypes and schizophrenia. *Biological Psychiatry*, *64*, 739–749. http://dx.doi.org/10.1016/j.biopsych.2008.05.015.

Endrass, T., Klawohn, J., Schuster, F., & Kathmann, N. (2008). Overactive performance monitoring in obsessive-compulsive disorder: ERP evidence from correct and erroneous reactions. *Neuropsychologia*, *46*, 1877–1887.

Falkenstein, M., Hohnsbein, J., Hoormann, J., & Blanke, L. (1991). Effects of crossmodal divided attention on late ERP components. II. Error processing in choice reaction tasks. *Electroencephalography and Clinical Neurophysiology*, *78*, 447–455.

Frank, M. J. (2006). Hold your horses: a dynamic computational role for the subthalamic nucleus in decision making. *Neural Networks: The Official Journal of the International Neural Network Society*, *19*, 1120–1136. http://dx.doi.org/10.1016/j.neunet.2006.03.006.

Frank, M. J., Doll, B. B., Oas-Terpstra, J., & Moreno, F. (2009). Prefrontal and striatal dopaminergic genes predict individual differences in exploration and exploitation. *Nature Neuroscience*, *12*, 1062–1068. http://dx.doi.org/10.1038/nn.2342.

Gehring, W. J., Goss, B., Coles, M. G. H., Meyer, D. E., & Donchin, E. (1993). A neural system for error detection and compensation. *Psychological Science*, *4*, 385–390.

Gurney, K., Humphries, M., Wood, R., Prescott, T. J., & Redgrave, P. (2004). Testing computational hypotheses of brain systems function: a case study with the basal ganglia. *Network*, *15*, 263–290.

Gurney, K., Prescott, T. J., & Redgrave, P. (2001). A computational model of action selection in the basal ganglia. II. Analysis and simulation of behaviour. *Biological Cybernetics*, *84*, 411–423.

Hazy, T. E., Frank, M. J., & O'Reilly, R. C. (2010). Neural mechanisms of acquired phasic dopamine responses in learning. *Neuroscience & Biobehavioral Reviews, 34*, 701–720. http://dx.doi.org/10.1016/j.neubiorev.2009.11.019.

Hikosaka, O. (1989). Role of basal ganglia in saccades. *Revue Neurologique (Paris), 145*, 580–586.

Hikosaka, O., & Wurtz, R. H. (1989). The basal ganglia. *Reviews of Oculomotor Research, 3*, 257–281.

Hoffmann, S., & Beste, C. (2015). A perspective on neural and cognitive mechanisms of error commission. *Frontiers in Behavioral Neuroscience, 9*. http://dx.doi.org/10.3389/fnbeh.2015.00050 Article 50.

Hoffmann, S., & Wascher, E. (2012). Spatial cueing modulates the monitoring of correct responses. *Neuroscience Letters, 506*, 225–228. http://dx.doi.org/10.1016/j.neulet.2011.11.011.

Holroyd, C. B., & Coles, M. G. (2002). The neural basis of human error processing: reinforcement learning, dopamine, and the error-related negativity. *Psychological Review, 109*, 679–709.

Humphries, M. D., Stewart, R. D., & Gurney, K. N. (2006). A physiologically plausible model of action selection and oscillatory activity in the basal ganglia. *Journal of Neuroscience, 26*, 12921–12942. http://dx.doi.org/10.1523/JNEUROSCI.3486-06.2006.

Hübner, R., Steinhauser, M., & Lehle, C. (2010). A dual-stage two-phase model of selective attention. *Psychological Review, 117*, 759–784. http://dx.doi.org/10.1037/a0019471.

Jocham, G., & Ullsperger, M. (2009). Neuropharmacology of performance monitoring. *Neuroscience & Biobehavioral Reviews, 33*, 48–60.

Jonides, J., Lewis, R. L., Nee, D. E., Lustig, C. A., Berman, M. G., & Moore, K. S. (2008). The mind and brain of short-term memory. *Annual Review of Psychology, 59*, 193–224. http://dx.doi.org/10.1146/annurev.psych.59.103006.093615.

Knudsen, E. I. (2007). Fundamental components of attention. *Annual Review of Neuroscience, 30*, 57–78.

Lawrence, N. S., Ross, T. J., Hoffmann, R., Garavan, H., & Stein, E. A. (2003). Multiple neuronal networks mediate sustained attention. *Journal of Cognitive Neuroscience, 15*, 1028–1038. http://dx.doi.org/10.1162/089892903770007416.

Makeig, S., Debener, S., Onton, J., & Delorme, A. (2004). Mining event-related brain dynamics. *Trends in Cognitive Sciences, 8*, 204–210. http://dx.doi.org/10.1016/j.tics.2004.03.008.

May, P. J., McHaffie, J. G., Stanford, T. R., Jiang, H., Costello, M. G., Coizet, V., et al. (2009). Tectonigral projections in the primate: a pathway for pre-attentive sensory input to midbrain dopaminergic neurons. *European Journal of Neuroscience, 29*, 575–587. http://dx.doi.org/10.1111/j.1460-9568.2008.06596.x.

McHaffie, J. G., Jiang, H., May, P. J., Coizet, V., Overton, P. G., Stein, B. E., et al. (2006). A direct projection from superior colliculus to substantia nigra pars compacta in the cat. *Neuroscience, 138*, 221–234. http://dx.doi.org/10.1016/j.neuroscience.2005.11.015.

Meckler, C., Allain, S., Carbonnell, L., Hasbroucq, T., Burle, B., & Vidal, F. (2011). Executive control and response expectancy: a Laplacian ERP study. *Psychophysiology, 48*, 303–311. http://dx.doi.org/10.1111/j.1469-8986.2010.01077.x.

Miller, E. K. (2000). The prefrontal cortex and cognitive control. *Nature Reviews Neuroscience, 1*, 59–65. http://dx.doi.org/10.1038/35036228.

Miller, E. K., & Cohen, J. D. (2001). An integrative theory of prefrontal cortex function. *Annual Review of Neuroscience, 24*, 167–202. http://dx.doi.org/10.1146/annurev.neuro.24.1.167.

O'Reilly, R. C., Herd, S. A., & Pauli, W. M. (2010). Computational models of cognitive control. *Current Opinion in Neurobiology, 20*, 257–261. http://dx.doi.org/10.1016/j.conb.2010.01.008.

Plenz, D. (2003). When inhibition goes incognito: feedback interaction between spiny projection neurons in striatal function. *Trends in Neuroscience, 26*, 436–443. http://dx.doi.org/10.1016/S0166-2236(03)00196-6.

Posner, M. I., & Petersen, S. E. (1990). The attention system of the human brain. *Annual Review of Neuroscience, 13*, 25–42. http://dx.doi.org/10.1146/annurev.ne.13.030190.000325.

Ratcliff, R. (1978). A theory of memory retrieval. *Psychological Review, 85*, 59–108.

Ratcliff, R. (1979). Group reaction time distributions and an analysis of distribution statistics. *Psychological Bulletin, 86*, 446–461.

Ratcliff, R. (1980). A note on modeling accumulation of information when the rate of accumulation changes over time. *Journal of Mathematical Psychology, 21*, 178–184.

Ratcliff, R. (2006). Modeling response signal and response time data. *Cognitive Psychology, 53*, 195–237.

Ratcliff, R. (2013). Parameter variability and distributional assumptions in the diffusion model. *Psychological Review, 120*, 281–292. http://dx.doi.org/10.1037/a0030775.

Ratcliff, R., & Frank, M. J. (2012). Reinforcement-based decision making in corticostriatal circuits: mutual constraints by neurocomputational and diffusion models. *Neural Computation, 24*, 1186–1229. http://dx.doi.org/10.1162/NECO_a_00270.

Redgrave, P., & Gurney, K. (2006). The short-latency dopamine signal: a role in discovering novel actions? *Nature Reviews Neuroscience, 7*, 967–975. http://dx.doi.org/10.1038/nrn2022.

Redgrave, P., Prescott, T. J., & Gurney, K. (1999). The basal ganglia: a vertebrate solution to the selection problem? *Neuroscience, 89*, 1009–1023.

Redgrave, P., Vautrelle, N., & Reynolds, J. N. J. (2011). Functional properties of the basal ganglia's re-entrant loop architecture: selection and reinforcement. *Neuroscience, 198*, 138–151. http://dx.doi.org/10.1016/j.neuroscience.2011.07.060.

Schneider, D., Beste, C., & Wascher, E. (2012). Attentional capture by irrelevant transients leads to perceptual errors in a competitive change detection task. *Frontiers in Psychology, 3*. http://dx.doi.org/10.3389/fpsyg.2012.00164.

Seamans, J. K., & Yang, C. R. (2004). The principal features and mechanisms of dopamine modulation in the prefrontal cortex. *Progress in Neurobiology, 74*, 1–58. http://dx.doi.org/10.1016/j.pneurobio.2004.05.006.

Silkis, I. (2000). The cortico-basal ganglia-thalamocortical circuit with synaptic plasticity. I. Modification rules for excitatory and inhibitory synapses in the striatum. *Biosystems, 57*, 187–196.

Stroop, J. R. (1935). Studies of interference in serial verbal reactions. *Journal of Experimental Psychology, 18*, 643–662. http://dx.doi.org/10.1037/h0054651.

Tomkins, A., Vasilaki, E., Beste, C., Gurney, K., & Humphries, M. D. (2013). Transient and steady-state selection in the striatal microcircuit. *Frontiers in Computational Neuroscience, 7*. Article 192. http://dx.doi.org/10.3389/fncom.2013.00192.

Vandekerckhove, J., & Tuerlinckx, F. (2007). Fitting the Ratcliff diffusion model to experimental data. *Psychonomic Bulletin & Review, 14*, 1011–1026.

Vidal, F., Hasbroucq, T., Grapperon, J., & Bonnet, M. (2000). Is the "error negativity" specific to errors? *Biological Psychology, 51*, 109–128.

Voss, A., Rothermund, K., & Voss, J. (2004). Interpreting the parameters of the diffusion model: an empirical validation. *Memory & Cognition, 32*, 1206–1220.

Voss, A., & Voss, J. (2007). Fast-dm: a free program for efficient diffusion model analysis. *Behavior Research Methods, 39*(4), 767–775. http://dx.doi.org/10.3758/BF03192967.

White, C. N., Ratcliff, R., & Starns, J. J. (2011). Diffusion models of the flanker task: discrete versus gradual attentional selection. *Cognitive Psychology, 63*, 210–238. http://dx.doi.org/10.1016/j.cogpsych.2011.08.001.

Wiecki, T. V., & Frank, M. J. (2013). A computational model of inhibitory control in frontal cortex and basal ganglia. *Psychological Review, 120*, 329–355. http://dx.doi.org/10.1037/a0031542.

# Chapter 10

# Performance and Error Monitoring: Causes and Consequences

### Sven Hoffmann
*Department of Performance Psychology, Institute of Psychology, German Sport University Cologne, Cologne, Germany*

---

*What is it that makes us conduct errors? In today's multimedia world, the tasks we have to conduct are becoming increasingly complex with respect to cognitive demands. Consider a simple situation like steering your car while attending to your navigation system: Obviously, you have to divide your attention between route guidance and observing the street just ahead. Further, immediate action might be necessary to adapt to the traffic situation. To be able to adapt, you have to not only control your actions but also allocate your attention.*

Both processes are fundamental to appropriate action and, thus, play an important role in sports, as well. In many sports situations, athletes can expect to face the demands of allocating attention and controlling action. Obviously, these situations are error prone. The detection of errors and the efficient adaptation or control of behavior are crucial with respect to performance in this regard. Consider the risk of injuries or the efficiency of training to improve performance. In this chapter, I describe the neurobiological mechanisms that are highly relevant to this applied context, since they are fundamental to controlled action (i.e., *cognitive control*).

This phenomenon might be familiar: You are initiating a movement, and in the same moment you think: "Oops!" However, obviously there exist several different error types. A distinction can be made between "mistakes" (e.g., not knowing the correct decision) and "slips" (the selected action is not what was intended). Another way to look at it is that there are basically two error types: errors due to cognitive overload and impulsive errors (i.e., slips, fast guessing; Reason, 1990). The latter is what this chapter is about: a situation leading to an inappropriate action selection.

Rabbitt showed as far back as 1966 that if subjects committed an error in a response-selection task, they detected this error immediately and adapted their strategy either by immediately correcting the error or by prolonging the response time in the subsequent trial. Indeed, this early study showed that following an error, the probability of another one is reduced (Rabbitt, 1966). This can be explained in terms of a phenomenon closely related to performance, the speed–accuracy trade-off. This finding led to the assumption of a general "response-monitoring" system. This was replicated later, and a neurophysiologic system was identified that seemed to be central to this mechanism (Gehring, Goss, Coles, Meyer, & Donchin, 1993; Hohnsbein, Falkenstein, Hoormann, & Blanke, 1991).

In the following sections, I describe the core theories of error monitoring followed by a discussion of how neurophysiological error monitoring correlates are being measured, as well as methodological issues that arise when doing so. Finally, I look at the predictive use of this error-monitoring system, that is, its use in the applied context.

## THEORIES OF ERROR PROCESSING

The first brain signal that was identified accompanying errors was found in studies utilizing the electroencephalogram (EEG). A sharp negative deflection was observed immediately following an erroneous response (about 40–100 ms following response onset) at fronto-central electrode positions (compare Figure 1). This signal was termed "error negativity" (Ne; Falkenstein, Hohnsbein, Hoormann, & Blanke, 1991) or "error-related negativity" (ERN; Gehring et al., 1993). Later, the Ne was shown to be a correlate of error monitoring and also an important correlate of behavioral adaptation in general.

There is much evidence from source localization (Dehaene, Posner, & Tucker, 1994) and functional magnetic resonance imaging (Debener et al., 2005; Mathalon, Whitfield, & Ford, 2003) studies that the neural generators of the Ne

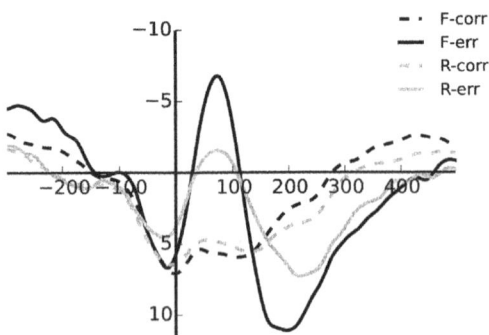

**FIGURE 1** Response-related averaged electroencephalogram (EEG) readings (y-axis in μV, negativity upward, x-axis=time in ms, time point zero indicates response onset) at location FCz of correct and erroneous responses in two tasks (F=flanker task; R=mental rotation; err=error; corr=correct). As an example, in the upper right, a topographic map of the EEG activity across the whole scalp at the time point of the maximum error negativity peak in a flanker task is shown. One can see the typical distribution across the scalp: a maximum at fronto-central electrode leads.

are located in the rostral part of the anterior cingulate cortex (ACC). This finding is underlined by the apparent involvement of the ACC in performance monitoring in general (Ridderinkhof, Ullsperger, Crone, & Nieuwenhuis, 2004).

Currently, several theories exist regarding the functional role of brain signals accompanying errors. I have been—and only can be—selective in those theories I have included in this discussion; however, those I describe here appear to be the most influential ones in the literature.

The mismatch hypothesis was one of the first hypotheses that aimed to explain the functional role of neurophysiological processes accompanying errors. The mismatch hypothesis assumed that the neural representations of initiated and/or demanded (re-) actions are compared. Accordingly, the Ne/ERN reflects a process of comparing the output of the motor system (i.e., an efference copy) with the plan of the response. Put more simply, a Ne is elicited when there is a mismatch between the efference copy and the action plan (Falkenstein et al., 1991; Gehring et al., 1993). As much more is now known about the neuronal architecture and function of the involved systems, technically the mismatch hypothesis is no longer up-to-date. However, the basic concept of a comparison of input and output is still, at least implicitly, present in newer models. Today, the most influential models seem to be the conflict model (Botvinick, Braver, Barch, Carter, & Cohen, 2001; van Veen, Cohen, Botvinick, Stenger, & Carter, 2001) and the reinforcement learning theory (RFL) of Holroyd and Coles (2002); the predicted-response–outcome (PRO) model (Alexander & Brown, 2011) is also gaining influence.

Botvinick et al. (2001) formulated the conflict monitoring theory because the mismatch hypothesis could not explain several empirical patterns. Whereas the mismatch hypothesis assumes that there is a comparator system that has access to information specifying which piece of information to be compared is correct (the system is some kind of "accuracy evaluator"), the conflict hypothesis circumvents this "homunculus" difficulty by proposing that the degree of "conflict" is central to monitoring. More specifically, the conflict theory assumes that in all situations in which two or more actions can be performed, a conflict between these response options emerges. Here, the term conflict refers to a temporal overlap of (pre-)activated response sets. On the functional level, this conflict indicates the need to increase (cognitive) control. With respect to the occurrence of errors, the conflict-monitoring theory assumes that errors (at least fast guesses or action slips) emerge from conflicts of nearly simultaneously activated response sets from which the erroneous response is nearly automatically activated. There is strong evidence for this model, at least for tasks that typically induce response conflict (Yeung, Botvinick, & Cohen, 2004). Indeed, fast guesses and reaction slips often occur in these types of tasks. In other tasks in which the stimulus–response mapping is (initially) unclear (e.g., probability learning tasks) and in tasks with high working memory load (e.g., mental rotation, Hoffmann, Labrenz, Themann, Wascher, & Beste, 2014), there is a minimum of response conflict. This is because if errors occur in these tasks, people either have no information about the correct solution or simply cannot maintain the correct stimulus–response mapping in working memory. However, conflict can even exist at the attentional and stimulus-processing levels, at least in situations that require efficient perceptual processing (see also Hoffmann, Chapter 9, and Schneider, Chapter 11).

The second influential theory about the processing of erroneous outcomes suggests that the dopamine (DA) system of the basal ganglia is the key system that mediates error processing. The RFL hypothesis (Holroyd & Coles, 2002) assumes that error signals are transferred by the mesencephalic DA system to the ACC and are used to train the ACC to optimize performance on the task at hand. In this theory, the ACC acts as a motor control filter (Holroyd and Coles termed this the "adaptive critic") that calculates what motor commands are issued to the motor system. Furthermore, various neural structures that project to the ACC, such as the dorsolateral cortex, the orbitofrontal cortex, the basal ganglia, and the amygdala, have access to motor execution (Holroyd & Coles, 2002). With respect to error detection and the relation to DA functions, the RFL theory is related to the temporal difference model of DA function (for details, see Suri, 2002; Suri & Schultz, 1998, 2001). However, the reinforcement learning does not exclude conflict mechanisms, nor vice versa (Holroyd, Yeung, Coles, & Cohen, 2005).

Of course, actions are not adapted only via the detection of and compensation for errors. Many actions require more subtle adaptation, for example, the evaluation of correct responses to increase the precision of the performance of a motor task and/or the evaluation of whether the outcome of an action is expected/appropriate. In fact, even following correct responses, a similar negativity can be observed, known as the "Ne-like" negativity (Vidal, Hasbroucq, Grapperon, & Bonnet, 2000). It is probable that the Ne-like wave is a reflection of the same process that underlies the Ne (Hoffmann & Falkenstein, 2010; Roger, Benar, Vidal, Hasbroucq, & Burle, 2010; Vidal et al., 2000). Additionally, the key source of these signals, the ACC, is involved not only in processes related to error detection and compensation but also more generally in processes related to response monitoring and learning of stimulus–response relations (Rushworth, Walton, Kennerley, & Bannerman, 2004). For example, if an event occurs that is unexpected or undesired (i.e., external unexpected/negative feedback), a so-called feedback-related negativity (FRN) can be observed (Miltner, Braun, & Coles, 1997). The sources of this FRN can be located close to those of the Ne (Ridderinkhof et al., 2004). The functional connection of Ne and FRN is well explained by the RFL hypothesis. It has been assumed that both reflect the adaptation of the involved neurobiological systems during the learning of stimulus–response contingencies (Holroyd, Pakzad-Vaezi, & Krigolson, 2008; Nieuwenhuis, Holroyd, Mol, & Coles, 2004). More specifically, Ne and FRN have been related to reinforcement learning and expectations (Ridderinkhof et al., 2004): According to the RFL theory, FRN and Ne indicate a negative reward prediction error, a signal that is elicited whenever the monitoring system has to adapt to its reward expectations (Bellebaum, Kobza, Thiele, & Daum, 2011). Accordingly, the amplitude of the Ne/FRN should be proportional to the size of the prediction error. From this, it can be predicted that the amplitude of the FRN should be dependent on the difference between the actual and the expected outcome of an action or event (Holroyd & Coles, 2002).

The third model of response monitoring that is gaining attention is the PRO model (Alexander & Brown, 2011), which focuses on the functional role of the medial prefrontal cortex (mPFC), or more specifically, the ACC, with respect to errors, error likelihood, conflict, reward valence, and punishment. This model is a probabilistic model of the mPFC. It assumes that the mPFC is mainly involved in learning and predicting the outcome of actions, regardless of the motivational saliency. Essentially, the model assumes that over the course of an experiment, the ACC learns a prediction of the possible responses related to a stimulus and their corresponding outcomes. The signal from which this relationship is learned consists of integrated response–outcome combinations, that is, prediction errors, which consist of unexpected outcomes and unexpected nonoccurrences (see also Alexander & Brown, 2011; Figure 1(A)). This model is important because it can make exact predictions about the functions of the PFC and learning progress. Additionally, the PRO model details how information is processed until a decision is reached (i.e., action selection and response execution). In sum, the PRO model suggests not only that the mPFC is the key region involved in cognitive control, but also that the mPFC is concerned with establishing predictions with respect to actions in general.

In addition to behavioral adaptation, there is a modulation on the level of stimulus processing following an error. It appears reasonable to focus attention on task-relevant features following an error if the error was due to inadequate stimulus processing. Indeed, it has been shown that rare events draw attention away from the processing of target stimuli (Notebaert et al., 2009). Furthermore, activity in the posterior mPFC predicts activity both in motor areas and in areas involved in visual processing (Danielmeier, Eichele, Forstmann, Tittgemeyer, & Ullsperger, 2011). Indeed, task-relevant information can be strengthened on the neural level via cognitive control mechanisms (see also Hoffmann, Chapter 9; Egner & Hirsch, 2005). Further, the Ne amplitude correlates with the amount of adjustment with respect to selective attention; that is, the congruency effect in a flanker task was modulated by Ne amplitude[1] (Maier, Yeung, & Steinhauser, 2011). It is likely that the ACC also solves conflicts by strengthening the correct stimulus–response relation

---

1. Congruency effect: In a standard flanker task, a central target occurs accompanied by lateral "flankers." These flankers can be "congruent," i.e., match the target and the corresponding response, or "incongruent," i.e., match another target and the corresponding response alternative. The congruency effect refers to an increase in the response time in the incongruent condition compared to the congruent condition.

(Burnod et al., 1999; Inoue et al., 2000). Accordingly, it has been hypothesized that cholinergic projections of the ACC contribute to executive control to inhibit distracting information (Sarter, Gehring, & Kozak, 2006).

Before turning to phenomena related to this response-monitoring system and some applications, it is important to consider a few methodological issues. This is so important because measuring the exact timing and temporal synchronization of experimental events and EEG is an especially critical issue when trying to quantify response-related EEG activity. In this case, there may be issues related to the hardware that are not immediately obvious. Therefore, the following section provides a short overview of common problems and practical advice on solving or controlling for them.

## METHODOLOGICAL ISSUES WHEN MEASURING RESPONSE-RELATED EEG ACTIVITY

When measuring response-related EEG activity, some initial methodological issues have to be considered, especially if one is interested in measuring error-related EEG activity (i.e., measuring the Ne), since there are only a few trials available, and thus, data quality is of particular importance. So let us first consider the basic procedure.

The Ne is quantified response related, and in most cases the participant has to indicate the desired response via a button press. At this time point, a marker (or trigger) is written to the EEG. During the off-line analysis, the EEG is segmented relative to this response trigger, and the response-related segments are averaged to estimate the Ne.[2] However, one problem is that a button (at least a mechanical one—the problem is not relevant if utilizing, e.g., photo sensors) requires physical pressure and has a predefined level threshold at which the electrical signal is sent. Thus, the signal for a button press is generated several milliseconds later than the response is actually initiated. This leads to a considerable latency shift in the estimated Ne. Furthermore, the steepness of the rise of force is related to the time point at which the button signal comes after the true initiated response. Obviously, if the rise of force is rather flat, the temporal difference between true onset and button press is larger than when there is a less steep rise of force. This might lead to problems in experimental designs employing response-related EEG. If this steepness is modulated by experimental manipulations (irrespective of whether it is a between-subjects or within-subject design), response-related EEG parameters such as the Ne cannot be estimated reliably. For example, if participants were unsure in one condition and sure in another, the difference between the two response-related potentials might be due to a higher variability of the Ne as a result of Ne latency jitter in one condition. Indeed, with respect to force, time-to-peak (the time from force onset to maximum force) increases with stimulus intensity and duration

---

2. For clarity, the description leaves out preprocessing, such as artifact handling or filtering.

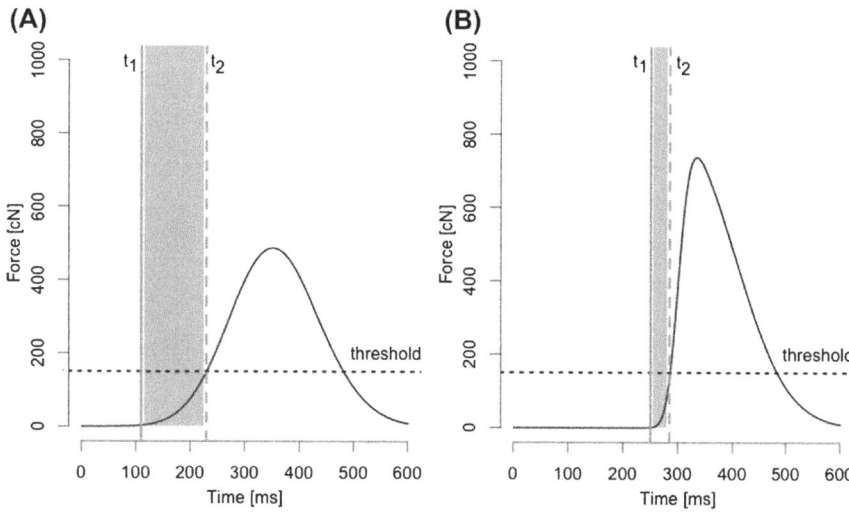

**FIGURE 2** For responses where the rise of force is not so steep (A), the temporal difference between the time point of the detected button press $t_2$ and the time point of true force onset $t_1$ is larger compared to a steep rise of force (A). Also, the level threshold for the button press (dotted line) overestimates the response time (B). Consider that the "shape" of force, that is, e.g., its skewness and kurtosis, determine how precisely level thresholds, that is, button presses, estimate the response time. *(Figure modified from Hoffmann, 2014.)*

(Ulrich, Rinkenauer, & Miller, 1998; Ulrich, Wing, & Rinkenauer, 1995). Figure 2 illustrates the discrepancy between level threshold and "true" response onset.

Looking at Figure 2, one can see that if using a threshold, it is possible that the true reaction time is overestimated. This has important implications for the Ne. To test whether the parameterization of the Ne if modulated by an over- or underestimation of the "true" response onset, we used a standard flanker paradigm for a two-alternative forced-choice task (Kopp, Rist, & Mattler, 1996). Response force was measured continuously and a "level trigger" was sent to the EEG device if force exceeded a predefined threshold. The true force inset was calculated off-line and a trigger was written at this time point. Then, the EEG was segmented relative to force onset or relative to the level trigger (which is a simulation of a button press). Finally, EEG parameters were estimated from both data sets (for details, see Hoffmann, 2014). Indeed, the Ne was decreased if the EEG was segmented relative to the level trigger. Furthermore, time–frequency estimates of the Ne were affected, too (Figure 3), such that theta power (4–7 Hz) was lower in segments where the parameters were related to the threshold level compared to segments where theta power was derived from force-onset-related EEG segments.

However, despite these methodological problems, the Ne is an ERP that can be measured very precisely. Its reliability, that is, its stability, is quite high compared to other ERPs (Olvet & Hajcak, 2009a,b) even on the single-trial level (Figure 4).

Figure 4 shows the concatenated single-trial traces of erroneous trials of every participant in a mental rotation paradigm (Hoffmann & Falkenstein, 2010; see

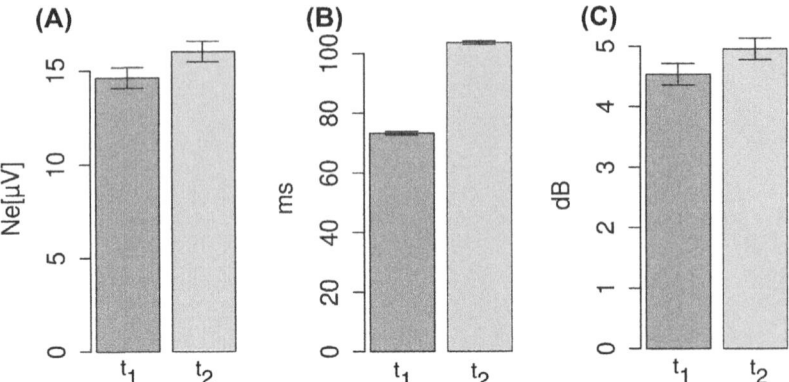

**FIGURE 3** The error negativity (Ne) peaks (A), Ne latencies (B), and theta power in decibels (C, 4–7 Hz) for level threshold (i.e., button presses) and true force onset for erroneous responses. The error bars indicate within-subject 95% confidence intervals. The time points of true force onset and the detected button press are indicated by $t_1$ and $t_2$, respectively. *(Figure modified from Hoffmann, 2014.)*

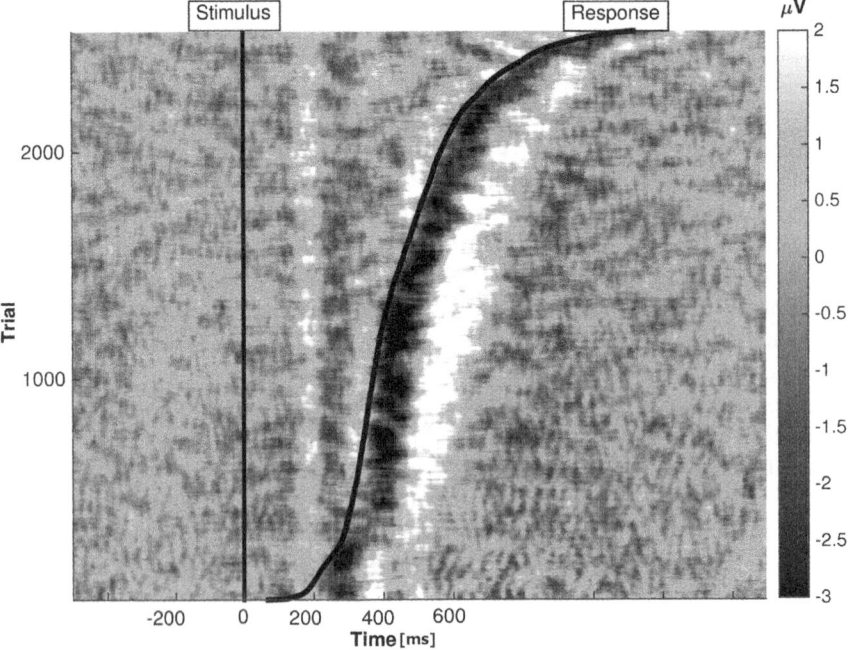

**FIGURE 4** Concatenated single-trial traces of the EEG at the electrode position FCz of erroneous trials of all participants performing a mental rotation task (cf. Hoffmann & Falkenstein, 2010). All trials were normalized prior to concatenation. The vertical line indicates stimulus onset; the sinusoid line indicates the response time. Trials were sorted according to response time. The plot was made using the EEGLAB toolbox (Delorme & Makeig, 2004). Consider the strong concordance of the single-trial error negativity (Ne) peak and response time.

also Hoffmann et al., 2014). Obviously, the Ne can be derived in most of the cases. Furthermore, utilizing sophisticated signal-processing methods, such as independent component analysis (Comon, 1994), can further increase the signal-to-noise ratio (Debener, Ullsperger, Siegel, & Engel, 2006; Hoffmann & Falkenstein, 2010; Jung et al., 2001). This approach is especially appealing because the ERP consists of temporally overlapping brain activities (Jung et al., 2001). This might be problematic especially when measuring response-related ERPs: The brain activity associated with the subject's response might partially overlap with brain activity accompanying stimulus processing or attention processes, especially in tasks producing very fast reaction times. Indeed, independent component analysis can be utilized to extract components representing the Ne as well as the correct-response negativity (CRN; e.g., Hoffmann & Falkenstein, 2010).

Another problem when measuring response-related EEG activity is how to quantify the peak of interest (which is a basic problem in ERP research, but especially relevant for response-related potentials). The amplitude of an ERP does not represent an absolute value because it cannot be measured relative to an electrically neutral reference. Therefore, the amplitude is measured relative to the mean activity in an interval of some neutral reference time point *before* the event of interest up to this event of interest (e.g., stimulus, response). From this interval, termed baseline, it is assumed that the activity is randomly distributed, whereas the activity following the event of interest is *event related*. This method of calculating the difference between amplitude and baseline is termed baseline-to-peak measurement. Another possibility is peak-to-peak measurement, where the amplitude difference between the peak of interest and the peak before is being calculated. Finally, one might measure the mean activity or area under the curve and calculate the difference between it and the baseline. All these approaches (and many more) have both advantages and disadvantages (for a detailed introduction and discussion, see Luck, 2014). The baseline problem is not a trivial one; there might well be some neural processes before the event of interest, such as processes related to preparation, that influence the baseline. In this case, a peak-to-peak measurement might be more appropriate. This, however, requires easily and clearly identifiable peaks, which is certainly not the case for every ERP. When peaks are not identifiable, the mean amplitude method might be more appropriate. What is the relevance of these issues for response-related potentials? Prior to the response there surely is activity related to stimulus processing and response preparation. Luckily, the Ne has a well-accentuated morphology and the peak-to-peak approach might be a good choice. However, this is not the case for every experiment. In some there is a clear Ne; in others, the shape of the curve is not so accentuated. In this case, measuring mean activity, that is, calculating the difference between mean post-response activity and pre-response activity might be a good choice, too. For an excellent introduction and discussion of the problem of peak picking and baseline, the reader should see Luck's (2014) "bible": *An Introduction to the Event-related Potential Technique*. In the next section, I consider how response-related potentials can be utilized in a more practical way, that is, their predictive value.

## PREDICTING ERRORS

In terms of the functional implementation of the Ne and its relation to the DA system, it might be that the Ne is predictive not only of behavioral adaptation, but also of inter-individual differences. Thus, the Ne might well be used as a dependent variable in a more practical way. In this section, I outline the predictive use of the Ne with respect to behavioral adaptation and then provide a short introduction on individual differences with respect to error monitoring. The latter is a research field that is rapidly evolving. Thus, this section focuses on the historically most relevant findings.

### Predicting Behavioral Adaptation

A core function of the process that is reflected in the Ne is the initiation and maintenance of cognitive control following erroneous or at least unexpected actions (Ridderinkhof et al., 2004). As already described in the Introduction, such a control system is indicated by a slowing of the response time compared to that in correct trials prior to an error (Rabbitt, 1966). Indeed, the assumed key correlate of such a system, the Ne, predicts on a single-trial level posterror response times (Debener et al., 2005). However, though this finding appears convincing at first glance, this relation has not been found consistently (Hajcak, McDonald, & Simons, 2003; Masaki, Falkenstein, Sturmer, Pinkpank, & Sommer, 2007), which might be because different experimental paradigms and measures were used. In some studies, post-error slowing was measured via the difference between the actual error trial and the following trial, whereas others quantified the difference between mean post-error response time and mean correct response time. Obviously, this could lead to completely different statistical inferences due to the distribution of reaction times. Furthermore, not every study used single-trial measures; most correlated mean reaction time with mean Ne.

Another aspect of cognitive control is the immediate correction of errors. Here, the Ne latency, that is, the time point when the Ne reaches its peak, is shorter for errors that are immediately corrected compared to "full" errors, that is, those that remain uncorrected (Carbonnell & Falkenstein, 2006; Hoffmann & Falkenstein, 2010). These results suggest that the Ne plays an important role in response monitoring (i.e., "online control"): When the error is detected relatively early (short Ne latency), there is a greater likelihood it can be suppressed (partial rather than overt error). What might be interesting in a more applied context would be to investigate if such control is involved in "online inhibition" of upcoming errors, that is, if the brain activity accompanying responses "foreshadows" errors. This should be the case if it is a reflection of online response monitoring.

After correct trials there is, as already mentioned, also a negativity with a morphology and source comparable to that of the Ne: the Ne-like wave (Vidal, Bonnet, & Macar, 1995; Vidal et al., 2000) or CRN (Ford, 1999). It has been suggested that the Ne and CRN reflect the same functional process (Vidal et al., 2000), that is, response monitoring.

Indeed, it has been shown by means of independent component analysis that the ERP following errors and correct responses can be modeled by a single factor/component (Hoffmann & Falkenstein, 2010; Roger et al., 2010). Interestingly, the CRN amplitude predicts the occurrence of an error in the following trial (Allain, Carbonnell, Falkenstein, Burle, & Vidal, 2004), which fits nicely with the effect of pre-error speeding and post-error slowing (e.g., Notebaert et al., 2009). When monitoring is low, it is likely the next trial will result in an error. In summary, it appears that the Ne and CRN might be used to predict short-term behavior.

To test this, in a recent study, Plewan, Falkenstein, Wascher, and Hoffmann (submitted for publication) investigated whether the Ne can be used to predict errors across different task sets and whether response brain activity that already occurs prior to errors can be used to predict errors. This is plausible if one assumes that CRN as well as the Ne reflect online response monitoring or control. To test this, a support vector machine approach was developed to classify response-related EEG activity reliably. Plewan et al. assumed that if the Ne/CRN reflects a generalizable mechanism, the Ne/CRN (i.e., reflection of response monitoring) of one task should predict the performance in another, cognitively different task. This was exactly what was found. If the support vector machine was trained in a standard flanker task, it could be used to predict errors in a mental rotation task with high sensitivity, and vice versa.

Besides this behavioral adjustment, it has recently been shown that response-monitoring processes affect processes that are related to perception, as well. Error-related posterior mPFC activity predicts activity in motor areas but also activity enhancement in areas encoding task-relevant stimulus features, as well as decreased activity in perceptual areas encoding distraction information (Danielmeier et al., 2011). Further, as already described, the Ne amplitude seems to be correlated with adjustments of selective attention such that if one commits an error, the resulting Ne is predictive for the degree of attentional conflict (Maier et al., 2011). These results indicate that the functional role of the aforementioned response monitoring functions is not restricted to motor/behavioral adaptation but also applies to adaptation of attention or perceptual processes (cf. Hoffmann, Chapter 9).

Can these findings be put to practical use? That is, can the Ne/CRN and other correlates of response monitoring be used to predict behavior? A relatively new application, coming from engineering science, is the brain–computer interface (BCI). The goal of BCIs is to identify and classify correlates of human cognition in online biophysical signals to communicate them to machine devices. The aim is to use these classifications in such a way that not only is a subject able to control the device, but the device is able to evaluate continually the state of the EEG signal and, thus, the "neural" state of the subject. As a result, BCIs can be used to train computers, but computers can also train BCIs. Several new implementations are promising.

For example, in clinical applications, it has been shown that patients suffering from amyotrophic lateral sclerosis can still control voluntary modulations of sensorimotor rhythms. These can be measured in the EEG and used to interact with a PC via a BCI (Kubler et al., 2005). Even patients suffering from locked-in syndrome can train to use EEG BCIs (Hinterberger, Kubler, Kaiser, Neumann, & Birbaumer, 2003).

Another application is the online utilization of the Ne to improve the performance of the BCI with respect to the choice of the correct consequence in a learning task or when a certain behavioral consequence is required. Imagine an experiment where you have to learn a predefined stimulus–response setting. In other words, you have to learn which response belongs to each stimulus that is provided on a screen. The stimuli appear sequentially, and on each trial you have to indicate your guess by pushing a button. This response is followed by feedback indicating whether your response was correct. A classic finding is that in such a situation, there is initially no Ne following the response because the "correct" response has not yet been learned. This is not the case for the feedback: The feedback is followed by FRN at frontocentral electrode leads at about 250 ms. It has been assumed that this FRN indicates whether the feedback was expected (Holroyd & Coles, 2002; Nieuwenhuis et al., 2002, 2004). Interestingly, the Ne and FRN show an inverse pattern during learning: Initially the Ne is almost absent, and the FRN is quite large in comparison. During the learning process, this pattern reverses, and at the end, the Ne is large during an error, and the FRN is small. This empirical finding supports the idea underpinning Holroyd and Coles's RFL theory.

Now how might this be applied? Researchers in the field of BCIs stumbled across this finding and adapted it to the needs of human–machine interaction systems. And indeed, there exists a kind of FRN that has been termed the interaction error potential or ErrP (Ferrez & Millan, 2008). This ErrP occurs in a BCI experiment when the subject responds correctly; that is, the subject's EEG activity should indicate a correct response, but the machine-learning algorithm makes an incorrect classification and provides an incorrect choice to the subject. Indeed, the sensitivity[3] in BCIs is often below 90%, and so many trials will be classified incorrectly. It has been shown that the FRN (or ErrP) following an erroneous choice of the personal computer, that is, machine-learning algorithm, might well be used to improve the performance of the BCI (Ferrez & Millan, 2008). However, within the framework of the RFL theory, the FRN can be used in BCIs to improve, for example, learning environments.

## Individual Differences and Error Monitoring

Finally, the amplitude of the Ne, because of its dependence on the DA system (and other transmitter systems, as well; see also Hoffmann, Chapter 9), is

---

3. The sensitivity is the number of correct positives/(number of correct positives + number of erroneous negatives).

correlated with several neurological and psychiatric diseases in which ACC activity is directly affected by DA influx (e.g., Parkinson's disease or Huntington's chorea; Beste, Saft, Andrich, Gold, & Falkenstein, 2006; Falkenstein, Willemssen, Hohnsbein, & Hielscher, 2006). This is also true for diseases where ACC activity is modulated by indirect DA modulation, that is, where the DA influx itself is modulated by another transmitter system (e.g., obsessive-compulsive disorder; Endrass et al., 2010). Further research is needed to test whether EEG correlates of response/error monitoring could provide added value to existing diagnostics tools, for example, with respect to states or outbreaks of diseases.

In another example, the Ne was decreased in manifest Huntington's disease compared to matched healthy controls (Beste et al., 2006) but, most importantly, in preclinical patients (i.e., patient not showing overt neuropsychological symptoms), the Ne was rather enhanced, and the degree of this enhancement was related to the estimated age of onset of the clinical manifestation of the disease, which was interpreted as an indicator of preclinical compensatory efforts (Beste et al., 2007). There is a large body of literature about the relation of the Ne to different pathologies. However, an ERP cannot be interpreted apart from the experiment and, thus, the functional context in which it was derived. Consider, that the term ERP means event-*related* potential, and thus, it reflects the fact that the ERP is temporally related to the event. Due to this, it cannot be interpreted in a causal fashion. However, if results are interpreted with caution, the investigation of error processes might be a promising approach to contribute to improving clinical assessments, such as diagnostic screening instruments, especially because the measurement of the Ne is highly reliable.

Given that the Ne is modulated by "hard- and soft-wired" neuronal mechanisms, it is possible that the Ne might be related to personality traits (for more on personality traits see Laborde & Mosely, Chapter 18), as well. Individuals showing higher values on a conscientiousness scale showed smaller motivation-dependent changes in the Ne (Pailing & Segalowitz, 2004). The Ne amplitude is correlated with risk taking, sensation seeking, and sensitivity to reward (Santesso & Segalowitz, 2009) as well as with negative affect and negative emotionality (Luu, Collins, & Tucker, 2000).

Furthermore, the assumption that the Ne, due to its relation to the DA system and other neurotransmitters expressions (Jocham & Ullsperger, 2009), is correlated with personality traits is supported by several clinical findings. Studies have shown that the Ne amplitude is modulated in borderline disorder (Ruchsow et al., 2006), psychopathy (Bresin, Finy, Sprague, & Verona, 2014; Hall, Bernat, & Patrick, 2007), alcohol abuse (Ridderinkhof et al., 2002), depression, obsessive-compulsive disorder, and general anxiety disorder (Weinberg, Kotov, & Proudfit, 2015). Though this is only a selection, it is easy to see that through its connection to several transmitter systems, the Ne can be used to study the cause of psychopathologies and perhaps as a diagnostic marker.

## CONCLUSION

In this chapter, I introduced the core concepts and theories of error and response monitoring or processing, discussed methodological concerns when measuring response-related EEG activity, and outlined potential predictive uses of the neural correlates of error monitoring. The aim of the chapter was to show that error-processing mechanisms are central to behavioral adaptation (Ne) but also to learning stimulus–response contingencies and to the process of detecting violations of expectations (FRN).

How is this related to sports? First, in sports, response monitoring or the monitoring of actions is one of the core mechanisms in the planning, execution, and evaluation of actions. For example, during motor learning, the concept of "closed loops" and their functional implementation has a close connection to the previously described neural mechanisms. Thus, the ACC, its functional role, and its connections (cf. Hoffmann, Chapter 9) seem to be crucial in this regard.

Second, besides this basic scientific value, error processing is important in risky sports; here error detection and rapid behavioral adaptation are crucial. The investigation of errors or mistakes provides the opportunity to investigate the reliability and dynamics of (action) systems in general (Nitsch, 2001). Furthermore, it is possible to detect the functional reasons why errors occur in given situations if an appropriate experimental design is chosen that operationalizes the mechanisms of interest. In other words, it might provide an answer to the question of "why and how we make something wrong" (Nitsch, 2001).

The investigation of errors might have practical implications for the optimization of education and training, as well, for example, by identifying error types and optimizing training with respect to the involved mechanisms. Consider the possibility to identify specific mechanisms related to different types of errors. Obviously, this is closely related to performance via the reduction of error probability. Another application might be the optimization of security management. Risks might be reduced and training conditions improved when the sources of errors (e.g., deficient attention or perception conflict) are known.

Finally, the neurophysiological models and methods can potentially be used to investigate learning processes or even predict if a learning technique will be successful. Until recently, it was quite difficult to measure these processes reliably in the field. Luckily, this has changed; advances in hardware (e.g., mobile EEG systems) and sophisticated data-processing methods for use in neuroscientific investigations have improved research by making it possible to study working environments or naturalistic movements in the field (De Vos & Debener, 2014; Kranczioch, Zich, Schierholz, & Sterr, 2014; Wascher, Heppner, & Hoffmann, 2014).

## REFERENCES

Alexander, W. H., & Brown, J. W. (2011). Medial prefrontal cortex as an action-outcome predictor. *Nature Neuroscience, 14,* 1338–1344. http://dx.doi.org/10.1038/nn.2921.

Allain, S., Carbonnell, L., Falkenstein, M., Burle, B., & Vidal, F. (2004). The modulation of the Ne-like wave on correct responses foreshadows errors. *Neuroscience Letters, 372,* 161–166. http://dx.doi.org/10.1016/j.neulet.2004.09.036.

Bellebaum, C., Kobza, S., Thiele, S., & Daum, I. (2011). Processing of expected and unexpected monetary performance outcomes in healthy older subjects. *Behavioral Neuroscience, 125,* 241–251. http://dx.doi.org/10.1037/a0022536.

Beste, C., Saft, C., Andrich, J., Gold, R., & Falkenstein, M. (2006). Error processing in Huntington's disease. *PLoS One, 1,* e86.

Beste, C., Saft, C., Yordanova, J., Andrich, J., Gold, R., Falkenstein, M., et al. (2007). Functional compensation or pathology in cortico-subcortical interactions in preclinical Huntington's disease? *Neuropsychologia, 45,* 2922–2930.

Botvinick, M. M., Braver, T. S., Barch, D. M., Carter, C. S., & Cohen, J. D. (2001). Conflict monitoring and cognitive control. *Psychological Review, 108,* 624–652.

Bresin, K., Finy, M. S., Sprague, J., & Verona, E. (2014). Response monitoring and adjustment: differential relations with psychopathic traits. *Journal of Abnormal Psychology, 123,* 634–649. http://dx.doi.org/10.1037/a0037229.

Burnod, Y., Baraduc, P., Battaglia-Mayer, A., Guigon, E., Koechlin, E., Ferraina, S., et al. (1999). Parieto-frontal coding of reaching: an integrated framework. *Experimental Brain Research, 129,* 325–346.

Carbonnell, L., & Falkenstein, M. (2006). Does the error negativity reflect the degree of response conflict? *Brain Research, 1095,* 124–130.

Comon, P. (1994). Independent component analysis—a new concept? *Signal Processing, 36,* 287–314.

Danielmeier, C., Eichele, T., Forstmann, B. U., Tittgemeyer, M., & Ullsperger, M. (2011). Posterior medial frontal cortex activity predicts post-error adaptations in task-related visual and motor areas. *Journal of Neuroscience, 31,* 1780–1789. http://dx.doi.org/10.1523/JNEUROSCI.4299-10.2011.

De Vos, M., & Debener, S. (2014). Mobile EEG: towards brain activity monitoring during natural action and cognition. *International Journal of Psychophysiology, 91,* 1–2. http://dx.doi.org/10.1016/j.ijpsycho.2013.10.008.

Debener, S., Ullsperger, M., Siegel, M., & Engel, A. K. (2006). Single-trial EEG-fMRI reveals the dynamics of cognitive function. *Trends in Cognitive Sciences, 10,* 558–563.

Debener, S., Ullsperger, M., Siegel, M., Fiehler, K., Cramon, D. Y. von, & Engel, A. K. (2005). Trial-by-trial coupling of concurrent electroencephalogram and functional magnetic resonance imaging identifies the dynamics of performance monitoring. *Journal of Neuroscience, 25,* 11730–11737.

Dehaene, S., Posner, M., & Tucker, D. M. (1994). Localization of a neural system for error-detection and compensation. *Psychological Science, 5,* 303–305.

Delorme, A., & Makeig, S. (2004). EEGLAB: an open source toolbox for analysis of single-trial EEG dynamics including independent component analysis. *Journal of Neuroscience Methods, 134,* 9–21.

Egner, T., & Hirsch, J. (2005). Cognitive control mechanisms resolve conflict through cortical amplification of task-relevant information. *Nature Neuroscience, 8,* 1784–1790. http://dx.doi.org/10.1038/nn1594.

Endrass, T., Schuermann, B., Kaufmann, C., Spielberg, R., Kniesche, R., & Kathmann, N. (2010). Performance monitoring and error significance in patients with obsessive-compulsive disorder. *Biological Psychology, 84*, 257–263. http://dx.doi.org/10.1016/j.biopsycho.2010.02.002.

Falkenstein, M., Hohnsbein, J., Hoormann, J., & Blanke, L. (1991). Effects of crossmodal divided attention on late ERP components. II. Error processing in choice reaction tasks. *Electroencephalography and Clinical Neurophysiology, 78*, 447–455.

Falkenstein, M., Willemssen, R., Hohnsbein, J., & Hielscher, H. (2006). Effects of stimulus-response compatibility in Parkinson's disease: a psychophysiological analysis. *Journal of Neural Transmission, 113*, 1449–1462. http://dx.doi.org/10.1007/s00702-005-0430-1.

Ferrez, P. W., & del R. Millan, J. (2008). Error-related EEG potentials generated during simulated brain-computer interaction. *IEEE Transactions on Biomedical Engineering, 55*, 923–929.

Ford, J. M. (1999). Schizophrenia: the broken P300 and beyond. *Psychophysiology, 36*, 667–682.

Gehring, W. J., Goss, B., Coles, M. G. H., Meyer, D. E., & Donchin, E. (1993). A neural system for error detection and compensation. *Psychological Science, 4*, 385–390.

Hajcak, G., McDonald, N., & Simons, R. F. (2003). To err is autonomic: error-related brain potentials, ANS activity, and post-error compensatory behavior. *Psychophysiology, 40*, 895–903.

Hall, J. R., Bernat, E. M., & Patrick, C. J. (2007). Externalizing psychopathology and the error-related negativity. *Psychological Science, 18*, 326–333.

Hinterberger, T., Kubler, A., Kaiser, J., Neumann, N., & Birbaumer, N. (2003). A brain-computer interface (BCI) for the locked-in: comparison of different EEG classifications for the thought translation device. *Clinical Neurophysiology, 114*, 416–425.

Hoffmann, S. (2014). Brief report: some remarks about the response relatedness of the error negativity. *Journal of Psychophysiology, 28*, 22–31. http://dx.doi.org/10.1027/0269-8803/a000108.

Hoffmann, S., & Falkenstein, M. (2010). Independent component analysis of erroneous and correct responses suggests online response control. *Human Brain Mapping, 31*, 1305–1315. http://dx.doi.org/10.1002/hbm.20937.

Hoffmann, S., Labrenz, F., Themann, M., Wascher, E., & Beste, C. (2014). Crosslinking EEG time–frequency decomposition and fMRI in error monitoring. *Brain Structure and Function, 219*, 595–605. http://dx.doi.org/10.1007/s00429-013-0521-y.

Hohnsbein, J., Falkenstein, M., Hoormann, J., & Blanke, L. (1991). Effects of crossmodal divided attention on late ERP components. I. Simple and choice reaction tasks. *Electroencephalography and Clinical Neurophysiology, 78*, 438–446.

Holroyd, C. B., & Coles, M. G. (2002). The neural basis of human error processing: reinforcement learning, dopamine, and the error-related negativity. *Psychological Review, 109*, 679–709.

Holroyd, C. B., Pakzad-Vaezi, K. L., & Krigolson, O. E. (2008). The feedback correct-related positivity: sensitivity of the event-related brain potential to unexpected positive feedback. *Psychophysiology, 45*, 688–697. http://dx.doi.org/10.1111/j.1469-8986.2008.00668.x.

Holroyd, C. B., Yeung, N., Coles, M. G. H., & Cohen, J. D. (2005). A mechanism for error detection in speeded response time tasks. *Journal of Experimental Psychology. General, 134*, 163–191. http://dx.doi.org/10.1037/0096-3445.134.2.163.

Inoue, K., Kawashima, R., Satoh, K., Kinomura, S., Sugiura, M., Goto, R., et al. (2000). A PET study of visuomotor learning under optical rotation. *Neuroimage, 11*, 505–516. http://dx.doi.org/10.1006/nimg.2000.0554.

Jocham, G., & Ullsperger, M. (2009). Neuropharmacology of performance monitoring. *Neuroscience and Biobehavioral Reviews, 33*, 48–60.

Jung, T. P., Makeig, S., McKeown, M. J., Bell, A. J., Lee, T. W., & Sejnowski, T. J. (2001). Imaging brain dynamics using independent component analysis. *Proceedings of the IEEE, 89*, 1107–1122.

Kopp, B., Rist, F., & Mattler, U. (1996). N200 in the flanker task as a neurobehavioral tool for investigating executive control. *Psychophysiology, 33*, 282–294. http://dx.doi.org/10.1111/j.1469-8986.1996.tb00425.x.

Kranczioch, C., Zich, C., Schierholz, I., & Sterr, A. (2014). Mobile EEG and its potential to promote the theory and application of imagery-based motor rehabilitation. *International Journal of Psychophysiology, 91*, 10–15. http://dx.doi.org/10.1016/j.ijpsycho.2013.10.004.

Kubler, A., Nijboer, F., Mellinger, J., Vaughan, T. M., Pawelzik, H., Schalk, G., et al. (2005). Patients with ALS can use sensorimotor rhythms to operate a brain-computer interface. *Neurology, 64*, 1775–1777.

Luck, S. J. (2014). *An introduction to the event-related potential technique*. Cambridge, MA: MIT Press.

Luu, P., Collins, P., & Tucker, D. M. (2000). Mood, personality, and self-monitoring: negative affect and emotionality in relation to frontal lobe mechanisms of error monitoring. *Journal of Experimental Psychology. General, 129*, 43–60.

Maier, M. E., Yeung, N., & Steinhauser, M. (2011). Error-related brain activity and adjustments of selective attention following errors. *Neuroimage, 56*, 2339–2347. http://dx.doi.org/10.1016/j.neuroimage.2011.03.083.

Masaki, H., Falkenstein, M., Sturmer, B., Pinkpank, T., & Sommer, W. (2007). Does the error negativity reflect response conflict strength? Evidence from a Simon task. *Psychophysiology, 44*, 579–585.

Mathalon, D. H., Whitfield, S. L., & Ford, J. M. (2003). Anatomy of an error: ERP and fMRI. *Biological Psychology, 64*, 119–141.

Miltner, W. H. R., Braun, C. H., & Coles, M. G. H. (1997). Event-related brain potentials following incorrect feedback in a time estimation task: evidence for a "generic" neural system for error detection. *Journal of Cognitive Neuroscience, 9*, 788–798.

Nieuwenhuis, S., Holroyd, C. B., Mol, N., & Coles, M. G. (2004). Reinforcement-related brain potentials from medial frontal cortex: origins and functional significance. *Neuroscience and Biobehavioral Reviews, 28*, 441–448. http://dx.doi.org/10.1016/j.neubiorev.2004.05.003.

Nieuwenhuis, S., Ridderinkhof, K. R., Talsma, D., Coles, M. G. H., Holroyd, C. B., Kok, A., et al. (2002). A computational account of altered error processing in older age: dopamine and the error-related negativity. *Cognitive, Affective, & Behavioral Neuroscience, 2*, 19–36.

Nitsch, J. (2001). Handlungsfehler im Sport: Theoretische Ausgangspunkte einer Forschungskonzeption. In *Handlungspsychologische Forschung für die Theorie und Praxis der Sportpsychologie* (pp. 65–121). Cologne, Germany: BPS.

Notebaert, W., Houtman, F., Opstal, F. V., Gevers, W., Fias, W., & Verguts, T. (2009). Post-error slowing: an orienting account. *Cognition, 111*, 275–279. http://dx.doi.org/10.1016/j.cognition.2009.02.002.

Olvet, D. M., & Hajcak, G. (2009a). Reliability of error-related brain activity. *Brain Research, 1284*, 89–99.

Olvet, D. M., & Hajcak, G. (2009b). The stability of error-related brain activity with increasing trials. *Psychophysiology, 46*, 957–961.

Pailing, P. E., & Segalowitz, S. J. (2004). The error-related negativity as a state and trait measure: motivation, personality, and ERPs in response to errors. *Psychophysiology, 41*, 84–95.

Plewan, T., Falkenstein, M., Wascher, E., &Hoffmann, S. *Classifying response accuracy across different task sets: a machine learning approach*, submitted for publication.

Rabbitt, P. M. (1966). Errors and error correction in choice reaction tasks. *Journal of Experimental Psychology, 71*, 264–272.

Reason, J. (1990). *Human error*. Cambridge England: Cambridge University Press.

Ridderinkhof, K. R., de Vlugt, Y., Bramlage, A., Spaan, M., Elton, M., Snel, J., et al. (2002). Alcohol consumption impairs detection of performance errors in mediofrontal cortex. *Science, 298*, 2209–2211. http://dx.doi.org/10.1126/science.1076929.

Ridderinkhof, K. R., Ullsperger, M., Crone, E. A., & Nieuwenhuis, S. (2004). The role of the medial frontal cortex in cognitive control. *Science, 306*, 443–447.

Roger, C., Benar, C. G., Vidal, F., Hasbroucq, T., & Burle, B. (2010). Rostral cingulate zone and correct response monitoring: ICA and source localization evidences for the unicity of correct- and error-negativities. *NeuroImage, 51*, 391–403. http://dx.doi.org/10.1016/j.neuroimage.2010.02.005.

Ruchsow, M., Walter, H., Buchheim, A., Martius, P., Spitzer, M., Kächele, H., et al. (2006). Electrophysiological correlates of error processing in borderline personality disorder. *Biological Psychology, 72*, 133–140. http://dx.doi.org/10.1016/j.biopsycho.2005.08.006.

Rushworth, M., Walton, M., Kennerley, S., & Bannerman, D. (2004). Action sets and decisions in the medial frontal cortex. *Trends in Cognitive Sciences, 8*, 410–417. http://dx.doi.org/10.1016/j.tics.2004.07.009.

Santesso, D. L., & Segalowitz, S. J. (2009). The error-related negativity is related to risk taking and empathy in young men. *Psychophysiology, 46*, 143–152.

Sarter, M., Gehring, W. J., & Kozak, R. (2006). More attention must be paid: the neurobiology of attentional effort. *Brain Research Reviews, 51*, 145–160.

Suri, R. E. (2002). TD models of reward predictive responses in dopamine neurons. *Neural Networks, 15*, 523–533.

Suri, R. E., & Schultz, W. (1998). Learning of sequential movements by neural network model with dopamine-like reinforcement signal. *Experimental Brain Research, 121*, 350–354.

Suri, R. E., & Schultz, W. (2001). Temporal difference model reproduces anticipatory neural activity. *Neural Computation, 13*, 841–862.

Ulrich, R., Rinkenauer, G., & Miller, J. (1998). Effects of stimulus duration and intensity on simple reaction time and response force. *Journal of Experimental Psychology: Human Perception and Performance, 24*, 915–928. http://dx.doi.org/10.1037/0096-1523.24.3.915.

Ulrich, R., Wing, A. M., & Rinkenauer, G. (1995). Amplitude and duration scaling of brief isometric force pulses. *Journal of Experimental Psychology: Human Perception and Performance, 21*, 1457–1472. http://dx.doi.org/10.1037/0096-1523.21.6.1457.

van Veen, V., Cohen, J. D., Botvinick, M. M., Stenger, V. A., & Carter, C. S. (2001). Anterior cingulate cortex, conflict monitoring, and levels of processing. *Neuroimage, 14*, 1302–1308.

Vidal, F., Bonnet, M., & Macar, F. (1995). Programming the duration of a motor sequence: role of the primary and supplementary motor areas in man. *Experimental Brain Research, 106*, 339–350.

Vidal, F., Hasbroucq, T., Grapperon, J., & Bonnet, M. (2000). Is the "error negativity" specific to errors? *Biological Psychology, 51*, 109–128.

Wascher, E., Heppner, H., & Hoffmann, S. (2014). Towards the measurement of event-related EEG activity in real-life working environments. *International Journal of Psychophysiology, 91*, 3–9. http://dx.doi.org/10.1016/j.ijpsycho.2013.10.006.

Weinberg, A., Kotov, R., & Proudfit, G. H. (2015). Neural indicators of error processing in generalized anxiety disorder, obsessive-compulsive disorder, and major depressive disorder. *Journal of Abnormal Psychology, 124*, 172–185. http://dx.doi.org/10.1037/abn0000019.

Yeung, N., Botvinick, M. M., & Cohen, J. D. (2004). The neural basis of error detection: conflict monitoring and the error-related negativity. *Psychological Review, 111*, 931–959.

Chapter 11

# Committing Errors as a Consequence of an Adverse Focus of Attention

### Daniel Schneider
*Leibniz Research Centre for Working Environment and Human Factors, Dortmund, Germany*

## "SERIOUS" ERRORS IN THE COURSE OF PERCEPTION

*"You cannot be serious!" For those interested in professional tennis, this quote is linked to a Wimbledon match in 1981, where player John McEnroe loudly challenged the referee's decision of calling his ball out. Considering that the referee in a tennis match obviously has an outstanding view on the tennis court, is highly trained in his profession, and is naturally unbiased with respect to the match outcome, it is difficult to imagine that the respective call was not based on a "serious" decision. Given his loud remark, it is also very likely that McEnroe did not just complain to get an advantage in the course of the game, but truly judged the ball in. However, this entails that either the referee or the professional tennis player made an error sometime between the perception of the ball hitting the ground and the behavioral response.*

Recent evidence actually shows that questioning the referee's decision in comparable situations is justified. Psychophysical studies revealed that the location of moving objects is erroneously perceived as being shifted in the direction of movement (De Valois & De Valois, 1991; Ramachandran & Anstis, 1990). This will lead to errors when a tennis ball bounces at the outer margin of a line, as its bouncing position will be overestimated. Whitney, Wurnitsch, Hontiveros, and Louie (2008) revealed that such error types are predominantly found in incorrect judgments of professional tennis referees. However, John McEnroe's decision might also have been based on an erroneous perceptual judgment, as his view to the other side of the court was impaired by the net on the field. Additionally, McEnroe might rather have attended the movement of his opponent to anticipate his next shot. This might have led to an inability to perceive the exact position of the ball when it hit the ground.

The examples clearly demonstrate that there are many different factors that can bias perception and lead to errors on the decision level. These factors can be related to external conditions (e.g., the partly concealed ball position) and to internal states (e.g., the overestimation of the position of moving objects). However, the adverse attentional focus as a reason for perceptual errors cannot be assigned entirely to either of both factors. Attention can be driven exogenously and may be captured by irrelevant events in our surrounding, leading to an inferior processing of relevant inputs. Furthermore, we might deliberately focus on certain incoming signals and miss other events that are potentially relevant in a given situation. The current chapter approaches these different types of attention-related errors in the course of perception and uses them to give an idea of the role of attention for transferring perceptual information into an adequate decision and performance.

## SELECTIVE ATTENTION AS BIASED COMPETITION

"Everyone knows what attention is. It is the taking possession by the mind, in clear and vivid form, of one out of what seem several simultaneously possible objects or trains of thought" (James, 1890/1950). This quote by William James describes that attention is a term that is commonly used in everyday life. We all remember situations in school when a teacher remarked, "Pay attention!" while we were rather listening to our seatmate or were absorbed in thought while observing some scene on the schoolyard. This suggests that "paying" attention improves the processing of a subset of information at the cost of further signals in our surroundings. To understand to what extent such a mechanism can be error-prone, it is important to know about the processes that underlie selective attention and their neural implementations. As this chapter focuses on perceptual errors in the visual domain, this first of all requires a description of the functional organization of the visual system.

Visual inputs are processed in the retina depending on their position in the visual field. Information proceeds from there to nuclei in the thalamus. The most relevant thalamic nucleus in the current context is the dorsal part of the lateral geniculate nucleus that projects information via the optic radiation to the primary visual cortex (striate cortex or V1). These projections to V1 are organized in a retinotopic way, meaning that visual signals are represented as a function of their position on the retina (Tootell, Silverman, Switkes, & De Valois, 1982). Thus, each neuron in V1 has distinct receptive fields and encodes information from only a confined fraction of the visual surroundings. The neurons' receptive fields increase as information proceeds from this early processing stage to higher order areas like V4, inferior temporal cortex, posterior parietal cortex, or prefrontal cortex. Consequently, the amount of information integrated within these receptive fields also increases (Desimone & Ungerleider, 1989). While early visual areas provide the fine spatial and feature analysis of the presented stimuli, this information is subsequently required to form position-invariant object representations in higher level areas like V4 (Serences & Yantis, 2006). However, what does this functional organization tell us about how attentional selection proceeds in the course of visual processing?

The biased competition account of visual selective attention postulates that selection is an emergent feature of competitive processing within and between visual areas (Desimone & Duncan, 1995). When multiple stimuli are present in our visual surroundings, they will compete for representation in the visual system, because each neuron is limited with respect to the information that can be processed. Given that the stimuli are not presented close together, the fine spatial and feature analysis in V1 will work in parallel for all presented items. However, as information proceeds from V1 to higher level extrastriate areas for object categorization and recognition (e.g., V4) and the respective receptive fields become larger, the proportion of neurons that are triggered by multiple stimuli increases. Therefore, the neuronal activity becomes ambiguous with respect to the coded information, and the search for relevant information within the visual scene is prone to fail. This is where competition-based attentional selection comes into play. Those stimuli that reach the strongest neural representation prevail in competitive processing within the visual system and are, therefore, represented at higher level processing stages that are responsible for further cognitive operations with the incoming information and the planning and initiation of goal-directed behavior.

This leads us to question what factors determine the strength of neural representations and promote the selection of certain visual inputs at the expense of others. First, V1 neurons encode visual stimuli as a function of their saliency referred to surrounding inputs (Knierim & van Essen, 1992). As the extrastriate areas subsequent to V1 are also mainly organized in a retinotopic way, this leads to the creation of a spatial map that represents visual information as a function of its saliency. However, if the most salient signal within the visual field would always win the competition and reach a representation in higher level areas, our behavior would simply be bottom-up according to the physically strongest signal in our surroundings. Therefore, competitive processing in the retinotopically arranged visual areas has to be biased toward relevant information. This bias is enabled by a competitive advantage for information that is contingent on attentional templates activated in working memory (Baddeley, 1986; Desimone & Duncan, 1995). Such attentional templates are composed of stimulus features like location, color, or shape and define the perceptual inputs that are relevant for achieving current behavioral goals (Duncan & Humphreys, 1989). Although the neural source related to these attentional templates is located in higher level prefrontal areas, the resulting top-down bias interacts with the competitive processing in early visual areas and can, therefore, result in the selection of otherwise noncompetitive signals (Desimone & Duncan, 1995; Miller, Erickson, & Desimone, 1996; Reynolds & Chelazzi, 2004).

Figure 1 shows a schematic depiction of the effect of a selective top-down bias in a simple visual task that requires the detection of a predefined shape. When two stimuli (e.g., diamond and triangle) are equally salient and their particular position was not given in advance (see Figure 1(A) and (B)), their initial neural representation is comparable irrespective of a selective attentional template for a certain shape. Only later competitive processing in extrastriate and higher level visual areas is biased in favor of the relevant signal. This leads to

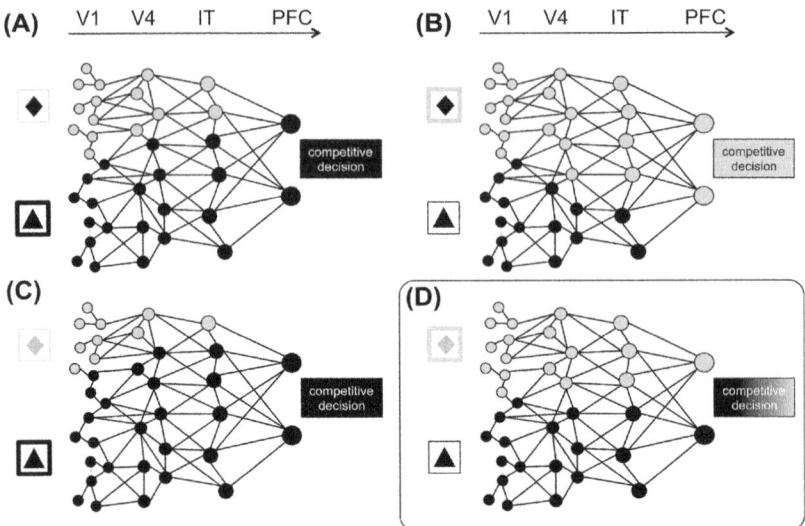

**FIGURE 1** An exemplary depiction of the biased competitive processing of two stimuli in visual areas *(modified from Serences & Yantis, 2006)*. The stimulus surrounded by the broader rectangle is "attended." The fill of the circles indicates which stimulus drives the response of a neuronal cluster in the specific areas from V1 to PFC. (A) and (B) show conditions with different attentional settings but equally salient stimuli. In (C), the more salient stimulus is attended, while (D) shows a condition with an attentional bias in favor of the less salient stimulus.

a predominant representation of the relevant stimulus in higher level executive areas that are responsible for the integration of incoming information and the current behavioral goal (Duncan, 2001). However, the stimuli may also differ with respect to their saliency leading to a representational advantage for the more salient input in early visual processing (see Figure 1(C) and (D)). This might cause an ambiguous representation of information in higher level visual areas when the irrelevant stimulus happens to be the more salient one (see Figure 1(D)). As a consequence, the irrelevant stimulus strongly interferes with the selection of relevant visual information and affects primary task performance.

There are numerous examples in everyday life for the impact of irrelevant information on goal-directed performance. For instance, the capture of attention toward bright billboards near the road or toward flashing alarm displays in the interior of the car might impair the perception of relevant information in a traffic scene. In experimental psychology, such an interfering effect of distractors is often studied by means of visual search paradigms. In the so-called additional singleton task, participants are instructed to search for a relevant stimulus among irrelevant stimuli. In a subset of trials, one of the irrelevant stimuli (the singleton) has a unique feature that is not related to the search task (e.g., Theeuwes, 1991, 1992). For example, additional singletons that were presented in an irrelevant color or were displayed with a certain delay compared to the remaining items caused an inferior search performance referred to conditions without a singleton. Performance in such visual search tasks is usually measured by means of modulations in response times or eye movement patterns that suggest an adverse capture of attention toward the irrelevant singleton position. This gives valuable information about our ability to ignore salient but irrelevant visual information and about the interaction of exogenous and endogenous attentional mechanisms. Yet, visual search tasks do not usually cause significant failures in the explicit report of intended information. This is why the current chapter focuses on two visual phenomena that are more appropriate for studying the attentional processes that are related to erroneous performance: change blindness and inattentional blindness.

## SELECTIVE ATTENTION AND CHANGE BLINDNESS

*Imagine you are sitting in a cinema and are watching a movie. In a certain scene, the leading actor of the movie is talking to a person that is not of central interest for the plot. Then a cut is shown during this conversation, and afterward, the leading actor continues to speak, but the person he or she has been speaking to has changed to another person looking only roughly similar to the first. This is a salient change in the movie scene, and of course, you will notice it; won't you?*

Psychological research has pointed out that we usually overestimate our ability to perceive changes to objects, scenes in a photograph, or movie scenes when vision is only shortly interrupted (see Simons, 2000; Simons & Levin, 1997). This effect is called "change blindness" and was shown to occur "within the blink of an eye." In an experiment by Grimes (1996), participants were presented photographs of natural scenes and were instructed to memorize the details of each photograph for a later testing phase. The spontaneous eye blinks of the observers were measured during the analysis of the photographs. Critically, changes of details within the photographs could either occur during an eye blink or during fixation (without an eye blink). Participants had to report these changes within the pictures independent of their primary task (memorizing details in the photographs). Interestingly, even very salient changes that were easily detected during fixation were often missed when presented during an eye blink (Grimes, 1996). There are different reasons why such blindness effects might occur.

Blindness might be based on processes triggered by the eye blink itself. However, comparable results can also be obtained by means of experimental manipulations (see Figure 2). Presenting a blank screen between two pictures for a short interval (see Figure 2(A)) or inserting so-called mud-splashes at the moment of the change (see Figure 2(B)) can also affect change detection (e.g., O'Regan, Rensink, & Clark, 1999; Simons, 1996). In these experimental variants, it was the participants' task to search for changes between stimulus displays. Nevertheless, these short interruptions of vision critically affected detection performance. This leads to a further exclusion of a possible reason for

**FIGURE 2** Typical visual paradigms for studying change blindness. Participants are instructed to report the change between the two photographs. Performance is indicated by the number of changes that occurred before the change was finally reported (i.e., a flicker paradigm). (A) depicts an experimental design with short gaps between two pictures. (B) shows a so-called mud-splashes paradigm where certain areas of each picture (never the area containing the change) are shortly presented with overlaying stimuli simultaneous to the change. The changing object (a famous skyscraper in San Francisco) is indicated by the dashed oval.

change blindness in the experiment by Grimes (1996). Change blindness is not exclusively due to an inability to divide attention between a primary and a secondary task. Visual interruptions lead to change blindness even when we exclusively focus attention on the detection of changes between stimulus displays or within motion scenes (Simons & Levin, 1997; Simons & Rensink, 2005).

Furthermore, Rensink and colleagues revealed that the ability to perceive changes depends on the relevance of the changing object in a picture (Rensink, O'Regan, & Clark, 1997). When an object was of central interest for the picture (e.g., a wedge of cheese on a buffet), participants' detection performance was increased compared to changes that were only of marginal interest for the picture's theme (e.g., changing the position of an object in the background). Additionally, spatial cues at the position of change reduced the blindness effect significantly (Scholl, 2000). Thus, both exogenous and endogenous attentional biases seem to support the processing of changes between disrupted visual scenes. Conclusions might go even further. Almost all studies on change blindness suggest that selective attention is inevitably required to see changes between stimulus displays (for review, see Simons & Rensink, 2005). When there is no short interruption of vision (e.g., by eye blinks, saccades, or blank screens between two displays), a change creates a motion transient that automatically draws attention to its location. For example, by waving our hand, we cause motion transients that automatically capture the attention of a person searching for us. Visual interruptions affect this motion transient and prevent the automatic orienting of attention to the location of change. Yet, at what stage of processing does selective attention promote change detection and transfer the perceptual representation of change into an adequate decision and performance that would otherwise be error-prone?

To answer this question, the paradigms used for investigating change blindness have to be distinguished with respect to the response type that is used to indicate change detection. In most change blindness paradigms, participants are required to report the mere presence of a change, for example, by button press. The explicit report requires awareness of the change and high degrees of change blindness are usually observed. Rensink and colleagues measured performance in such paradigms by counting the change repetitions that are required for detection when two different images appear in fast succession (i.e., a "flicker paradigm," see Rensink et al., 1997). It is also possible to present changing and non-changing pictures only once before a response is required (i.e., a "forced choice detection paradigm," see Simons, 1996). This provides more alternatives for performance measures as it also allows for analyzing error rates and signal detection parameters. Furthermore, participants can be instructed to guess the location of change in paradigms that exclusively include changing visual displays. This allows measuring the forced choice report of unaware changes. Fernandez-Duque and Thornton (2000) revealed that even when participants were unaware of a change between two stimulus displays, their performance in guessing the location of change was above chance. Interestingly, there were no effects of selective attentional orienting to unaware visual changes. The findings point toward the notion that selective attention is associated with the explicit report of change but might not be involved in early processing stages that are unrelated to aware perception. This point will be treated further in the "selective attention and visual awareness" section of this chapter.

As the change blindness effect is very robust and we usually overestimate our ability to perceive changes, it also provides valuable information about the mechanisms that are involved in everyday perception of our visual surrounding. Therefore, change blindness paradigms are valuable tools for applied research. Car driving is a typical task involving the "looked but did not see" errors that are also reported by participants in a change blindness paradigm. It, thus, stands to reason that perceptual errors in driving are frequently related to a failure to perceive changes in critical visual scenes. For example, Caird and colleagues used a modified flicker paradigm with pictures of intersections (Caird, Edwards, Creaser, & Horrey, 2005). Younger and older adults watched the flickering pictures and were instructed to decide whether it was save to complete a certain driving maneuver at the intersection. The changing objects were relevant for the driving situation (e.g., pedestrians, vehicles, signs, or traffic control devices). Although all participants revealed a high amount of decision errors, younger adults outperformed the older participants. The results suggest that perceptual errors related to change blindness might be responsible for a large amount of decision errors in perceptually complex surroundings like intersections. This should especially have negative effects when the perceptual and cognitive abilities of the driver have declined with age.

Furthermore, change blindness paradigms can be used to determine unconscious biases on information processing. As already mentioned, the time required for detecting a changing object in a flicker paradigm relates to its relevance for the presented scenes (Rensink et al., 1997). This allows for figuring out which objects in a visual scene are most relevant for the observer. Accordingly, Jones and colleagues investigated the relevance of such a diagnostic tool in clinical contexts (Jones, Jones, Smith, & Copley, 2003). Social users of alcohol and cannabis were divided in heavy social users, light social users, and nonusers referred to each substance. Although no participants reported substance-related problems, the authors found out that heavy social users were faster in detecting changing objects related to the substance than light social users or nonusers. This high-level attentional bias toward the substance might result in a problematic future consumption behavior. Therefore, a diagnosis of such an unconscious bias on information processing might have a preventive effect regarding the risk to develop an addiction.

## PERCEPTUAL ERRORS UNDER CONDITIONS OF INATTENTION

The change blindness effect is a perceptual error based on a failure to perceive change when vision is shortly interrupted, even when attention is focused on the search for changes in the visual surroundings. As already mentioned, change blindness might contribute to a variety of decision errors in complex visual surroundings. However, the errors we encounter the most in everyday life distinguish themselves from change blindness with respect to the perceptual and attentional mechanisms involved. *Imagine you are walking through a shopping center and are looking for a certain store or article you want to buy. Meanwhile, a good friend of yours is walking past you and looking directly at you for starting a conversation. However, you do not even notice this and walk on by. The same might happen to you when you are walking down the street buried deep in thought. This might seem strange, but nearly everybody should have encountered comparable situations. What type of error is responsible for such behavior?*

The examples describe situations in which you focused attention inwards or on desired visual information (e.g., the billboard of a certain store). Meanwhile, new information that is potentially relevant for you does not reach awareness. This phenomenon is referred to as "inattentional blindness." Early experiments on inattentional blindness were conducted in the field of aviation psychology. For example, Fischer and colleagues tested commercial aircraft pilots in simulated landing situations with and without using a so-called Head-Up Display (HUD) (Fischer, Haines, & Price, 1980). The HUD projects visual information into an observer's viewing direction (e.g., by means of a display mounted in the helmet) and, thus, prevents fixating away from the critical visual information to check the instruments on the console.

However, pilots had slowed down responses to other aircrafts presented on the runway during landing when using the HUD. Occasionally, they even revealed a complete blindness referred to such a hazardous situation when attention was focused on the HUD. In the same way, pilots often missed information on the HUD when instructed to focus on the visual information outside the cockpit. Thus, attending either to the HUD or to the information straight ahead out the windshield leads to a blindness referred to unexpected visual events outside the attentional focus. Considering modern technologies like head-mounted displays in glasses that are available commercially, these findings highlight the importance of investigating the mechanisms underlying perceptual errors resulting from "inattention."

Heightened interest in psychological experiments on inattentional blindness was caused by a series of studies by Mack and Rock in the 1990s (see Mack & Rock, 1998). Participants saw a cross, consisting of two lines of different lengths for a period of 200 ms and had to judge which line was longer. In a subset of trials, the cross was accompanied by an irrelevant stimulus presented within 2.3° visual angle from fixation. Although it was a suprathreshold stimulus and it was presented for 200 ms, about 25% of the participants failed to report it. These results suggest that blindness occurred because participants did not expect the irrelevant event and focused attention on the primary task. In a further condition, the task-relevant cross and the unexpected irrelevant stimulus changed locations. While the irrelevant stimulus was presented at fixation, the cross was presented away from fixation. Interestingly, about 75% of participants failed to report the unexpected event under these circumstances. This does not sound reasonable. How can it be that our ability to perceive an unexpected suprathreshold stimulus is impaired when it is presented at the point of fixation, especially since fixation is the point of most detailed vision under normal conditions and is usually believed to coincide with the focus of attention? The results can only be explained by an inhibitory effect of attention. The increased blindness referred to the original experimental condition suggests that shifting attention away from fixation leads to an additional inhibition of processing at the fixated location (Mack & Rock, 1998).

The findings of Mack and Rock (1998) were as much interesting as unexpected and are in line with the results obtained by earlier work on perceptual errors while using an HUD. However, both experimental types involve very special perceptual situations. Although the unexpected stimuli used in the experiments by Mack and Rock (1998) were displayed above perceptual threshold, results cannot be transferred to real-world situations because the stimuli were highly artificial and static. The flight simulator experiments involved the perception of visual information on separate artificial plains in visual space (a 2D-level HUD display and a simulated 3D situation). An investigation of inattentional blindness in everyday perceptual surroundings, therefore, requires psychological experiments outside of such laboratory settings. Simons and colleagues developed a promising and well-known approach regarding this (see Simons & Chabris, 1999). Participants viewed video scenes that showed six basketball players of two different teams (black and white team). They were instructed either to count all passes of the black/white team (easy condition) or to separately count the aerial and bounce passes (hard condition). The critical events in these experimental session occurred when more than half of the video sequence was over. Either a woman wearing an opened umbrella or a woman wearing a whole-body gorilla costume walked through the scene while the other actors kept playing basketball. After viewing the video sequence, the participants in both the hard and easy conditions were asked to report the number of passes their attended team (black vs. white) completed. Afterward, they were asked several questions concerning the unexpected event (umbrella woman or gorilla). Surprisingly, about half of the participants missed the very salient but unexpected visual event in the video scenes. The blindness effect was strongest for participants performing the hard task. Furthermore, observers were more likely to miss the unexpected event when it was dissimilar with respect to their attentional setting (e.g., black gorilla while focusing on the white team). The experiments clearly show that objects that are not attended often fail to reach awareness. This unaware state of perception occurs although the unexpected events clearly "walk through" the spatial focus of attention. The experiments by Simons and Chabris (1999) clearly show that the inattentional blindness phenomenon is not restricted to the laboratory but is also evident in naturalistic and dynamic visual scenes.

Both the error types revealed in change blindness and inattentional blindness experiments contradict our assumptions about the way we perceive our visual surroundings. We typically believe that we see most of the salient and potentially relevant visual information in front of us. This belief is erroneous, even under conditions we meet every day. Furthermore, both visual phenomena indicate that attention shapes the way we perceive and interact with our perceptual surroundings. However, the aspects of attention that give information about change blindness and inattentional blindness differ. More specifically, inattentional blindness can only occur when a visual event is unexpected and attentional resources are engaged on other information within a visual scene. Thus, dividing attentional resources between multiple tasks is usually sufficient to perceive visual information. Only without knowledge about future visual information (i.e., no attentional resources are provided for this information) can blindness occur even for highly salient and moving objects (Simons & Chabris, 1999). As for change blindness, this is a more robust error in the course of perception. It can occur when we are expecting changes in a visual scene and are actively searching for them. Change detection requires transferring the pre-change information into a short-term store. Representations in short-term memory allow for the comparison of the stored information with the new visual inputs and the aware access to the result of the comparison process. Dividing attention between multiple stimuli is clearly not sufficient to enable the correct concurrence of these operations. Change detection rather requires an exogenous and/or endogenous shift of attention toward the location of a changing object. Based on the change blindness phenomenon, the following sections of this chapter describe how research in the field of cognitive neuroscience can help to further distinguish the role of selective attention for aware perception and goal-directed performance.

## Electrophysiological Evidence for the Role of Selective Attention in Change Detection

The prior sections of this chapter revealed that attention is involved in the way we perceive, access, and report relevant perceptual inputs. First, studies revealed that directing attention toward a subset of the presented visual information can bias competitive processing in visual areas and promote the representation of relevant information in higher level cognitive areas (Desimone & Duncan, 1995; Reynolds & Chelazzi, 2004). This line of research was mainly based on single-cell recordings in the behaving monkey (Chelazzi, Duncan, Miller, & Desimone, 1998; Chelazzi, Miller, Duncan, & Desimone, 1993, 2001). The biased competition account of visual selective attention postulates that only information that is represented in higher level areas can reach awareness and be available for report (Everling, Tinsley, Gaffan, & Duncan, 2002). This corresponds to findings from human testing showing that selective attention is critically related to the explicit report of change between two visual scenes. Noninvasive techniques for imaging the neural processes underlying cognitive operations in humans contributed to further bringing together these lines of research.

A prominent method for studying the neural mechanisms involved in selective attention and aware perception is the measurement of event-related potentials (ERPs) of the electroencephalogram (EEG). As its name implies, an ERP is the electrophysiological response measured at electrodes placed on the scalp that is related to a sensory, cognitive, or motor event. The ERP technique implies that all neural activity that is not related to a certain stimulus or event is random and can, therefore, be cancelled out by averaging across many trials in an experiment. The resulting ERP waveform consists of multiple positive and negative voltage deflections that are typically differentiated by means of their temporal occurrence and spatial distribution on the scalp. Referred to the study of perceptual errors and selective attention, ERPs can, for example, be used to investigate the neural processes that differ between conditions of blindness and detection of visual changes. An investigation in this regard was conducted by Eimer and Mazza (2005). Participants were presented with four faces (two on the left and right side of fixation, respectively) in two successive visual displays. In two-thirds of the trials, a face on the left or right side of fixation changed between the two displays, while no change occurred on the remaining trials. When a change was detected, the ERP time-locked to the second display revealed an increased negativity at posterior electrodes contralateral compared to ipsilateral to the changed face starting at about 220 ms subsequent to stimulus presentation. Based on its negative polarity, temporal characteristic, and spatial scalp distribution, this ERP component was referred to as N2pc (N2 posterior contralateral; see Luck & Hillyard, 1994a, 1994b). Corresponding to the role of selective attention in the detection of change, N2pc is associated with the selection of relevant visual information surrounded by distracting signals (Eimer, 1995; Luck & Hillyard, 1994a, 1994b) and was shown to originate both from the amplification of relevant information and suppression of irrelevant signals in the course of visual processing (Hickey, Di Lollo, & McDonald, 2009).

Based on these results, further studies concentrated on N2pc as a tool for investigating the relation of selective attention, aware perception, and goal-directed performance. For example, Busch and colleagues used ERPs as correlates of the neural mechanisms involved in the mere detection and identification of change (Busch, Frund, & Herrmann, 2010). Participants were presented two visual displays in fast succession containing 16 different objects arranged in a $4\times4$ matrix. In half of the trials, one object changed between the displays. Participants were subsequently asked if they detected any change. If this was affirmed on a change trial, participants were also asked to identify the modified object in the second display. Interestingly, only trials with fully identified changes revealed an N2pc and a prior lateralized positivity at posterior electrodes. This early asymmetry might be a correlate of the motion transient elicited by the change (Schneider, Hoffmann, & Wascher, 2014). Additionally, the N2pc effect clearly supports the notion that selective attention is associated with higher level identification of the changing object. However, the kind of "gut feeling" or sensing that a change appeared emerges at a pre-attentive processing stage. This corresponds to the results by Fernandez-Duque and Thornton (2000) mentioned earlier in this chapter showing that guessing the location of change without change awareness is not associated with selective attention.

Schneider and Wascher (2013) conducted a further ERP study on the neural mechanisms underlying the detection of change. Participants were instructed to respond to the location (left vs. right side of fixation) of a relevant change of luminance of a rectangular stimulus that was occasionally accompanied by a distracting change of orientation of the same or contralateral stimulus. Higher rates of localization errors and increased response times were observed in the contralateral distractor condition compared to the remaining stimulus conditions featuring a relevant luminance change. On ERP level, different posterior patterns were revealed for correct and erroneous localization trials in this contralateral distractor condition (see Figure 3). For correct trials, an early increased negativity ipsilateral to the relevant luminance change (i.e., contralateral to irrelevant orientation change) was observed prior to the increased contralateral negativity starting at about 200ms subsequent to change presentation (i.e., N2pc). This suggests that while attention was initially driven toward the irrelevant orientation change, it was re-allocated to the target stimulus. Furthermore, a sustained posterior contralateral negativity (SPCN; see Robitaille & Jolicoeur, 2006) followed N2pc. This component was associated with the sustained higher level representation of relevant visual information in short-term or working memory

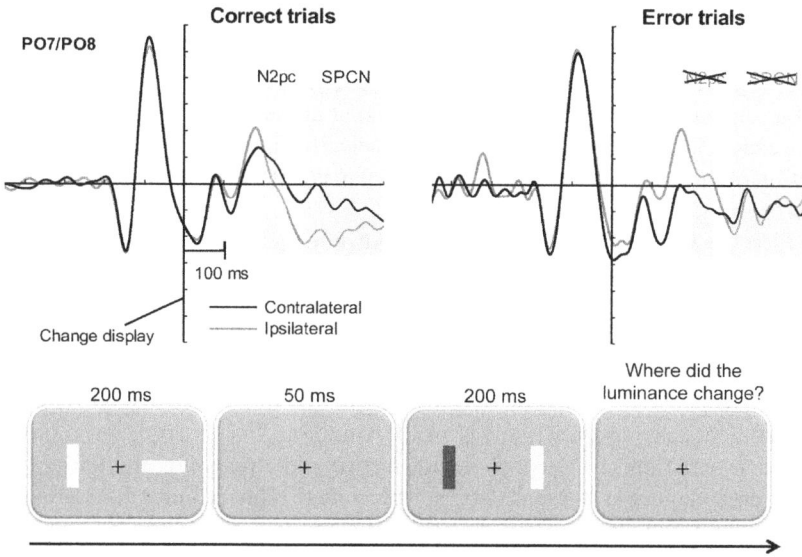

**FIGURE 3** ERP results from a change blindness experiment (*modified from Schneider & Wascher, 2013*). Participants were instructed to respond to the location (left vs. right) of a luminance change occurring between two displays shortly interrupted by a blank display. The condition shown in the ERP data contained an irrelevant change of orientation simultaneous to the relevant luminance change (contralateral distractor condition). For correct localization trials, an N2pc with higher negativity contralateral compared to ipsilateral to the luminance change was shown (see shaded timewindow). Subsequent to N2pc, this posterior contralateral effect was sustained for about 200 ms (i.e., sustained posterior contralateral negativity or SPCN). On the contrary, a strong asymmetry in favor of the contralateral distractor was revealed for error trials, and no subsequent N2pc or SPCN effects were shown.

required for further cognitive operations with the stimulus material. Compared to this ERP pattern, a large contralateral negativity in favor of the irrelevant orientation change appeared for error trials and neither N2pc nor SPCN were shown (see Figure 3). These results suggest that erroneous decisions on the basis of the stimulus material were associated with the distraction of attention by an irrelevant feature change and a failure to re-allocate attention and create a higher level representation of the critical visual information (Schneider & Wascher, 2013). Again, this supports the critical role of selective attention for establishing an aware percept of stimulus change by drawing a connection between selection and higher level visual representations.

## SELECTIVE ATTENTION AND VISUAL AWARENESS

It was shown that studies with different experimental approaches consistently pointed toward a prominent role of selective attention for aware perception. Findings on the basis of single-cell recordings, EEG, and also other imaging techniques in human testing (e.g., functional magnetic resonance imaging) suggest that attentional selection is involved in visual awareness and, therefore, is a prerequisite for the goal-directed interaction with our visual surroundings. This might lead to the assumption that attention and awareness are two sides of the same coin. Accordingly, all stimuli that are attended become aware and awareness cannot be established without selective attention. However, the last section of this chapter will show that this view is not in accordance with current neuroscientific research.

First of all, it is important to point out our meaning of visual awareness and to recapitulate why this is important for the current purposes. The interaction with our environment starts with the sensory experience of the information surrounding us. In order to perform in a goal-directed way, we have to filter the sensory information that is currently relevant and, in turn, create a certain motoric response. This response can be based on an aware decision regarding the different response alternatives in the situation. For example, in a change detection paradigm, we have to decide between the response alternatives that a certain change was presented or not presented. This requires having access to the neural representation of change and using this access for the control of behavior. Therefore, what is meant here by "awareness" was also labeled access awareness (Block, 1996; Lamme, 2003). Selective attention might, therefore, be involved in enabling the aware access to visual information that would otherwise not be available for cognitive and behavioral control.

This connection between selective attention and the ability to have aware access to sensory information was already mentioned in the sections concerning the role of attention in change detection and change blindness. However, it can be made clearer by means of a simple illustration. *Think about sitting in the kitchen and learning for your next exam at the university. If you are really focused on this task, you will obviously fade down all other incoming information from your surroundings. For instance, you will no longer perceive the loud ticking of the clock that you usually find so annoying when trying to take a nap. However, when this ticking suddenly stops you will certainly recognize that something has just changed in your surroundings.* This is because the sound change drew attention, and the missing clock ticking reached awareness. In other words, you suddenly have access to what changed in your surrounding and might, therefore, decide to change your clock's batteries. Apparently, selective attention plays an important role in transferring sensory inputs to a stage that enables aware access to perform higher level decisional operations. Yet, this still brings up the question: what is the neural process that directly determines if awareness can evolve?

Important progress in this respect was achieved by neuroscientific studies that concentrated on the direction of information processing in the visual system. It is important to mention at this point that visual information processing is not exclusively directed in a unidirectional way from the lowest areas in the occipital cortex to the highest visual areas in the frontal cortex. Visual information processing rather consists both of a feed-forward processing stream and iterative or reentrant connections between and within visual areas (Lamme & Roelfsema, 2000). When signals are first represented in primary visual cortex, they are rapidly transferred in a feed-forward way to higher level visual areas. As already mentioned, the receptive fields along this processing stream become larger and, therefore, the visual processing advances from the analysis of very confined features to more global patterns and representations of whole objects. By about 120 ms subsequent to the presentation of visual information, the feed-forward representations are present in all cortical processing areas (see Lamme & Roelfsema, 2000). When this processing stream reaches a certain area, a reentrant signal is automatically sent back to lower visual areas.

Several studies suggested that only the reentrant signals but not feed-forward processing is associated with aware perception. For example, a study by Pascual-Leone and Walsh (2001) used transcranial magnetic stimulation (TMS) to affect visual processing. Usually, when TMS is applied to the motion selective areas in the visual cortex (area "middle temporal" or area MT/V5), this can create a motion sensation. However, this motion sensation is not created when TMS disrupts processing in early sensory areas subsequent to the stimulation of area MT. As area MT is a higher level extrastriate visual area, this suggests that the disruption of reentrant connections with the earlier visual areas caused the blindness referred to the motion sensation. Backward masking is another way to affect the reentrant interactions in the course of visual processing. This experimental technique implies the presentation of a second stimulus shortly (i.e., <100 ms) after a first target stimulus. This typically leads to an invisibility of the target stimulus. In an experiment by Lamme, Zipser, and Spekreijse (2002), macaque monkeys had to respond to the presence of a figure in a certain background pattern. The study was based on results indicating that the selective coding of the orientation of lines proceeds very early in primary visual cortex (V1), whereas the segregation of whole figures from background is a later process that was assumed to be based on reentrant interactions with higher level extrastriate visual areas (Lamme, Rodriguez–Rodriguez, & Spekreijse, 1999). It was shown that when backward masking led to an inability to detect the target stimulus, it selectively affected the figure-ground signals in V1 but had no impact on earlier orientation segregation.

Thus, there are two factors that are related to the ability to have an aware access to information in our visual surrounding, selective attention and reentrant or recurrent processing between visual areas. The question remains whether these factors interact in creating awareness and, if so, how this interaction proceeds. According to Lamme (2003), reentrant processing determines if visual awareness can evolve after all, because the interconnection of visual areas is required to integrate low-level features (e.g., local feature detection like line orientations) and higher level features (e.g., object representations). Thus, reentrant processing can be seen as the most likely neural correlate of awareness. Aware access to information is created when the recurrent interactions are so widespread as to include higher level visual areas, because at this stage the information is set into the context of current behavioral goals and needs (Duncan, 2001; Lamme, 2003). This is a prerequisite for decisional processes on the basis of the visual inputs and enables an adequate interaction with our visual surrounding. Importantly, selective attention is the key factor here, because attending to certain inputs leads to a competitive advantage and enables the higher level visual representations of relevant information.

## CONCLUSION

This section was concerned with the role of attention for transferring perceptual information into an adequate decision and performance. Based on change blindness and inattentional blindness phenomena as examples for errors that can occur in the course of visual processing, it was shown that selective attention is associated with the ability to have an aware access to visual inputs and perform decisional processes based on this. The inability to report a change when attention was not directed at its specific location is a good example. Without the access to the representation of change, there is no basis for higher level decisional processes leading to a failure in report. However, as it was shown that guessing the location of change in such a situation still works above chance level (Fernandez-Duque & Thornton, 2000), it is still possible to respond adequately to the change signal without any aware access. This might be achieved because the reentrant neuronal interactions that are responsible for the processing of information between visual areas occur regardless of attention and lead to an integrated visual representation independent from access and report. Research from the field of cognitive neuroscience contributed to the further understanding of how attention shapes perception and behavioral performance. It was shown that selectively attending to certain information in our visual surroundings leads to a competitive bias that can enable the representation of relevant information in higher level processing areas even when it is less salient than surrounding signals (Desimone & Duncan, 1995). This higher level representation is important because by means of reentrant interactions with lower areas, it allows for setting the incoming signals into the context of current behavioral goals and needs. This is an essential prerequisite of access awareness and the goal-directed interaction with our environment.

## REFERENCES

Baddeley, A. D. (1986). *Working memory*. Oxford: Oxford University Press.
Block, N. (1996). How can we find the neural correlate of consciousness? *Trends in Neurosciences*, *19*(11), 456–459.
Busch, N. A., Frund, I., & Herrmann, C. S. (2010). Electrophysiological evidence for different types of change detection and change blindness. *Journal of Cognitive Neuroscience*, *22*(8), 1852–1869. http://dx.doi.org/10.1162/jocn.2009.21294.
Caird, J. K., Edwards, C. J., Creaser, J. I., & Horrey, W. J. (2005). Older driver failures of attention at intersections: using change blindness methods to assess turn decision accuracy. *Human Factors*, *47*(2), 235–249.

Chelazzi, L., Duncan, J., Miller, E. K., & Desimone, R. (1998). Responses of neurons in inferior temporal cortex during memory-guided visual search. *Journal of Neurophysiology, 80*(6), 2918–2940.

Chelazzi, L., Miller, E. K., Duncan, J., & Desimone, R. (1993). A neural basis for visual search in inferior temporal cortex. *Nature, 363*(6427), 345–347. http://dx.doi.org/10.1038/363345a0.

Chelazzi, L., Miller, E. K., Duncan, J., & Desimone, R. (2001). Responses of neurons in macaque area V4 during memory-guided visual search. *Cerebral Cortex, 11*(8), 761–772.

De Valois, R. L., & De Valois, K. K. (1991). Vernier acuity with stationary moving Gabors. *Vision Research, 31*(9), 1619–1626.

Desimone, R., & Duncan, J. (1995). Neural mechanisms of selective visual attention. *Annual Review of Neuroscience, 18*, 193–222. http://dx.doi.org/10.1146/annurev.ne.18.030195.001205.

Desimone, R., & Ungerleider, L. G. (1989). Neural mechanisms of visual processing in monkeys. In F. Boller & J. Grafman (Eds.), *Handbook of neuropsychology* (pp. 267–299). New York: Elsevier.

Duncan, J. (2001). An adaptive coding model of neural function in prefrontal cortex. *Nature Reviews Neuroscience, 2*(11), 820–829. http://dx.doi.org/10.1038/35097575.

Duncan, J., & Humphreys, G. W. (1989). Visual search and stimulus similarity. *Psychological Review, 96*(3), 433–458.

Eimer, M. (1995). Event-related potential correlates of transient attention shifts to color and location. *Biological Psychology, 41*(2), 167–182.

Eimer, M., & Mazza, V. (2005). Electrophysiological correlates of change detection. *Psychophysiology, 42*(3), 328–342. http://dx.doi.org/10.1111/j.1469-8986.2005.00285.x.

Everling, S., Tinsley, C. J., Gaffan, D., & Duncan, J. (2002). Filtering of neural signals by focused attention in the monkey prefrontal cortex. *Nature Neuroscience, 5*(7), 671–676. http://dx.doi.org/10.1038/nn874.

Fernandez-Duque, D., & Thornton, I. M. (2000). Change detection without awareness: do explicit reports underestimate the representation of change in the visual system. *Visual Cognition, 7*, 324–344.

Fischer, E., Haines, R. F., & Price, T. A. (1980). *Cognitive issues in head-up displays (NASA Technical Paper 1711)*. Moffett Field, CA: NASA Ames Research Center.

Grimes, J. (1996). On the failure to detect changes in scenes across saccades. In K. Akins (Ed.), *Perception (Vancouver studies in cognitive science)* (Vol. 2) (pp. 89–110). New York: Oxford University Press.

Hickey, C., Di Lollo, V., & McDonald, J. J. (2009). Electrophysiological indices of target and distractor processing in visual search. *Journal of Cognitive Neuroscienc, 21*(4), 760–775. http://dx.doi.org/10.1162/jocn.2009.21039.

James, W. (1890/1950). *Principles of psychology* (Vol. 2). New York: Dover.

Jones, B. T., Jones, B. C., Smith, H., & Copley, N. (2003). A flicker paradigm for inducing change blindness reveals alcohol and cannabis information processing biases in social users. *Addiction, 98*(2), 235–244.

Knierim, J. J., & van Essen, D. C. (1992). Neuronal responses to static texture patterns in area V1 of the alert macaque monkey. *Journal of Neurophysiology, 67*(4), 961–980.

Lamme, V. A. (2003). Why visual attention and awareness are different. *Trends in Cognitive Science, 7*(1), 12–18.

Lamme, V. A., Rodriguez-Rodriguez, V., & Spekreijse, H. (1999). Separate processing dynamics for texture elements, boundaries and surfaces in primary visual cortex of the macaque monkey. *Cerebral Cortex, 9*(4), 406–413.

Lamme, V. A., & Roelfsema, P. R. (2000). The distinct modes of vision offered by feedforward and recurrent processing. *Trends in Neurosciences, 23*(11), 571–579.

Lamme, V. A., Zipser, K., & Spekreijse, H. (2002). Masking interrupts figure-ground signals in V1. *Journal of Cognitive Neuroscienc, 14*(7), 1044–1053. http://dx.doi.org/10.1162/089892902320474490.

Luck, S. J., & Hillyard, S. A. (1994a). Electrophysiological correlates of feature analysis during visual search. *Psychophysiology, 31*(3), 291–308.

Luck, S. J., & Hillyard, S. A. (1994b). Spatial filtering during visual search: evidence from human electrophysiology. *Journal of Experimental Psychology: Human Perception and Performance, 20*(5), 1000–1014.

Mack, A., & Rock, I. (1998). *Inattentional blindness*. Cambridge, MA: MIT Press.

Miller, E. K., Erickson, C. A., & Desimone, R. (1996). Neural mechanisms of visual working memory in prefrontal cortex of the macaque. *Journal of Neuroscience, 16*(16), 5154–5167.

O'Regan, J. K., Rensink, R. A., & Clark, J. J. (1999). Change-blindness as a result of 'mudsplashes'. *Nature, 398*(6722), 34. http://dx.doi.org/10.1038/17953.

Pascual-Leone, A., & Walsh, V. (2001). Fast backprojections from the motion to the primary visual area necessary for visual awareness. *Science, 292*(5516), 510–512. http://dx.doi.org/10.1126/science.1057099.

Ramachandran, V. S., & Anstis, S. M. (1990). Illusory displacement of equiluminous kinetic edges. *Perception, 19*(5), 611–616.

Rensink, R. A., O'Regan, J. K., & Clark, J. J. (1997). To see or not to see: the need for attention to perceive changes in scenes. *Psychological Science, 8*, 368–373.

Reynolds, J. H., & Chelazzi, L. (2004). Attentional modulation of visual processing. *Annual Review of Neuroscience, 27*, 611–647. http://dx.doi.org/10.1146/annurev.neuro.26.041002.131039.

Robitaille, N., & Jolicoeur, P. (2006). Fundamental properties of the N2pc as an index of spatial attention: effects of masking. *Canadian Journal of Experimental Psychology, 60*(2), 101–111.

Schneider, D., Hoffmann, S., & Wascher, E. (2014). Sustained posterior contralateral activity indicates re-entrant target processing in visual change detection: an EEG study. *Frontiers in Human Neuroscience, 8*, 247. http://dx.doi.org/10.3389/fnhum.2014.00247.

Schneider, D., & Wascher, E. (2013). Mechanisms of target localization in visual change detection: an interplay of gating and filtering. *Behavioural Brain Research, 256*, 311–319. http://dx.doi.org/10.1016/j.bbr.2013.08.046.

Scholl, B. J. (2000). Attenuated change blindness for exogenously attended items in a flicker paradigm. *Visual Cognition, 7*, 377–396.

Serences, J. T., & Yantis, S. (2006). Selective visual attention and perceptual coherence. *Trends in Cognitive Science, 10*(1), 38–45. http://dx.doi.org/10.1016/j.tics.2005.11.008.

Simons, D. J. (1996). In sight, out of mind: when object representations fail. *Psychological Science, 7*, 301–305.

Simons, D. J. (2000). Current approaches to change blindness. *Visual Cognition, 7*, 1–15.

Simons, D. J., & Chabris, C. F. (1999). Gorillas in our midst: sustained inattentional blindness for dynamic events. *Perception, 28*(9), 1059–1074.

Simons, D. J., & Levin, D. T. (1997). Change blindness. *Trends in Cognitive Science, 1*(7), 261–267. http://dx.doi.org/10.1016/S1364-6613(97)01080-2.

Simons, D. J., & Rensink, R. A. (2005). Change blindness: past, present, and future. *Trends in Cognitive Science, 9*(1), 16–20. http://dx.doi.org/10.1016/j.tics.2004.11.006.

Theeuwes, J. (1991). Cross-dimensional perceptual selectivity. *Perception & Psychophysics, 50*(2), 184–193.

Theeuwes, J. (1992). Perceptual selectivity for color and form. *Perception & Psychophysics, 51*(6), 599–606.

Tootell, R. B., Silverman, M. S., Switkes, E., & De Valois, R. L. (1982). Deoxyglucose analysis of retinotopic organization in primate striate cortex. *Science, 218*(4575), 902–904.

Whitney, D., Wurnitsch, N., Hontiveros, B., & Louie, E. (2008). Perceptual mislocalization of bouncing balls by professional tennis referees. *Current Biology, 18*(20), 947–949. http://dx.doi.org/10.1016/j.cub.2008.08.021.

Chapter 12

# Lifestyle and Interventions for Improving Cognitive Performance in Older Adults

### Patrick D. Gajewski[1], Michael Falkenstein[1,2]
[1]*Leibniz Research Centre for Working Environment and Human Factors (IfADo), Dortmund, Germany;* [2]*Institute for Working, Learning and Aging (ALA), Bochum, Germany*

The present review of the current literature provides a short review on the impact of lifestyle as well as physical and cognitive training on cognitive performance in healthy, elderly subjects. Special emphasis is devoted to different formats of cognitive training from computer-based to more natural formats.

Findings from numerous epidemiologic and clinical studies suggest that biological, behavioral, social, environmental, and lifestyle factors influence cognition in older age and may even reduce cognitive decline (Plassman, Williams, Burke, Holsinger, & Benjamin, 2010). Biological factors have a clear and large impact on cognition, which is even reflected in functional brain activity (Raz, 2000). Unfortunately, those factors cannot be directly influenced in later life. However, lifestyle factors such as diet, social interaction, occupation, physical activity, and leisure cognitive activity can be well affected also in advanced age. In a comprehensive meta-analysis, Plassman et al. (2010) reviewed the influence of biological and lifestyle factors in 127 observational and some intervention studies. Factors with positive influence on cognition in older adults were nutrition, challenging occupation, leisure activity, and cognitive and physical activity. In a more recent meta-analysis of 247 studies, Beydoun et al. (2014) reviewed the influence of various factors on cognition in older adults. In addition to the factors identified by Plassman et al., they found specific nutritional factors such as caffeine and high educational attainment positively influencing cognition and reducing the incidence of dementia.

Many nutrition studies propose a Mediterranean diet with fish, olive oil, green vegetables, and red wine to promote cognition and preventing dementia in older adults (Scarmeas, Stern, Mayeux, & Luchsinger, 2006). More recent reviews confirmed a protective effect of unsaturated fatty acids (e.g., omega-3 fatty acids from oily sea fish) and micronutrients (such as polyphenols) from vegetables and fruits against brain aging (e.g., Barberger-Gateau, 2014). Moderate alcohol drinking, in particular red wine, is probably a protective factor against cognitive decline, while high consumption (≥36 g/day) is associated with faster decline in all cognitive domains compared with low consumption (Sabia et al., 2014). Coffee appears to improve cognition in both animals and humans. In humans, drinking coffee attenuated the cognitive decline in elderly men over 10 years, with three cups per day leading to the least decline (van Gelder et al., 2007). As to negative habits, smoking causes cognitive decline and loss of gray matter tissue in the brain over time (Almeida et al., 2011).

Chronic **stress** impairs various cognitive functions, particularly in older age (Lupien, Maheu, Tu, Fiocco, & Schramek, 2007). In a study with 811 older subjects, psychological stress had an inverse (U-shaped) association with cognition (Peters et al., 2010). This increased vulnerability may be due to the fact that during aging, neurons within the PFC become less resilient to stress (McEwen & Morrison, 2013).

**Occupation** and occupational environment have a pronounced influence on cognitive performance. The more complex the demand characteristics of the work-setting, the greater the potential for maintenance or even enhancement of cognitive functioning that may compensate age-related declines (e.g., Marquié et al., 2010; Schooler, Mulatu, & Oates, 1999). Potter, Helms, and Plassman (2008) observed a beneficial effect of demanding work on cognition in a sample of about 1000 older workers across a time range of seven years, whereas tedious physical job demands tended to impair cognition. Gajewski et al. (2010) showed specific performance deficits in older workers with long-lasting repetitive work, compared to older workers with flexible work demands. These deficits were seen only for a task with high working memory demand, but not for easier tasks.

**Physical activity** not only improves motor behavior and cardiovascular fitness, but appears to enhance brain function and cognitive performance and protects against neurodegenerative diseases (cf. Kramer & Erickson, 2007 for a review). In a recent study, Desjardins-Crépeau et al. (2014) showed that high physical fitness was associated with greater processing speed and better executive functions, whereas memory performance was hardly improved. The mechanisms of the beneficial effects of physical fitness are still under investigation. Higher physical fitness is associated with greater gray matter volume in the prefrontal cortex and hippocampus (Erickson, Leckie, & Weinstein, 2014), which are known to deteriorate with increasing age. Gajewski and Falkenstein (2015, in press) showed that seniors, who have been physically active during their entire lifespan, performed better in two executive tasks (Stroop and task switching) than carefully matched sedentary seniors. This was accompanied by enhanced electrical activity over fronto-central brain areas, which are known to support executive functions. As in the study of Desjardins-Crépeau, memory performance did not differ between physically active and passive seniors in the Gajewski and Falkenstein study. Hence, it appears that not all but specific executive functions are improved in older subjects by lifelong physical activity.

## COGNITIVE LEISURE ACTIVITY

In animal research, it has been consistently shown that environmental enrichment improves cognition (Milgram, Siwak-Tapp, Araujo, & Head, 2006). In humans, many observational studies have found that challenging leisure activities such as reading and playing board games or musical instruments may lead to slower cognitive decline in healthy elderly persons and reduce the risk of dementia (Singh-Manoux, Richards, & Marmot, 2003; Verghese et al., 2003). Plassman et al. (2010) found positive effects on cognition in older adults due to cognitive activity and training, but no consistent influence of social engagement. Hence, cognitively challenging leisure activities appear to have a positive effect on cognition in older people, whereas simple social interactions may be less efficient. Therefore, it seems that the degree of cognitive load, and hence of cognitive training (CT), is the crucial factor in lifestyle activities. In future studies, the assessment of leisure activities and also the cognitive test batteries and the definition of cognitive parameters should be standardized (Wang, Xu, & Pei, 2012).

### Cognitive Training: Intervention Studies

The goal of CT for older adults is to improve their performance on cognitive skills that usually deteriorate with age and which are crucial for everyday life performance (Zelinski, 2009). CT can be administered in various formats ranging from formal (PC- or game-based) to more natural interventions. In most CT studies, test-like tasks or games were trained via PC (computerized cognitive training; CCT) repeatedly and for an extended time. Reijnders, van Heugten, and van Boxtel (2013) reviewed the CCT studies published between 2007 and 2012. The methodological quality of the intervention studies differed widely and was low on average. The results show evidence that cognitive training can be effective in improving various aspects of objective cognitive functioning: memory performance, executive functioning, processing speed, attention, fluid intelligence, and subjective cognitive performance.

A more recent meta-analysis of CCT analyzing nearly 5000 cognitively healthy older adults states that CCT slightly increased overall cognitive function in healthy older adults (Lampit, Hallock, & Valenzuela, 2014; see also Burch, 2014 for critical comment). However, efficacy varies across cognitive domains and is largely determined by design choices.

The largest CCT trial conducted so far was the Advanced Cognitive Training for Independent and Vital Elderly (ACTIVE) (Willis et al., 2006). About 3000 elderly persons were trained for 10 sessions in memory, reasoning, or speed of processing. The training groups (compared to noncontact controls) improved cognition specific to the abilities trained. Transfer is a critical issue in all CCT literature (Green, Strobach, & Schubert, 2014). The crucial question is whether training leads not only to gains in the trained task but also to gains in those functions that are targeted by the training (near transfer) to gains in unrelated functions (far transfer) or even to everyday life performance. Thus, an important issue in the CCT literature is which cognitive domains should be trained to enhance transfer. In an influential study, Jaeggi, Buschkuehl, Jonides, and Perrig (2008) showed that the extended training of working memory by a very difficult dual $n$-back task led to an increase of fluid intelligence in young adults, even though the test was entirely different from the trained task. More recently, the group trained older adults with an adaptive verbal $n$-back task for either 10 or 20 sessions (Stepankova et al., 2014). The training groups showed a dose-dependent improvement in non-trained measures of working memory and visuospatial skills compared to a control group. Thus, the results show far transfer in older subjects when working memory was trained with an $n$-back task. Karbach and Kray (2009) showed that task switching training not only leads to near transfer but also to far transfer to other cognitive domains in young and older adults. In a recent meta-analysis, Karbach and Verhaeghen (2014) showed evidence for near transfer of executive and working memory training in older adults. Also, far transfer was consistently observed, but to a smaller degree. In summary, as to training contents, working memory and task switching training appear to enhance the chance for far transfer.

Although the literature, in general, shows good evidence for transfer effects of CCT to psychometric tests, effects on everyday performance are rarely reported (Reijnders et al., 2013). In the ACTIVE study, reasoning training resulted in less functional decline in self-reported instrumental activities of daily life. Edwards, Delahunt, and Mahncke (2009) found that a divided spatial attention training (UFOV) delayed driving cessation among older drivers. Caserta, Young, and Janelle (2007) showed that a perceptual-cognitive skills training improved the match performance of senior tennis players. Finally, the training of simple PC-based tasks can improve driving performance in older adults in a driving simulator (Cassavaugh & Kramer, 2009).

Long-term effects of CCT have also been rarely assessed. In the ACTIVE study, a very long-term follow-up assessment was conducted (Rebok et al., 2014). Ten training sessions as well as four booster sessions 11 and 35 months after initial training were administered. All interventions but the memory intervention maintained their effects at 10 years.

Cognitive training has been shown to lead to changes in neuronal structure and function, which are the basis of the observed improvements in performance (e.g., Lustig, Shah, Seidler, & Reuter-Lorenz, 2009). Those changes include volume changes in relevant structures (Engvig et al., 2012), increases in cerebral blood flow and white matter integrity (Chapman, Aslan, Spence, Hart, et al., 2013), increased frontal and parietal activity in functional MRI (Olesen, Westerberg, & Klingberg, 2004), and enhanced amplitude of event-related brain potentials (ERP) (Gajewski & Falkenstein, 2012; Wild-Wall, Falkenstein, & Gajewski, 2012).

Even though the overall evidence for benefits of cognitive training in older people is positive, at least for near transfer, there are some studies and reviews that show only modest or no benefit of CCT on young and elderly subjects (Lampit et al., 2014). Shipstead, Redick, & Engle (2012) mentioned several problems in CT studies, such as using single transfer tasks and non-contact control groups. Using more strict criteria, Redick et al. (2013) found no transfer to any of the cognitive ability tests despite improvements on both trained tasks in young adults. Fifty-four seniors were trained by van Muijden, Band, and Hommel (2012) with different CCT tasks online over seven weeks. An active control group answered quiz questions online. A large test battery covering executive functions, attention, and reasoning assessed transfer. The CCT group showed larger improvement of inhibition and reasoning than the control group, while the other domains were not differentially improved. Such rather negative outcomes may be due to motivational problems and the format used. In the van Muijden study, many trainees found the training to be not motivating. Hence, a most important issue for future research concerns the structure (contents, duration, timing, booster sessions, etc.), the design of the active control groups, and the motivational incentive of a CCT (Green & Bavelier, 2008). In addition, the pre-training cognitive state and the educational status must be considered.

Green and Bavelier (2008) and Green et al. (2014) proposed several rules for a successful training, namely care for high motivation and optimum arousal, slow increase of difficulty, immediate or delayed feedback, and high variability. They stated that multifaceted and variable training causes slower learning but better retention. Moreover, they argued that stimuli and tasks in different contexts lead to the extraction of more specific features and rules than static contexts. Lustig et al. (2009) likewise stressed the importance of variable training for success and transfer. They claimed that for successful training, a neuronal and, hence, functional overlap of training and testing is necessary, i.e., CCT is only effective when the targeted cognitive functions are implicitly trained. More recently, Fissler, Küster, Schlee, and Kolassa (2013) proposed that variable training rather than the training of one fixed task leads to better transfer. Thus, more variable and multifaceted trainings are promising in promoting stable learning and transfer.

In their "Dortmund three-training study," Gajewski and Falkenstein (2012) combined selected paper and PC-based tasks from commercial products to a variable multidomain training. The CCT provided immediate feedback and adaptive enhancement of difficulty. The training was administered to about 40 healthy seniors by a trainer who provided individual supervision. Three other groups received a physical or a relaxation or no training. Extensive paper- and PC-based psychometric tests were given before and after the training and to the same dates for the passive group. During the PC-based tests, the electroencephalogram (EEG) was recorded and event-related potentials (ERPs) were computed that reflect specific cognitive processes. Training duration was the same for the three interventions (two sessions of 90 min per week for 16 weeks). The results showed primarily accuracy improvements of different cognitive functions in the CT group vs. the other groups in both paper-and-pencil-based and PC-based tests.

Moreover, the ERPs showed occasional enhancements of components related to attention, decision, and error detection exclusively for the CT group.

In a further study (Gajewski & Falkenstein, 2011), 120 middle-aged assembly-line workers (40–65) were trained with supervised multitask/multidomain CT for three months (two sessions of 90 min per week). The training was evaluated with paper-based tests and a memory-based task-switching task (TST) administered via PC, while the EEG was recorded. Compared to a waiting control group, the trainees improved performance in a number of psychometric paper and pencil tests. Trainees with initially lower performance benefited most. In the TST, a strong decrease of error rates was seen. In the ERP, data correlates of response selection and error detection increased. These results persisted until three months after training.

From the beginning of the cognitive training literature more "natural" interventions were also examined regarding their effect on cognition in older persons.

A first approach is the use of video games. They combine variable training, which targets cognitive functions with more realistic game-like scenarios, whereby likely enhancing motivation. In their review, Bavelier and Davidson (2013) stated that video games improve visual attention, discrimination, and mental rotation, that is, implicitly trained functions. Several studies found effects of video game training on cognitive functions also in older individuals. Basak, Boot, Voss, and Kramer (2008) trained older adults in a strategy video game for 23.5 h. The trainees improved more than controls in executive functions and visuospatial skills. Individual differences in changes in game performance were correlated with improvements in task switching. More recently, Anguera et al. (2013) had older adults play a multitasking video game. The trainees reduced multitasking costs compared to both an active and a non-contact control group, with gains persisting for six months. Furthermore, the training reduced age-related deficits in EEG measures of cognitive control. Toril, Reales, and Ballesteros (2014) reviewed studies between 1986 and 2013 and stated that video game training produces positive effects on several cognitive functions in older subjects. The magnitude of the training effect was moderated by factors such as the age of the trainees and the duration of the intervention.

PC- and video game–based interventions usually lack social interaction and are not well accepted by older people. Fissler et al. (2013) proposed a more familiar training approach for seniors with highly variable card and board games, which train certain cognitive functions ("overlapping variability" framework). The older trainees completed 30 training hours in 15 sessions. The results show enhanced executive functions in the gaming compared to a passive control group. Even more natural multimodal interventions are learning to dance or to play a musical instrument. As to dancing, Kattenstroth, Kalisch, Holt, Tegenthoff, and Dinse (2013) found that six months of dance training enhances postural, sensorimotor, and cognitive performance in elderly.

In their SYNAPSE project, Park et al. (2014) trained older adults new complex skills. Participants learned either to quilt, digital photography, or engaged in both activities for three months. In the active groups, episodic memory was enhanced relative to two active control groups. As in the review of Plassman et al. (2010), the Park study found no substantial effect of sustained social activity on cognition in older people. In a subsequent study, older adults who were computer novices were trained to use a tablet and associated software for three months (Chan, Haber, Drew, & Park, 2014). The tablet group was compared with two separate controls: (1) a placebo group that engaged in passive tasks requiring little new learning and (2) a social group that had regular social interaction but no active skill acquisition. The tablet group showed greater improvements in episodic memory and processing speed but not mental control or visuospatial processing compared to both controls.

Stine-Morrow, Parisi, Morrow, and Park (2008) administered cognitive training in a naturalistic setting. Their training included frequent exposure to ill-defined problems and a collaborative and competitive context. This setting is thought to train many cognitive abilities. Relative to controls, the trained participants showed improvement in speed of processing, inductive reasoning, and divergent thinking in the absence of explicit training of those functions.

In summary, there is some evidence for beneficial effects of cognitive training on cognitive performance on older adults. However, CCT programs widely differ in quality and effectivity. In addition, CCT should be adequately shaped to improve cognition. First, the training should be adaptive but with shallow or self-administered increase of difficulty. Second, feedback should be frequently given either by a program or by a teacher. Third, the CCT should have a sufficient length of at least 10 to 15 sessions. Fourth, because far transfer is scarce, each targeted cognitive domain should be trained. Fifth, and most importantly, the training should induce fun to motivate seniors to continue the training without external guidance. The two last requirements can, to our opinion, best be induced by training several cognitive functions and by variable tasks targeting one single function. Hence game-like scenarios or complex leisure activities such as unspecific PC training and language learning appear to be a good platform particularly for older adults. However, because there are also reports of no transfer even with working memory training, the factors that influence the success of CT in seniors should be carefully investigated in further studies. In particular, complex everyday activity trainings (such as language learning) should be further evaluated in future research.

## Physical Training: Intervention Studies

As mentioned previously, observational studies reveal a robust and dose-dependent relationship between physical exercise and cognitive performance in older adults (e.g., Lautenschlager & Almeida, 2006). However, given the observational nature of these studies, causation cannot be established. Therefore, numerous physical intervention studies have been conducted with older adults. One example is the study of Albinet, Boucard, Bouquet, and Audiffren (2010), who trained sedentary elderly with an aerobic exercise or a stretching program three times a week for 12 weeks. Executive control was measured with the Wisconsin Card Sorting Test (WCST). Only the participants in the aerobic group improved their performance on the WCST.

Colcombe and Kramer (2003) were the first to conduct a meta-analytical study of the effect of physical training on cognition in healthy older adults in randomized controlled trials. Eighteen studies with active control groups were included. A moderate effect size (0.48) for physical training was obtained. The largest positive effects were observed for executive control processes, which usually showed a substantial age-related decline. Improvements in controlled, spatial, and speed tasks are smaller. The effects were slightly larger when aerobic training was combined with strength and flexibility training. In addition, the session duration of at least 30 min and the total duration of at least six months appear to be necessary. Studies that included more women showed larger fitness training benefits than studies with fewer women.

However, some reviews are more pessimistic. For example, van Uffelen, Chin, Paw, Hopman-Rock, and van Mechelen (2008) reviewed the evidence of exercise programs on cognition in older adults with and without cognitive decline in 23 selected training studies. Though some studies found positive effects, the majority of the studies did not find any effect. The small number of included studies, lack of high-quality studies, and the large variability in study populations, exercise protocols, and outcome measures, complicate interpretation of the results. More high-quality trials are needed to assess the effects of different types of exercise on cognitive function in older adults with and without cognitive decline. A meta-analysis conducted by Smith et al. (2010) including randomized controlled studies from 1966 until 2009 with a total number of 2049 participants showed modest improvement in attention and processing speed, executive function, whereas memory effects were less consistent. Also, a recent review of Kirk-Sanchez and McGough (2014) stated that many studies demonstrate positive effects of exercise on cognitive performance, whereas others show minimal to no effect. Hence, the moderator variables that influence success or failure have to be explored in much more detail in the future.

Physical training exerts various effects on the brain. Gomez-Pinilla and Hillman (2013) proposed that exercise exerts its effects on cognition by affecting molecular events related to the management of energy metabolism and synaptic plasticity. Chapman, Aslan, Spence, Defina, et al. (2013) found increased cerebral blood flow (CBF) and a related increase of memory performance in older adults after 12 weeks physical training. Although the effects of physical activity on the brain are often widespread, prefrontal and hippocampal areas appear to be more influenced than other areas of the brain. In a trial with 120 older adults, Erickson et al. (2011) showed that aerobic training vs. active control increases the size of the hippocampus, which was correlated with improvements in spatial memory. An increased hippocampal volume was associated with greater serum levels of BDNF, a mediator of neurogenesis in the dentate gyrus.

Several recent studies investigated different types of physical training. Voelcker-Rehage, Godde, and Staudinger (2011) compared a 12-month cardiovascular or coordination training in older adults. Changes in brain activation were investigated by functional magnetic resonance imaging (fMRI). Both groups improved in executive functioning and perceptual speed. The fMRI revealed unspecific changes for both groups in prefrontal areas and also training-specific changes. In a follow-up study, the group showed that motor fitness due to coordination training led to an increase of subcortical brain areas relevant for motor control (Niemann, Godde, Staudinger, & Voelcker-Rehage, 2014).

Berryman et al. (2014) compared the effects of different short-term (eight-week) physical interventions in 51 older adults. All groups showed similar improvements in cognition, with maximum effect on inhibition, an executive function. Forte et al. (2013) trained 42 older adults for three months in either coordination or progressive resistance training. Inhibitory control improved after the intervention, independent of the training type. These results confirm that different types of physical training benefits executive function (in this case inhibition).

Chapman, Aslan, Spence, Defina, et al. (2013) administered short-term aerobic exercise for three sessions per week 1 h each for 12 weeks. Despite the short training, the trainees (compared to a passive control group) showed memory improvements and enhancements of cerebral blood perfusion.

These data suggest that even shorter term aerobic exercise can improve cognition in sedentary older adults, and that different regimes show both similar effects (on executive functions) but also specific effects on cognition.

In their aforementioned study, Gajewski and Falkenstein (2012) compared the effects of a combined physical training (strength and aerobic) to those of two other active groups (cognitive training and relaxation) and a non-contact group on sedentary seniors. A battery of psychometric tests was administered before and after the training. The effects of the cognitive training were largest (see above). However, the physical training also yielded specific effects on cognition, for example, on processing speed, some executive functions, and some aspects of memory. The improvements mainly affected the speed of performance, while the quality of performance was less improved, inversely to the effects of cognitive training. Also, larger improvements were seen for the younger trainees (65–70 years). Hence, physical training may improve the performance of elderly in everyday situations where speed rather than accuracy is relevant, such as braking in traffic situations.

All in all, physical as well as cognitive training appear to have a positive impact on cognition in older adults. Because both interventions act via different mechanisms, their combination may maximize the effect on cognition (Thom & Clare, 2011). However, by now, there are very few combination studies. Fabre, Chamari, Mucci, Massé-Biron, and Préfaut (2002) provided the first evidence that a combination of CCT and exercise yields larger effects on cognition in older adults than either training alone. Theill, Schumacher, Adelsberger, Martin, and Jäncke (2013) randomized 63 older adults to either a single working memory training, a simultaneous working memory and cardiovascular training, or no training. They found larger improvements in an executive control task for both training groups vs. the control group. In addition, the simultaneous training showed better gains in some cognitive functions than the single cognitive training. Further studies are, hence, highly needed that carefully explore and adjust the parameters of the two approaches to yield an optimum effect on cognition.

As with cognitive training, motivation has to be maximized to yield high compliance of older trainees. Game-like and virtual-reality enhanced types of exercise ("exergaming") are, hence, promising. Anderson-Hanley et al. (2012) investigated the effect of stationary cycling with virtual-reality tours ("cybercycle") on executive function compared to traditional exercise. They found that cybercycling improved executive functions and enhanced BDNF more than traditional exercise. Chao, Scherer, and Montgomery (2014) reviewed the few studies using the Nintendo Wii™ exergames on cognition, physical function, and psychosocial outcomes in older adults. Indeed, positive effects on physical function, cognition, and quality of life were found.

## CONCLUSION

Concerning environment and lifestyle, the main factors that show robust effects on cognition in advanced age are education, occupation, nutrition, and regular physical and cognitive activity. Physical activity appears to specifically improve executive functions such as interference processing and task switching. Cognitively challenging leisure activities such as reading, playing board games or musical instruments, and dancing appear to improve cognition in older people, while a possible impact of social activities may depend on their complexity.

In line with the lifestyle literature, the main interventions with positive effects on cognition in older age assessed in randomized, controlled trials are cognitive and physical training. Formal cognitive training is usually administered via PC-supported tasks or video games. Such trainings certainly improve the trained task, but they also appear to transfer to the implicitly trained cognitive functions (near transfer). Far transfer to untrained domains is less frequent and weaker. Training of working memory and executive functions (in particular, task switching) appears to have some potential for far transfer. Transfer to everyday life cognitive performance has only been rarely reported. However, because there are also negative outcomes with no specific effects, the parameters of a cognitive training that lead to success have to be further explored. By now, it seems that more variable and multifaceted trainings are promising in promoting stable improvements of cognition. Most importantly, cognitive training should induce fun to motivate the trainees to continue training without external guidance. More natural and cognitively challenging activities, which exercise multiple cognitive abilities such as learning quilting or working with new PC software, also lead to improvements of cognition in older adults, as measured with non-trained cognitive tests.

The majority of randomized, controlled physical training studies show improvements in cognition in older adults. As in the observational (cross-sectional) studies, executive functions are mostly affected. Different types of physical training such as aerobic exercise vs. coordination training exert both specific and general effects on brain activity and functions and on cognitive performance. Because vigorous coordination training includes the effects of both training types, it is recommended. Studies comparing the effects of short-term physical with cognitive training showed less impact due to physical training. Moreover, physical training mainly affected the speed of performance, while quality of performance was less improved. In contrast, cognitive training improved accuracy more than speed. Hence, combination studies of cognitive and physical training appear to be promising but are extremely rare. By now, it has been shown that simultaneous cognitive and physical training showed better gains in some cognitive functions than single cognitive training. Taking into consideration the literature, we propose that an optimum approach to improve cognition in older age is the combination of proper nutrition, cognitively challenging lifestyle with multiple and variable activities, and vigorous physical coordination training. Frequent and challenging dancing coupled with gorgeous dinners including a Mediterranean diet is a good example, which covers most of those ingredients that promote health and cognition in older age.

## REFERENCES

Albinet, C. T., Boucard, G., Bouquet, C. A., & Audiffren, M. (2010). Increased heart rate variability and executive performance after aerobic training in the elderly. *European Journal of Applied Physiology, 109*(4), 617–624.

Almeida, O. P., Garrido, G. J., Alfonso, H., Hulse, G., Lautenschlager, N. T., Hankey, G. J., et al. (2011). 24-month effect of smoking cessation on cognitive function and brain structure in later life. *NeuroImage, 55*(4), 1480–1489.

Anderson-Hanley, C., Arciero, P. J., Brickman, A. M., Nimon, J. P., Okuma, N., Westen, S. C., et al. (2012). Exergaming and older adult cognition: a cluster randomized clinical trial. *American Journal of Preventive Medicine, 42*(2), 109–119.

Anguera, J. A., Boccanfuso, J., Rintoul, J. L., Al-Hashimi, O., Faraji, F., Janowich, J., et al. (2013). Video game training enhances cognitive control in older adults. *Nature, 501*(7465), 97–101.

Barberger-Gateau, P. (2014). Nutrition and brain aging: how can we move ahead. *European Journal of Clinical Nutrition, 68*(11), 1245–1249. http://dx.doi.org/10.1038/ejcn.2014.177.

Basak, C., Boot, W. R., Voss, M. W., & Kramer, A. F. (2008). Can training in a real-time strategy video game attenuate cognitive decline in older adults? *Psychology and Aging, 23*(4), 765–777.

Bavelier, D., & Davidson, R. J. (2013). Brain training: games to do you good. *Nature, 494*(7438), 425–426.

Berryman, N., Bherer, L., Nadeau, S., Lauzière, S., Lehr, L., Bobeuf, F., et al. (2014). Multiple roads lead to Rome: combined high-intensity aerobic and strength training vs. gross motor activities leads to equivalent improvement in executive functions in a cohort of healthy older adults. *Age, 36*(5), 9710.

Beydoun, M. A., Beydoun, H. A., Gamaldo, A. A., Teel, A., Zonderman, A. B., & Wang, Y. (2014). Epidemiologic studies of modifiable factors associated with cognition and dementia: systematic review and meta-analysis. *BMC Public Health, 14*, 643.

Burch, D. (2014). What could computerized brain training learn from evidence-based medicine? *PLoS Medicine, 11*, 11.

Caserta, R. J., Young, J., & Janelle, C. M. (2007). Old dogs, new tricks: training the perceptual skills of senior tennis players. *Journal of Sport & Exercise Psychology, 29*(4), 479–497.

Cassavaugh, N., & Kramer, A. F. (2009). Transfer of computer-based cognitive training to simulated driving in older adults. *Applied Ergonomy, 40*(5), 943–952.

Chan, M. Y., Haber, S., Drew, L. M., & Park, D. C. Training older adults to use tablet computers: does it enhance cognitive function? *The Gerontologist.* http://dx.doi.org/10.1093/geront/gnu057. [Epub ahead of print]

Chao, Y. Y., Scherer, Y. K., & Montgomery, C. A. Effects of using Nintendo Wii™ exergames in older adults: a review of the literature. *Journal of Aging and Health, 27*(3), 379–402. http://dx.doi.org/10.1177/0898264314551171.

Chapman, S. B., Aslan, S., Spence, J. S., Defina, L. F., Keebler, M. W., Didehbani, N., et al. (2013). Shorter term aerobic exercise improves brain, cognition, and cardiovascular fitness in aging. *Frontiers in Aging Neuroscience, 5*, 75.

Chapman, S. B., Aslan, S., Spence, J. S., Hart, J. J., Jr., Bartz, E. K., Didehbani, N., et al. Neural mechanisms of brain plasticity with complex cognitive training in healthy seniors. *Cerebral Cortex, 25*(2), 396–405. http://dx.doi.org/10.1093/cercor/bht234.

Colcombe, S., & Kramer, A. F. (2003). Fitness effects on the cognitive function of older adults: a meta-analytic study. *Psychological Science, 14*(2), 125–130.

Desjardins-Crépeau, L., Berryman, N., Vu, T. T., Villalpando, J. M., Kergoat, M. J., Li, K. Z., et al. (2014). Physical functioning is associated with processing speed and executive functions in community-dwelling older adults. *Journals of Gerontology Series B: Psychological Sciences and Social Sciences, 69*(6), 837–844.

Edwards, J. D., Delahunt, P. B., & Mahncke, H. W. (2009). Cognitive speed of processing training delays driving cessation. *Journals of Gerontology Series B: Biological Sciences and Medical Sciences, 64*(12), 1262–1267.

Engvig, A., Fjell, A. M., Westlye, L. T., Skaane, N. V., Sundseth, Ø., & Walhovd, K. B. (2012). Hippocampal subfield volumes correlate with memory training benefit in subjective memory impairment. *NeuroImage, 61*(1), 188–194.

Erickson, K. I., Leckie, R. L., & Weinstein, A. M. (2014). Physical activity, fitness, and gray matter volume. *Neurobiology of Aging, 35*(Suppl. 2), 20–28.

Erickson, K. I., Voss, M. W., Prakash, R. S., Basak, C., Szabo, A., Chaddock, L., et al. (2011). Exercise training increases size of hippocampus and improves memory. *Proceedings of the National Academy of Sciences of the United States of America, 108*(7), 3017–3022.

Fabre, C., Chamari, K., Mucci, P., Massé-Biron, J., & Préfaut, C. (2002). Improvement of cognitive function by mental and/or individualized aerobic training in healthy elderly subjects. *International Journal of Sports Medicine, 23*(6), 415–421.

Fissler, P., Küster, O., Schlee, W., & Kolassa, I. T. (2013). Novelty interventions to enhance broad cognitive abilities and prevent dementia: synergistic approaches for the facilitation of positive plastic change. *Progress in Brain Research, 207*, 403–434.

Forte, R., Boreham, C. A., Leite, J. C., De Vito, G., Brennan, L., Gibney, E. R., et al. (2013). Enhancing cognitive functioning in the elderly: multicomponent vs. resistance training. *Clinical Interventions in Aging, 8*, 19–27.

Gajewski, P. D., & Falkenstein, M. (2011). Neurocognition of aging in working environments. *Journal of Labour Market Research, 44*(4), 307–320.

Gajewski, P. D., & Falkenstein, M. (2012). Training-induced improvement of response selection and error detection in aging assessed by task switching: effects of cognitive, physical and relaxation training. *Frontiers in Human Neuroscience, 6*, 130.

Gajewski, P. D., & Falkenstein, M. (in press). Long-term physical activity is associated with lower distractibility in a Stroop interference task in older age: ERP evidence. Brain & Cognition.

Gajewski, P. D., & Falkenstein, M. (2015). Lifelong physical activity and executive functions in older age assessed by memory based task switching. *Neuropsychologia, 73*, 195–207.

Gajewski, P. D., Wild-Wall, N., Schapkin, S. A., Erdmann, U., Freude, G., & Falkenstein, M. (2010). Effects of aging and job demands on cognitive flexibility assessed by task switching. *Biological Psychology, 85*(2), 187–199.

van Gelder, B. M., Buijsse, B., Tijhuis, M., Kalmijn, S., Giampaoli, S., Nissinen, A., et al. (2007). Coffee consumption is inversely associated with cognitive decline in elderly European men: the FINE study. *European Journal of Clinical Nutrition, 61*(2), 226–232.

Gomez-Pinilla, F., & Hillman, C. (2013). The influence of exercise on cognitive abilities. *Comprehensive Physiology, 3*(1), 403–428.

Green, C. S., & Bavelier, D. (2008). Exercising your brain: a review of human brain plasticity and training-induced learning. *Psychology and Aging, 23*(4), 692–701.

Green, C. S., Strobach, T., & Schubert, T. (2014). Methodological standards in training and transfer experiments. *Psychological Research, 78*(6), 756–772.

Jaeggi, S. M., Buschkuehl, M., Jonides, J., & Perrig, W. J. (2008). Improving fluid intelligence with training on working memory. *Proceedings of the National Academy of Sciences of the United States of America, 105*(19), 6829–6833.

Karbach, J., & Kray, J. (2009). How useful is executive control training? Age differences in near and far transfer of task-switching training. *Developmental Science, 12*(6), 978–990.

Karbach, J., & Verhaeghen, P. (2014). Making working memory work: a meta-analysis of executive-control and working memory training in older adults. *Psychological Science, 25*(11), 2027–2037.

Kattenstroth, J. C., Kalisch, T., Holt, S., Tegenthoff, M., & Dinse, H. R. (2013). Six months of dance intervention enhances postural, sensorimotor, and cognitive performance in elderly without affecting cardio-respiratory functions. *Frontiers in Aging Neuroscience, 26*, 5.

Kirk-Sanchez, N. J., & McGough, E. L. (2014). Physical exercise and cognitive performance in the elderly: current perspectives. *Clinical Interventions in Aging, 9*, 51–62.

Kramer, A. F., & Erickson, K. I. (2007). Capitalizing on cortical plasticity: influence of physical activity on cognition and brain function. *Trends in Cognitive Science, 11*(8), 342–348.

Lampit, A., Hallock, H., & Valenzuela, M. J. (2014). Computerized cognitive training in cognitively healthy older adults: a systematic review and meta-analysis of effect modifiers. *PLoS Medicine, 11*, 11.

Lautenschlager, N. T., & Almeida, O. P. (2006). Physical activity and cognition in old age. *Current Opinion in Psychiatry, 19*(2), 190–193.

Lupien, S. J., Maheu, F., Tu, M., Fiocco, A., & Schramek, T. E. (2007). The effects of stress and stress hormones on human cognition: implications for the field of brain and cognition. *Brain and Cognition, 65*(3), 209–237.

Lustig, C., Shah, P., Seidler, R., & Reuter-Lorenz, P. A. (2009). Aging, training, and the brain: a review and future directions. *Neuropsychological Review, 19*, 504–522.

Marquié, J. C., Duarte, L. R., Bessieres, P., Dalm, C., Gentil, C., & Ruidavets, B. (2010). Higher mental stimulation at work is associated with improved cognitive functioning in both young and older workers. *Ergonomics, 53*(11), 1287–1301.

McEwen, B. S., & Morrison, J. H. (2013). The brain on stress: vulnerability and plasticity of the prefrontal cortex over the life course. *Neuron, 79*(1), 16–29.

Milgram, N. W., Siwak-Tapp, C. T., Araujo, J., & Head, E. (2006). Neuroprotective effects of cognitive enrichment. *Ageing Research Reviews, 5*(3), 354–369.

van Muijden, J., Band, G. P., & Hommel, B. (2012). Online games training aging brains: limited transfer to cognitive control functions. *Frontiers in Human Neuroscience, 6*, 221.

Niemann, C., Godde, B., Staudinger, U. M., & Voelcker-Rehage, C. (2014). Exercise-induced changes in basal ganglia volume and cognition in older adults. *Neuroscience, 281*, 147–163.

Olesen, P. J., Westerberg, H., & Klingberg, T. (2004). Increased prefrontal and parietal activity after training of working memory. *Nature Neuroscience, 7*(1), 75–79.

Park, D. C., Lodi-Smith, J., Drew, L., Haber, S., Hebrank, A., Bischof, G. N., et al. (2014). The impact of sustained engagement on cognitive function in older adults: the Synapse Project. *Psychological Science, 25*(1), 103–112.

Peters, J. L., Weisskopf, M. G., Spiro, A., III, Schwartz, J., Sparrow, D., Nie, H., et al. (2010). Interaction of stress, lead burden, and age on cognition in older men: the VA Normative Aging Study. *Environmental Health Perspectives, 118*(4), 505–510.

Plassman, B. L., Williams, J. W., Jr., Burke, J. R., Holsinger, T., & Benjamin, S. (2010). Systematic review: factors associated with risk for and possible prevention of cognitive decline in later life. *Annals of Internal Medicine, 153*(3), 182–193.

Potter, G. G., Helms, M. J., & Plassman, B. L. (2008). Associations of job demands and intelligence with cognitive performance among men in late life. *Neurology, 70*, 1803–1808.

Raz, N. (2000). Aging of the brain and its impact on cognitive performance: integration of structural and functional findings. In F. I. M. Craik & T. A. Salthouse (Eds.), *Handbook of aging and cognition* (2nd ed.) (pp. 1–90). Mahwah, NJ: Erlbaum.

Rebok, G. W., Ball, K., Guey, L. T., Jones, R. N., Kim, H. Y., King, J. W., et al. (2014). ACTIVE Study Group. Ten-year effects of the advanced cognitive training for independent and vital elderly cognitive training trial on cognition and everyday functioning in older adults. *Journal of the American Geriatrics Society, 62*(1), 16–24.

Redick, T. S., Shipstead, Z., Harrison, T. L., Hicks, K. L., Fried, D. E., Hambrick, D. Z., et al. (2013). No evidence of intelligence improvement after working memory training: a randomized, placebo-controlled study. *Journal of Experimental Psychology, General, 142*(2), 359–379.

Reijnders, J., van Heugten, C., & van Boxtel, M. (2013). Cognitive interventions in healthy older adults and people with mild cognitive impairment: a systematic review. *Ageing Research Reviews, 12*(1), 263–275.

Sabia, S., Elbaz, A., Britton, A., Bell, S., Dugravot, A., Shipley, M., et al. (2014). Alcohol consumption and cognitive decline in early old age. *Neurology, 82*(4), 332–339.

Scarmeas, N., Stern, Y., Mayeux, R., & Luchsinger, J. A. (2006). Mediterranean diet, Alzheimer disease, and vascular mediation. *Archives of Neurology, 63*(12), 1709–1717.

Schooler, C., Mulatu, M. S., & Oates, G. (1999). The continuing effects of substantively complex work on the intellectual functioning of older workers. *Psychology and Aging, 14*(3), 483–506.

Shipstead, Z., Redick, T. S., & Engle, R. W. (2012). Is working memory training effective? *Psychological Bulletin, 138*(4), 628–654. http://dx.doi.org/10.1037/a0027473. Epub 2012 Mar 12.

Singh-Manoux, A., Richards, M., & Marmot, M. (2003). Leisure activities and cognitive function in middle age: evidence from the Whitehall II study. *Journal of Epidemiology and Community Health, 57*, 907–913.

Smith, P. J., Blumenthal, J. A., Hoffman, B. M., Cooper, H., Strauman, T. A., Welsh-Bohmer, K., et al. (2010). Aerobic exercise and neurocognitive performance: a meta-analytic review of randomized controlled trials. *Psychosomatic Medicine, 72*, 239–252.

Stepankova, H., Lukavsky, J., Buschkuehl, M., Kopecek, M., Ripova, D., & Jaeggi, S. M. (2014). The malleability of working memory and visuospatial skills: a randomized controlled study in older adults. *Developmental Psychology, 50*(4), 1049–1059.

Stine-Morrow, E. A., Parisi, J. M., Morrow, D. G., & Park, D. C. (2008). The effects of an engaged lifestyle on cognitive vitality: a field experiment. *Psychology and Aging, 23*(4), 778–786.

Theill, N., Schumacher, V., Adelsberger, R., Martin, M., & Jäncke, L. (2013). Effects of simultaneously performed cognitive and physical training in older adults. *BMC Neuroscience, 14*, 103.

Thom, J. M., & Clare, L. (2011). Rationale for combined exercise and cognition-focused interventions to improve functional independence in people with dementia. *Gerontology, 57*(3), 265–275.

Toril, P., Reales, J. M., & Ballesteros, S. (2014). Video game training enhances cognition of older adults: a meta-analytic study. *Psychology and Aging, 29*(3), 706–716.

van Uffelen, J. G., Chin, A., Paw, M. J., Hopman-Rock, M., & van Mechelen, W. (2008). The effects of exercise on cognition in older adults with and without cognitive decline: a systematic review. *Clinical Journal of Sport Medicine, 18*(6), 486–500.

Verghese, J., Lipton, R. B., Katz, M. J., Hall, C. B., Derby, C. A., Kuslansky, G., et al. (2003). Leisure activities and the risk of dementia in the elderly. *New England Journal of Medicine, 348*(25), 2508–2516.

Voelcker-Rehage, C., Godde, B., & Staudinger, U. M. (2011). Cardiovascular and coordination training differentially improve cognitive performance and neural processing in older adults. *Frontiers in Human Neuroscience, 5*, 26.

Wang, H. X., Xu, W., & Pei, J. J. (2012). Leisure activities, cognition and dementia. *Biochimica et Biophysica Acta (BBA), 1822*(3), 482–491.

Wild-Wall, N., Falkenstein, M., & Gajewski, P. D. (2012). Neural correlates of changes in a visual search task due to cognitive training. *Neural Plasticity, 2012*, 529057.

Willis, S. L., Tennstedt, S. L., Marsiske, M., Ball, K., Elias, J., & Koepke, K. M. (2006). ACTIVE Study Group. Long-term effects of cognitive training on everyday functional outcomes in older adults. *Journal of the American Medical Association (JAMA), 296*(23), 2805–2814.

Zelinski, E. M. (2009). Far transfer in cognitive training of older adults. *Restorative Neurology and Neuroscience, 27*(5), 455–471.

# Section D

# Self-Other Perceptions and Other Performance Phenomena of Perception-Action Interactions

13. Bridging the Gap between Perception and Action: An Overview (Alexandra Pizzera)    207
14. Capturing Motion for Enhancing Performance: An Embodied Cognition Perspective on Sports and the Performing Arts (Vassilis Sevdalis and Clemens Wöllner)    223
15. Auditory Action Perception (Christian Kennel and Alexandra Pizzera)    235
16. Visual Perception in Expert Action (Rita de Oliveira)    253

## Overview

The phenomenon that our own body is activated while perceiving others in motion can be explained by addressing perception–action interactions. This section aims to bridge the gap between perception and action by focusing on different senses as well as the type of task and movement (Chapter 13). Chapters 14 and 15 start by explaining how humans perceive and control movements by including visual as well as acoustic information, provided by one's own body. Because different task demands may result in different perception–action couplings, Chapter 16 gives an overview of how the perception–action link is represented in fine-motor and complex movements.

Chapter 13

# Bridging the Gap between Perception and Action: An Overview

### Alexandra Pizzera[1,2]
[1]*Department of Performance Psychology, German Sport University Cologne, Cologne, Germany;*
[2]*Institute of Sports and Sports Sciences, Heidelberg University, Heidelberg, Germany*

> A gymnast is performing a handstand on the balance beam. Several people are observing this move, such as other gymnasts, coaches, spectators, and judges. What exactly happens within these spectators' own bodies while they are observing the gymnast? If the gymnast is currently at risk of falling to one side of the beam, quite often we see athletes, coaches, and spectators lean to the opposite side, even though they are fully aware that they will not be able to influence the balance of the gymnast.

The phenomenon that one's own body is engaged while perceiving others in motion can be explained by addressing perception–action interactions. This section explores how the gap between perception and action is bridged. The following chapters focus on different senses, such as visual action perception (Sevdalis & Wöllner, Chapter 14), acoustic action perception (Kennel & Pizzera, Chapter 15), as well as the type of task and movement (De Oliveira, Chapter 16).

Several researchers from different disciplines have examined the link between perception and action. These disciplines include social cognition (Goldman & de Vignemont, 2009; Niedenthal, Barsalou, Winkielman, Krauth-Gruber, & Ric, 2005), economics (Oullier & Basso, 2010), cognitive neuroscience (Decety & Grèzes, 1999; Rizzolatti, Fadiga, Gallese, & Fogassi, 1996), artificial intelligence (Brooks, 1986, 1991), philosophy (Hurley, 1998), linguistics and memory (Barsalou, 1999), linguistics and sport (Beilock, Lyons, Mattarella-Micke, Nusbaum, & Small, 2008), and sport psychology (Cañal-Bruland, van der Kamp, & van Kesteren, 2010; Dosseville, Laborde, & Raab, 2011; Pizzera & Raab, 2012a; Renden, Kerstens, Oudejans, & Cañal-Bruland, 2014). This section focuses on perception–action links in a highly ecologically valid environment, namely, sports. Wilson (2002) suggested that "the mind must be understood in the context of its relationship to a physical

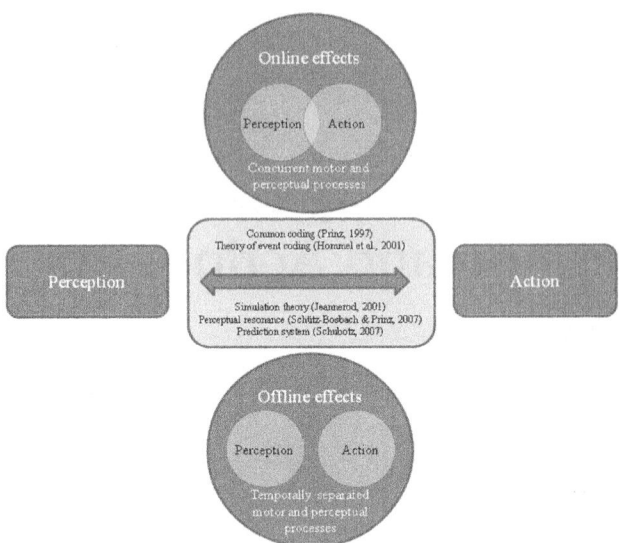

**FIGURE 1** The perception–action link, subdivided into online and offline effects.

body that interacts with the world" (p. 625). In the domain of sports, there is an opportunity to take this statement quite literally and experimentally test perception–action links with movement experts. In addition, this quote highlights that to understand the link between perception and action, I focus on the perceiver's own body, not the body of others or the situation or environment in which the body is acting.

The current chapter covers two general approaches that have been used to examine the bidirectional link between perception and action (see Hecht, Vogt, & Prinz, 2001): One looks at how perception affects action and the other looks at how action affects perception. I describe and explain this bidirectional link using several theoretical frameworks. In addition, I review and discuss the approaches with regard to online and offline effects (Schütz-Bosbach & Prinz, 2007). Online effects refer to perceptual and motor processes occurring simultaneously; offline effects describe temporally separated perception and action processes. For a graphical overview of the perception–action link, subdivided into online and offline effects, and the different theoretical frameworks reviewed and discussed in this chapter, see Figure 1.

## HOW DOES PERCEPTION AFFECT ACTION?

The first approach describes how a person learns a new movement, such as the gymnast learning the handstand on the balance beam. This can be done by observing other people perform a handstand, also known as observational learning, imitation learning, or modeling. If the handstand on the balance beam has already been learned (i.e., the skill is within the gymnast's repertoire), the gymnast might want to improve his or her performance of the handstand. For such performance enhancement, elite gymnasts have been shown to use observation (as opposed to observational learning; see Ste-Marie et al., 2012, for a discussion of the different terms). This observation helps gymnasts improve self-assessment, increase performance of technical execution, and increase use of imagery and visual perceptions (Hars & Calmels, 2007). For this transfer of observation (perception) to actual execution of the movement (action) to work, some form of communication between perceptual and motor processes is required. It is assumed that for this communication to take place, internal cognitive or purely visual mechanisms are activated during the perception of the action that can also be used during action generation (Bandura, 1986; Gray, Neisser, Shapiro, & Kouns, 1991). Numerous studies have been conducted in the area of observational learning (see Hodges & Williams, 2007, special issue of the *Journal of Sports Sciences* on observational learning, and a review by Ste-Marie et al., 2012).

Although they share the main message that there is a strong overlap between processes involved in motor execution, observation, and imitation, studies have differed in their focus with respect to developmental aspects, characteristics of the model (self, novice, expert, nonbiological, point-light display[1]), underlying mechanisms of observational learning or observation, and learning phases and action goals (movement outcome vs movement quality). For instance, Ashford, Davids, and Bennett (2007) suggested that there are developmental effects that influence observational learning as a result of differences in individuals' intrinsic dynamics and, hence, the ability to replicate an observed skill; adults focus more on movement dynamics (such as form), whereas children show a greater tendency to achieve movement outcome goals. The link between perception and action also seems to be different depending on the characteristics of the model, that is, on who is perceived. In an intervention study, children showed better performance after viewing themselves perform an adaptive behavior (self-modeling) as opposed to viewing themselves at their current skill level (self-observation) or not using observational learning (Clark & Ste-Marie, 2007). Similarly, other intervention studies showed positive effects for self-observation videos (Onate et al., 2005; Van Wieringen, Emmen, Bootsma, Hoogesteger, & Whiting, 1989). Thus, model similarity seems to be an important contributor to the effectiveness of modeling, confirming the close link between perceptual and motor processes. Therefore, observation of a model similar or even identical to oneself leads to even larger overlaps of these processes. Yet, to maximize learning and enhance performance, observation of a novice model alone, even

---

1. A point-light display (PLD) describes a procedure in which point-light dots are attached to the key joints of a moving body. Structural and contextual information is removed as the viewer sees only the dots, which reflect the motion being performed.

if the model is very similar to the observer (or if the performance is very similar to the observer's performance at the beginning of the learning phase), might not suffice. In fact, a study in which participants learned to hit barriers in a clockwise motion revealed that the observation of both a novice and an expert model (mixed observation) resulted in better learning than the observation of one of the models alone (Andrieux & Proteau, 2014). Similarly, it was shown that observable information can improve performance compared to individual learning, both with a skilled and unskilled demonstrator.

A general assumption when learning a new movement is that the learner is required to learn to control his or her movement through the coordination of various body parts. The crucial information taken up during observation seems to be the relative motion. If this relative motion is the important source of information for perception, then PLDs should be just as effective for acquiring a coordination pattern, since all non-relevant information is removed (Runeson, 1984). In a study with bowling action, adults confirmed this assumption in that their performance did not differ when viewing PLDs compared to video sequences (Hayes, Hodges, Scott, Horn, & Williams, 2007). However, children's performance was poorer after viewing PLDs as opposed to video sequences. It seems that the effectiveness of demonstrations and the question of what is perceived and used for movement reproduction need to be judged with respect to the age of the learner but also relative to the task and task context. Indeed, studies using complex movements such as ballet skills (Rodrigues, Ferracioli, & Denardi, 2010) or basketball dribbling (Romack & Briggs, 1995) revealed superiority of PLDs compared to video displays.

Another way to examine the link from perception to action is to study imagery. It is assumed that similar to observational learning or modeling, imagery uses models that extract relevant information and form cognitive representations for later action execution (Feltz & Landers, 1983; McCullagh & Weiss, 2001). Macuga and Frey (2012) showed that there is similar neural involvement for observation and use of imagery in sports. The only difference, according to Holmes and Calmels (2008), is that observations use external stimuli such as live demonstrations or video (a bottom-up and percept-driven process), whereas imagery requires individuals to create an internal image based on their own past experience and memory (internal stimulus; a top-down, knowledge-driven process). Ram, Riggs, Skaling, Landers, and McCullagh (2007) tested this assumption using a free-weight squat lift and a stabilometer balancing task. The groups that received modeling showed a more appropriate movement form and outcome in acquisition and retention than imagery condition groups. However, with imagery, performance was still more effective than in a control group. The authors argued that presenting an external stimulus through observation is more effective in enhancing skill acquisition. For a more detailed overview of the effects of imagery on motor performance, see also Kalicinski, Thomas and Lobinger (Chapter 8).

The behavioral studies on imitation learning previously described can be complemented by neurocognitive research. Several neuroimaging studies have examined the neural network that is engaged during motor execution, simulation, imagery, verbalization, and observation. These studies also relied on the concept of shared motor representations, as suggested by psychological experiments. A meta-analysis by Grèzes and Decety (2001) showed that there are common activation sites supporting the idea of a functional equivalence between action execution, simulation, and observation. These experiments originated from a study performed with monkeys, in which researchers detected that a particular set of neurons, that is, mirror neurons (MNs), that discharged both when the monkey observed meaningful hand movements made by the experimenter and when executing the movement himself (Di Pellegrino, Fadiga, Fogassi, Gallese, & Rizzolatti, 1992). The authors showed that the activity of these MNs "represented" the observed action, suggesting that this motor "representation" might be the basis for understanding motor events. There is some evidence that such a neurological network might exist in humans, as well (Binkofski et al., 2000; Buccino et al., 2001; Iacoboni et al., 1999). A neuroimaging study using position emission tomography found brain areas similar to those found in monkeys during action perception (i.e., grasping movements; Rizzolatti, Fadiga, Matelli, et al., 1996). However, further studies revealed that activation of brain areas associated with the planning and generation of actions occurred only if the intention of the observation was to imitate the perceived action (Decety et al., 1997). Further electroencephalographic research (Calmels et al., 2006) and brain-imaging studies (Buccino et al., 2001; Grèzes, Armony, Rowe, & Passingham, 2003) also supported the proposal that MNs relate to the observation–execution matching system in humans.

Yet, the MN theory of action understanding has been questioned. Specifically, problems have been formulated that there is no direct evidence in monkeys that MNs support action understanding. For instance, action understanding can be achieved through other non-mirroring mechanisms; action understanding and action production dissociate; damages to the inferior frontal gyrus are not correlated with action understanding deficits; and the generalization of the MN system to speech recognition fails on empirical grounds (for a discussion, see Hickok, 2008; Steinhorst & Funke, 2014). Nevertheless, neurophysiological studies revealed a common neural network for action observation and action production that includes the premotor cortex, inferior parietal lobule, superior temporal sulcus, supplementary motor area (SMA), cingulated gyrus, and cerebellum. This network is activated both when performing and observing others perform that same motor task and was termed the "action observation network" (Kilner, Neal, Weiskopf, Friston, & Frith, 2009; Oosterhof, Wiggett, Diedrichsen, Tipper, & Downing, 2010). In a study by Buccino et al. (2004), non-guitarists were scanned while imitating unfamiliar guitar chords. Using an event-related functional magnetic resonance imaging (fMRI) paradigm, the three stages of action observation, motor preparation, and execution were analyzed separately. The results showed an overall activation of the MN system throughout the three stages, while the left dorsolateral prefrontal cortex was active only during motor preparation. In addition, activations in the MN system were stronger during observing with the aim of imitating compared to observing without purpose. This latter aspect of perception (observation) for action was also confirmed in a study using gymnastics movements (Zentgraf et al., 2005). Participants instructed to observe sequential gymnastics movements with the intention of imitating showed a stronger activation in the MN system compared with participants instructed to judge the quality of performance. In addition, the two instructions activated the SMA in different ways, in that the pre-SMA (part of the SMA) was more strongly activated for the judging instruction and the more caudal SMA proper for the imitation instruction. The authors suggested that the requirements of transforming the gymnastic movements to body-centered coordinates explained this finding.

With regard to action consequences that seem to play a role in the perception–action link, Bellebaum and Colosio (2014) found that although action–outcome associations can be learned similarly well by actively performing a movement or by observing it, the mechanisms involved differ. For instance, participants in an active learning group showed ERP (event-related potential) patterns that were characterized by learning-related increases in ERN (error-related negativity) and decreases in FRN (feedback-related negativity)[2] amplitudes following erroneous choices and the respective negative feedback, whereas in the observational learners, the oFRN and oERN (observational FRN and ERN) decreased relative to the FRN in active learners and were not modulated by learning. Overall, neuroimaging studies have revealed high overlaps of active brain regions during the simulation and execution of actions, shown by fMRI, magnetoencephalography, positron emission tomography, TMS, or electromyography (Kosslyn, Ganis, & Thompson, 2001; Lotze & Halsband, 2006; Munzert & Zentgraf, 2009). In sum, the neural processes involved in observational learning are influenced by instructions and prior experience, and strong overlaps exist with regard to processes involved in motor execution, observation, and imitation (Vogt & Thomaschke, 2007).

Applying these findings to the sports domain, researchers have shown that motor experts show a special perceptual sensitivity to actions. For instance, in an fMRI study with expert dancers, Calvo-Merino, Glaser, Grèzes, Passingham, and Haggard (2005) revealed increased brain activity in the classic MN regions among expert dancers viewing movements they had previously performed themselves (classical ballet moves) compared to movements they had not (capoeira moves). Thus, areas in the brains of the participants responded to the stimuli (dance moves) in a way that depended on the observer's specific

---

2. ERN and FRN are components of ERP that reflect neural activity in the anterior cingulate cortex and are associated with negative prediction errors and, hence, signaling events that are "worse than expected."

motor expertise, although all groups saw the same stimuli. In a follow-up study, even pure motor experience effects (no influence of visual experience) were demonstrated, as male and female dancers responded even more to movements from their own motor repertoire as opposed to opposite-gender moves, for which they had only visual experience (Calvo-Merino, Grèzes, Glaser, Passingham, & Haggard, 2006).

The transfer from perception to action has been investigated for years, and the preceding paragraphs have summed up some of the research conducted in this area. For more detailed information, the reviews and empirical articles that have been cited are recommended. The transfer from action to perception, in contrast, has only recently been receiving more attention in research. This approach is addressed in the following section, and different possibilities for how to investigate this link in sports are shown.

## HOW DOES ACTION AFFECT PERCEPTION?

The second approach to understanding the bidirectional link between action and perception investigates if the opposite is also true, that is, if our own movements can influence the way we perceive these movements. Similar to the transfer from perception to action, transfer in the opposite direction (from action to perception) also assumes a strong relation between perceptual and motor processes. For example, according to the theoretical framework of common coding (Prinz, 1997), the final stages of perception (perceptual processing of stimuli) and the initial stages of action control (action planning or initiation) share a domain of coding, such that planned actions and perceived events are represented in the same format. Following this framework, changes in these codes due to motor learning should also be reflected in corresponding changes in perceptual skills. Applying this to the introductory example, a change in the gymnast's skill level in performing the handstand ("motor code") might also lead to a change in the gymnast perceiving that skill ("perceptual code").

The theory of event coding (TEC; Hommel, Müsseler, Aschersleben, & Prinz, 2001) extended Prinz's (1997) common-coding approach. According to TEC, cognitive representations of events, such as perception or planning of actions, subserve not only representational functions but also action-related functions. Hommel et al. (2001) proposed that stimulus representations underlying perception and action representations underlying action planning are coded and stored together in a common representational medium, not separately. Hommel et al. pointed out that TEC is limited to the late stages of perception and the early cognitive antecedents of action. The core concept of TEC is the event code, which consists of the codes that represent the distal features of an event (feature codes). These feature codes have two functions: first, to register sensory input from various sensory systems and second, to modulate the activities of various motor systems. They are, therefore, not specific to a particular stimulus or response. A gymnast may receive visual and kinesthetic input

(input from the two sensory systems), which may affect her arm and leg movements (two motor systems). The integration of information of multiple sources is realized by the distal feature codes, which enable a much more complex and complete use of information than proximal codes, which are restricted to one sensory channel. The term distal refers to distal attributes of the perceived and/or produced event, as opposed to proximal effects on the sensory surface or muscular innervations or patterns of muscular innervations.

An interesting approach that addresses the action–perception link from both a psychological and a neuroscientific perspective is Jeannerod's (2001) simulation theory, which proposes that the motor system is part of a simulation network that is activated under various conditions related to action, either self-intended or during action perception. The simulation theory predicts that there is a neural network that is similarly activated during the state when an action is simulated and/or executed. Referring to those mental states that involve action content and brain activity simulating the action during the same, executed action, Jeannerod created the term "S-states" (simulation states). He devised a taxonomy of behaviorally defined S-states that include different types of S-states, such as intended action, imagined action, prospective action judgments, perceptually based judgments, perceptually based decisions, observation of graspable objects, observation of actions performed by others, and action in dreams. For instance, during an S-state, an individual may simulate an action while perceptually judging it, planning/intending an action, or observing another individual performing an action. The prerequisite for the simulation theory is the activation of the motor system during these different types of S-states, giving them their action content. Jeannerod (2001) described a number of studies that support such activation from a neurophysiological perspective. For instance, studies indicated activations of the primary motor cortex, the pre-motor cortex, and associative cortical areas (for an overview, see Jeannerod, 2001). This idea of action simulation has received more and more attention in current research; yet, many aspects of the exact simulative process remain unclear, from an empirical, computational, and neural perspective (for an interesting review and discussion on this aspect, see Pezzulo, Candidi, Dindo, & Barca, 2013).

Schütz-Bosbach and Prinz (2007) have termed the link between action and perception "perceptual resonance," suggesting that action production primes perception in a way that causes observers to be selectively sensitive to actions that share features with their own actions. The authors speculated further that humans might be able to perceive and understand only those actions of others that they can do themselves. Contradictory to this view, a new framework was put forward, according to which the term "motor system" might be better changed to "prediction system" (Schubotz, 2007). Derived from the observation that humans are quite accurate in predicting the outcomes of inanimate events (such as waves rolling), for which there is no motor experience, this framework suggests that the prediction of events that we cannot reproduce is achieved by the aid of sensorimotor-driven forward models. Schubotz (2007) proposed that the predictive accounts of the sensorimotor system can be generalized from bodily action to event perception by using the audiomotor or visuomotor system in a simulation mode to represent and predict inanimate action, without proprioceptive (e.g., the sense of the relative position of the body parts) or other interoceptive (e.g., equilibrioceptive or sense of balance) information. This framework points out an additional aspect that other frameworks pay less attention to and that TEC (Hommel et al., 2001) describes only briefly, namely, that perception is not limited to the visual domain, and multimodality may be useful.

Humans can perceive their environment by using different senses. With regard to Schubotz (2007), gymnasts might use exteroceptive information such as hearing their own footsteps or seeing their own legs during skill execution. In addition, they might use interoceptive senses such as kinesthetic information (e.g., feeling if the knees are bent) to control their movements. In recent years, research has been increasingly addressing the aspect of multimodality by including the use of different senses. For instance, in a transcranial magnetic stimulation (TMS) study, subjects were instructed to perceive unimodal visual, unimodal auditory, or multimodal stimuli of a simple hand action (Alaeerts, Swinnen, & Wenderoth, 2009). Incongruent multimodal stimuli were included to test for multimodal interaction in the human motor system. A selective response increase in the primary motor cortex was observed for the congruent multimodal condition when compared to the unimodal and incongruent multimodal condition. The authors concluded that these results support the idea of shared action representations in the human motor system, additionally evoked in a modality-dependent manner. This aspect is addressed further by Kennel and Pizzera (Chapter 15). The above-described approaches and studies concerning the bidirectional link between perception and action show that there is a need to refine this link with regard to the temporal overlap of perceptual and motor processes.

## ONLINE AND OFFLINE EFFECTS OF THE PERCEPTION–ACTION LINK

One approach to temporal or perceptual refining suggests that the perception–action link can be subdivided into online and offline effects (Schütz-Bosbach & Prinz, 2007). With regard to online effects from perception to action, the gymnast may perceive another gymnast performing a skill while executing a skill herself. Such short-term effects of perception on action can be termed visuo-motor priming. Motor-visual priming effects (Vogt & Thomaschke, 2007) describe situations in which execution affects visual processing. For instance, the gymnast might execute a skill and concurrently perceive (see or hear) cues related to her own body that she can use for controlling the skill. For both priming directions, action execution can either hinder or facilitate perceptual processes and vice versa. With regard to negative online effects, in a study in which participants were asked to make sinusoidal arm movements, it was shown that when they observed an incongruent movement, this interfered with their own movement execution (Bouquet, Gaurier, Shipley, Toussaint, & Blandin, 2007). Facilitative online effects have been shown in a study in which participants were asked to turn a knob either to the left or to the right while observing a perceptually bistable apparent rotation of an object (Wohlschläger, 2000). The perceived direction of the object was biased in the direction of the produced knob-turning direction. This aspect of online effects is further addressed by Kennel and Pizzera (Chapter 15), in which a more applied perspective with respect to auditory online effects is presented. For a more detailed overview of different online effects, see Schütz-Bosbach and Prinz (2007).

An explanation of interference effects can be found in the TEC model. Although stating that there is a common representational domain for perception and action, it may not be possible for low-level processes of action planning and perception to access high-level action processes at the same time. Therefore, if these processes share the same codes, performance declines may be observed. To be more specific, if a cognitive code is currently being "used" by action processes, this code may be less available for perception. However, this approach does not explain facilitative effects. In contrast, Heyes (2001, 2005) explained such effects on the basis of learned associations. This so-called associative account of imitative phenomena states that experience forms the link between planned actions and perceived events, assuming distinct representations for the two. Therefore, both matching and non-matching sensory and motor representations can be associated in the same way, regulated by the frequency of exposure. Since non-matching representations occur less often and individuals have acquired stronger associations between matching representations (due to environmental contingencies in favor of forming matches), interference as well as facilitative effects can be explained. This associative learning approach was even used to explain the link between online perception and action processes when it comes to estimating slants of hills, distances, and even monetary values. For instance, one research program (Proffitt, 2006) revealed that participants estimated hills as steeper when they wore a heavy backpack or after a 1-h run. However, this was only evident in the verbal and visual assessments of the participants, and not when they were asked to indicate a visually guided haptic measure of the slant of the hill (adjusting a palm board by feel, matching the board's felt orientation to the hill's incline). With regard to judgments of importance, studies revealed that holding a heavy clipboard increases judgments of monetary value, showing the embodiment of abstract concepts such as importance (Jostmann, Lakens, & Schubert, 2009). Such effects are explained on the basis of associative learning in that people have learned to associate the experience of weight with increased expenditure of physical or mental effort.

With regard to offline effects, a gymnast might train the handstand for several weeks, months, or years; this, in turn, might influence how this gymnast perceives the execution of this skill (temporally separated, hence, not concurrent with action execution) by another gymnast. This effect is especially important for sports judges and can be linked to the type of S-state (see simulation theory) that includes perceptually based judgments and decisions. Considering a judge who is watching the gymnast perform a handstand on the beam, the judge might perceive the quality of this handstand differently depending on whether the judge has been able to perform the handstand him- or herself. With regard to facilitative or hindering effects of this action–perception transfer, such previous motor experiences could lead to better or worse perceptual judgments. Indeed, such beneficial effects have been found in gymnastics judges who were asked to judge gymnasts performing a skill on the balance beam (Pizzera, 2012). The judges who could perform the skill themselves were more accurate in their judgments than judges who could not perform the skill. However, this effect seems to be highly task or sport dependent. Specifically, the performance of sports officials, which is primarily based on perceptual judgments, has been shown to be differentially predicted by the sports officials' prior motor experiences as athletes (Pizzera & Raab, 2012b). For example, the performance of soccer and team handball referees did not relate to referees' prior experiences as players, whereas performance of ice hockey referees and trampoline judges did. These two studies show that these links between action and perception can be understood differently in terms of the specificity of action (what component of action is meant), which can influence perception or perceptual judgments. The gymnastics judge, for example, can have specific motor competencies with a certain skill, whereas a soccer referee might have a higher amount of general motor experience due to several years of having played soccer in the past.

## CONCLUSION AND FUTURE RESEARCH

In this chapter, I reviewed the literature on the bidirectional perception–action link. I described research showing both online or offline effects of this link. The aim of the current chapter was to present findings that focus on the perception–action links in a highly ecologically valid environment, namely, sports. The overview revealed that there are contradictory results, and many questions remain open for future studies to address. For instance, theoretical frameworks such as TEC might have different predictions for online effects when addressing fine-motor versus complex skills. However, this might also be true when comparing online and offline effects in general. Can the same theoretical frameworks explain the interaction between perception and action when motor and perceptual processes are running concurrently as opposed to being temporally separated? With respect to effects of action on perception, questions remain as to what, exactly, is affected on the level of perceptual processing. Do the effects take place on the level of actual perception or, if perceptual judgments are examined, on the level of information integration or memory processes? Alternatively, is one level affected more than the other and/or is this dependent on the task at hand? Future studies should address such questions and further investigate the bidirectional link to advance this interesting field of research.

Each of the following chapters of this section will cover one of the aspects described above. Chapter 14 (Sevdalis & Wöllner) will address the visual sense, by focusing on how motion can be captured, how individuals from sports and the performing arts visually perceive action, and how this action perception is related to the perceiver's own motor competencies or experiences. Chapter 15 (Kennel & Pizzera) will add the auditory sense, describing how people perceive and control movements by means of acoustic information provided by their own bodies. In addition, this chapter will differentiate between natural and artificial movement sounds, because different task demands may result in different perception–action couplings. Finally, Chapter 16 (De Oliveira) gives an overview of how the perception–action link is represented in fine-motor and complex tasks, thereby addressing the question of how the perception of complex movements is tuned by one's own body and vice versa. This chapter will provide additional approaches to the perception–action link, such as the Gibsonian approach that states that perception and action are functionally linked and that only their coordination allows for adaptive behavior (Gibson, 1979).

## REFERENCES

Alaeerts, K., Swinnen, S. P., & Wenderoth, N. (2009). Interaction of sound and sight during action perception: evidence for shared modality-dependent action representations. *Neuropsychologia, 47*, 2593–2599.

Andrieux, M., & Proteau, L. (2014). Mixed observation favors motor learning through better estimation of the model's performance. *Experimental Brain Research, 232*, 3121–3132. http://dx.doi.org/10.1007/s00221-014-4000-3.

Ashford, D., Davids, K., & Bennett, S. (2007). Developmental effects influencing observational modelling: a meta-analysis. *Journal of Sports Sciences, 25*, 547–558.

Bandura, A. (1986). *Social foundations of thought and action: A social cognitive theory*. Englewood Cliffs, NJ: Prentice-Hall.

Barsalou, L. W. (1999). Perceptual symbol systems. *Behavioral and Brain Sciences, 22*, 577–660.

Beilock, S. L., Lyons, I. M., Mattarella-Micke, A., Nusbaum, H. C., & Small, S. L. (2008). Sports experience changes the neural processing of action language. *Proceedings of the National Academy of Sciences of the United States of America, 105*, 13269–13273.

Bellebaum, C., & Colosio, M. (2014). From feedback- to response-based performance monitoring in active and observational learning. *Journal of Cognitive Neuroscience, 26*, 2111–2127. http://dx.doi.org/10.1162/jocn_a_00612.

Binkofski, F., Amunts, K., Stephan, K. M., Posse, S., Schormann, T., Freund, H. J., et al. (2000). Broca's region subserves imagery of motion: a combined cytoarchitectonic and fMRI study. *Human Brain Mapping, 11*, 273–285.

Bouquet, C. A., Gaurier, V., Shipley, T., Toussaint, L., & Blandin, Y. (2007). Influence of the perception of biological or non-biological motion on movement execution. *Journal of Sports Sciences, 25*, 519–530.

Brooks, R. (1986). A robust layered control system for a mobile robot. *Journal of Robotics & Automation, 2*, 14–23.

Brooks, R. (1991). New approaches to robotics. *Science, 253*, 1227–1232.

Buccino, G., Binkofski, F., Fink, G. R., Fadiga, L., Fogassi, L., Gallese, V., et al. (2001). Action observation activates premotor and parietal areas in a somatotopic manner: an fMRI study. *European Journal of Neuroscience, 13*, 400–404.

Buccino, G., Vogt, S., Ritzl, A., Fink, G. R., Zilles, K., Freund, H.-J., et al. (2004). Neural circuits underlying imitation learning of hand actions: an event-related fMRI study. *Neuron, 42*, 323–334.

Calmels, C., Holmes, P., Jarry, G., Hars, M., Lopez, E., Paillard, A., et al. (2006). Variability of EEG synchronization prior to, and during, observation and execution of a sequential finger movement. *Human Brain Mapping, 27*, 251–266.

Calvo-Merino, B., Glaser, D. E., Grèzes, J., Passingham, R. E., & Haggard, P. (2005). Action observation and acquired motor skills: an fMRI study with expert dancers. *Cerebral Cortex, 15*, 1243–1249.

Calvo-Merino, B., Grèzes, J., Glaser, D. E., Passingham, R. E., & Haggard, P. (2006). Seeing or doing? Influence of visual and motor familiarity in action observation. *Current Biology, 16*, 1905–1910.

Cañal-Bruland, R., van der Kamp, J., & van Kesteren, J. (2010). An examination of motor and perceptual contributions to the recognition of deception from others' actions. *Human Movement Science, 29*, 94–102.

Clark, S., & Ste-Marie, D. (2007). The impact of self-as-a-model intervention on children's self-regulation of learning and swimming performance. *Journal of Sports Sciences, 25*, 557–586.

Decety, J., & Grèzes, J. (1999). Neural mechanisms subserving the perception of human actions. *Trends in Cognitive Sciences, 3*, 172–178.

Decety, J., Grèzes, J., Costes, N., Perani, D., Jeannerod, M., Procyk, E., et al. (1997). Brain activity during observation of actions. Influence of action content and subject's strategy. *Brain, 120*, 1763–1777.

Di Pellegrino, G., Fadiga, L., Fogassi, L., Gallese, V., & Rizzolatti, G. (1992). Understanding motor events: a neurophysiological study. *Experimental Brain Research, 91*, 176–180.

Dosseville, F., Laborde, S., & Raab, M. (2011). Contextual and personal motor experience effects in judo referees' decisions. *The Sport Psychologist, 25*, 67–81.

Feltz, D. L., & Landers, D. M. (1983). The effects of mental practice on motor skill learning and performance: a meta-analysis. *Journal of Sport Psychology, 5*, 25–57.

Gibson, J. J. (1979). *The ecological approach to visual perception*. Boston, MA: Houghton Mifflin.

Goldman, A., & de Vignemont, F. (2009). Is social cognition embodied? *Trends in Cognitive Sciences, 13*, 154–159.

Gray, J. T., Neisser, U., Shapiro, B. A., & Kouns, S. (1991). Observational learning of ballet sequences: the role of kinematic information. *Ecological Psychology, 3*, 121–134.

Grèzes, J., Armony, J. L., Rowe, J., & Passingham, R. E. (2003). Activations related to "mirror" and "canonical" neurons in the human brain: an fMRI study. *NeuroImage, 18*, 928–937.

Grèzes, J., & Decety, J. (2001). Functional anatomy of execution, mental simulation, observation, and verb generation of actions: a meta-analysis. *Human Brain Mapping, 12*, 1–19.

Hars, M., & Calmels, C. (2007). Observation of elite gymnastic performance: processes and perceived functions of observation. *Psychology of Sport and Exercise, 8*, 337–354.

Hayes, S. J., Hodges, N. J., Scott, M. A., Horn, R. R., & Williams, A. M. (2007). The efficacy of demonstrations in teaching children an unfamiliar movement skill: the effects of object-oriented actions and point-light demonstrations. *Journal of Sports Sciences, 25*, 559–575.

Hecht, H., Vogt, S., & Prinz, W. (2001). Motor learning enhances perceptual judgment: a case for action-perception transfer. *Psychological Research, 65*, 3–14.

Heyes, C. (2001). Causes and consequences of imitation. *Trends in Cognitive Sciences, 5*, 253–261.

Heyes, C. (2005). Imitation by association. In S. Hurley, & N. Chater (Eds.), *Perspectives on imitation: From neuroscience to social science* (Vol. 1) (pp. 157–176). Cambridge, MA: MIT Press.

Hickok, G. (2008). Eight problems for the mirror neuron theory of action understanding in monkeys and humans. *Journal of Cognitive Neuroscience, 21*, 1229–1243.

Hodges, N. J., & Williams, A. M. (Eds.), (2007). Observational learning [Special Issue]. *Journal of Sports Sciences, 25*(5).

Holmes, P., & Calmels, C. (2008). A neuroscientific review of imagery and observation use in sport. *Journal of Motor Behavior, 40*, 433–445.

Hommel, B., Müsseler, J., Aschersleben, G., & Prinz, W. (2001). The theory of event coding (TEC): a framework for perception and action planning. *Behavioral and Brain Sciences, 24*, 849–937.

Hurley, S. L. (1998). *Consciousness in action*. Cambridge, MA: Harvard University Press.

Iacoboni, M., Woods, R. P., Brass, M., Bekkering, H., Mazziotta, J. C., & Rizzolatti, G. (1999). Cortical mechanisms of human imitation. *Science, 286*, 2526–2528.

Jeannerod, M. (2001). Neural simulation of action: a unifying mechanism for motor cognition. *NeuroImage, 14*, 103–109.

Jostmann, N. B., Lakens, D., & Schubert, T. W. (2009). Weight as an embodiment of importance. *Psychological Science, 20*, 1169–1174.

Kilner, J. M., Neal, A., Weiskopf, N., Friston, K. J., & Frith, C. D. (2009). Evidence of mirror neurons in human inferior frontal gyrus. *Journal of Neuroscience, 29*, 10153–10159.

Kosslyn, S. M., Ganis, G., & Thompson, W. L. (2001). Neural foundations of imagery. *Nature Reviews Neuroscience, 2*, 635–642.

Lotze, M., & Halsband, U. (2006). Motor imagery. *Journal of Physiology (Paris), 99*, 386–395.

Macuga, K. L., & Frey, S. H. (2012). Neural representations involved in observed, imagined, and imitated actions are dissociable and hierarchically organized. *NeuroImage, 59*, 2798–2807.

McCullagh, P., & Weiss, M. R. (2001). Modeling: considerations for motor skill performance and psychological responses. In R. N. Singer, H. A. Hausenblas, & C. M. Janelle (Eds.), *Handbook of sport psychology* (2nd ed.) (pp. 205–238). New York: Wiley.

Munzert, J., & Zentgraf, K. (2009). Motor imagery and its implications for understanding the motor system. In M. Raab, J. Johnson, & H. Heekeren (Eds.), *Progress in brain research Mind and motion: The bidirectional link between thought and action: Vol. 174.* (pp. 219–229). Amsterdam, The Netherlands: Elsevier Press.

Niedenthal, P. M., Barsalou, L. W., Winkielman, P., Krauth-Gruber, S., & Ric, F. (2005). Embodiment in attitudes, social perception, emotion. *Personality and Social Psychology Review, 9*, 184–211.

Onate, J. A., Guskiewic, K. M., Marshall, S. W., Giuliani, C., Yu, B., & Garrett, W. E. (2005). Instruction of jump-landing technique using videotape feedback altering lower extremity motion patterns. *American Journal of Sports Medicine, 33*, 831–842.

Oosterhof, N. N., Wiggett, A. J., Diedrichsen, J., Tipper, S. P., & Downing, P. E. (2010). Surface-based information mapping reveals crossmodal vision–action representations in human parietal and occipitotemporal cortex. *Journal of Neurophysiology, 104*, 1077–1089.

Oullier, O., & Basso, F. (2010). Embodied economics: how bodily information shapes the social coordination dynamics of decision-making. *Philosophical Transactions of the Royal Society of London, Series B: Biological Sciences, 365*, 291–301.

Pezzulo, G., Candidi, M., Dindo, H., & Barca, L. (2013). Action simulation in the human brain: twelve questions. *New Ideas in Psychology, 31*, 270–290.

Pizzera, A. (2012). Gymnastic judges benefit from their own motor experience as gymnasts. *Research Quarterly for Exercise and Sport, 83*, 603–607.

Pizzera, A., & Raab, M. (2012a). Does motor or visual experience enhance the detection of deceptive movements in football? *International Journal of Sports Science & Coaching, 7*, 269–283.

Pizzera, A., & Raab, M. (2012b). Perceptual judgments of sports officials are influenced by their motor and visual experience. *Journal of Applied Sport Psychology, 24*, 59–72.

Prinz, W. (1997). Perception and action planning. *European Journal of Cognitive Psychology, 9*, 129–154.

Proffitt, D. R. (2006). Embodied perception and the economy of action. *Perspectives on Psychological Science, 1*, 110–122.

Ram, N., Riggs, S. M., Skaling, S., Landers, D. M., & McCullagh, P. (2007). A comparison of modelling and imagery in the acquisition and retention of motor skills. *Journal of Sports Sciences, 25*, 587–597.

Renden, P. G., Kerstens, S., Oudejans, R., & Cañal-Bruland, R. (2014). Foul or dive? Motor contributions to judging ambiguous foul situations in football. *European Journal of Sport Science, 14*(S1), S221–S227.

Rizzolatti, G., Fadiga, L., Gallese, V., & Fogassi, L. (1996). Premotor cortex and the recognition of motor actions. *Cognitive Brain Research, 3*, 131–141.

Rizzolatti, G., Fadiga, L., Matelli, M., Bettinardi, V., Paulesu, E., Perani, D., et al. (1996). Localization of grasp representations in humans: by PET: 1. Observation versus execution. *Experimental Brain Research, 111*, 246–252.

Rodrigues, S. T., Ferracioli, M. C., & Denardi, R. A. (2010). Learning a complex motor skill from video and point-light demonstrations. *Perceptual & Motor Skills, 111*, 307–323.

Romack, J. L., & Briggs, R. S. (1995). Relevant visual information in a demonstration and its use in learning how to dribble a ball. *Journal of Sport and Exercise Psychology, 17*, S79.

Runeson, S. (1984). Perception of biological motion: the KSD principle and the implications of a distal versus proximal approach. In G. Jansson, S. Bergstom, & W. Epstein (Eds.), *Perceiving events and objects* (pp. 383–405). Hillsdale, NJ: Erlbaum.

Schubotz, R. I. (2007). Prediction of external events with our motor system: towards a new framework. *Trends in Cognitive Sciences, 11*, 211–218.

Schütz-Bosbach, S., & Prinz, W. (2007). Perceptual resonance: action-induced modulation of perception. *Trends in Cognitive Science, 11*, 349–355.

Ste-Marie, D., Law, B., Rymal, A. M., Jenny, O., Hall, C., & McCullagh, P. (2012). Observation intervention for motor skill learning and performance: an applied model for the use of observation. *International Review of Sport and Exercise Psychology, 5*, 145–176.

Steinhorst, A., & Funke, J. (2014). Mirror neuron activity is no proof for action understanding. *Frontiers in Human Neuroscience, 8*, 1–4.

Van Wieringen, P. C. W., Emmen, H. H., Bootsma, R. J., Hoogesteger, M., & Whiting, H. T. A. (1989). The effect of video-feedback on the learning of the tennis service by intermediate players. *Journal of Sports Sciences, 7*, 153–162.

Vogt, S., & Thomaschke, R. (2007). From visuo-motor interactions to imitation learning: behavioral and brain imaging studies. *Journal of Sports Sciences, 25*, 497–517.

Wilson, M. (2002). Six views of embodied cognition. *Psychonomic Bulletin and Review, 9*, 625–636.

Wohlschläger, A. (2000). Visual motion priming by invisible actions. *Vision Research, 40*, 925–930.

Zentgraf, K., Stark, R., Reiser, M., Künzell, S., Schienle, A., Kirsch, P., et al. (2005). Differential activation of pre-SMA and SMA proper during action observation: effects of instructions. *NeuroImage, 26*, 662–672.

Chapter 14

# Capturing Motion for Enhancing Performance: An Embodied Cognition Perspective on Sports and the Performing Arts

### Vassilis Sevdalis[1], Clemens Wöllner[2]
[1]Department of Performance Psychology, Institute of Psychology, German Sport University Cologne, Cologne, Germany; [2]Institute of Systematic Musicology, University of Hamburg, Hamburg, Germany

> *During learning, a young musician has to find an optimal body posture in relation to his/her instrument to produce sounds effectively and without strain; a dancer needs to develop synchronization skills to match the movements of a partner. During performance, an athlete has to be able to "read" incidental variations of the game and position his/her body appropriately to score a goal. Early stages of learning require boosting the motivation and pleasure of the young trainee; professional performance requires a sustained amount of learning musical materials or sustaining high-pressure situations where each action may result in a win or a loss of the game.*

Playing music, dancing, and practicing sports essentially entail some forms of bodily and cognitive engagement. These forms vary throughout learning processes and in performance situations. Ultimately, the goal of engagement with sports and the performing arts (e.g., music, dance) is to achieve an optimal outcome that meets certain standards compatible with one's level of training. This goal is achieved through long-term embodied practices: these practices require complex corporeal activities that combine high motor and cognitive demands. Applied motor performance domains, such as sports and the performing arts (i.e., music and dance), share considerable similarities regarding the engagement of the human body and the cognitive demands (e.g., perception and learning) imposed on the performing individuals. This chapter aims to shed light on the contributions of the body in guiding performance in perceptual and learning tasks, and it outlines how sensorimotor activities enable the long-term development of individual and interpersonal performance competences.

## OVERVIEW OF THEORETICAL APPROACHES

Recent advances in cognitive science have underlined the potential of bodily functions in shaping cognitive processes (Grafton, 2009; Wilson, 2002), a conception that is often referred to as embodied cognition in scientific literature. Although different definitions of what qualifies as "embodied" have been suggested (e.g., Goldman & de Vignemont, 2009; Wilson & Golonka, 2013), most of the embodied cognition approaches converge to the fact that bodily contributions are fundamentally potent in altering cognitive functions (Schubert & Semin, 2009), especially those that concern perception and learning of body movements (Sevdalis & Keller, 2011a). A fundamental premise of the mechanism of embodied cognition is that individuals rely on their own sensorimotor experience when perceiving and learning actions (Herwig, Beisert, & Prinz, 2013; Prinz, 1990). For instance, while observing someone performing an action, observers can recruit some of their own sensorimotor resources as if they were performing the action themselves. This mapping of observed movements onto one's own sensory-motor system is known as action simulation (Jeannerod, 2006; see also Pizzera, Chapter 13 and Kennel & Pizzera, Chapter 15 in this book). The direct consequence of this premise is that alterations in sensorimotor experience that occur incidentally (e.g., by everyday experience) or deliberately (e.g., through training) have the potential to influence the processing that accompanies the observation of actions: this is done by (re-) establishing and updating (internal models of) the relations of the body and the environment in which the actions take place.

Quite some time ago, some of these concepts had already been discussed as "ideomotor theory," postulating strong links between action and perception (Lotze, 1852). Ideomotor learning occurs when people become aware that certain action consequences are caused by their movements. Focusing on action outcomes and exerting control over them may, thus, enhance motor learning (see also Koch, Keller, & Prinz, 2004). Similar ideas were developed by Theodor Lipps (1883, 1903), who stated that human interactions are largely based on inner co-sensations. Perceiving someone else moving may resonate with one's own kinaesthetic representation of these movements, eventually leading to empathic responses between the observer and the performer (for a recent review on empathy in sports and the performing arts, see Sevdalis & Raab, 2014). Interestingly, in addition to the perception of actions as such, even the perception of action outcomes, such as expressive variations in sound patterns, may trigger motor resonance in the observer (for a review, see Sevdalis & Keller, 2014).

Further pioneering work highlighted the importance of the environment and its properties in shaping an organism's behavior, especially with regard to visual perception (Gibson, 1979). Within an ecological framework, the environment specifies affordances to the organism, as possibilities for action, which the organism then perceives and acts upon. The task of the organism is to pick up information from the environment by detecting features that are relevant for dealing with a specific situation. Later approaches extended the approach of Gibson to more interpersonal/social settings: apart from an individual organism's relation with its environments, detecting features of other individuals can shape an organism's interactions with them (Marsh, Richardson, Baron, & Schmidt, 2006; McArthur & Baron, 1983). The presence of others and the interaction with them can also shape an individual's action possibilities. However, only recently theoretical work on joint action has been conducted in ecologically valid environments, such as in music and sports (D'Ausilio, Novembre, Fadiga, & Keller, 2015; Novembre & Keller, 2014; also Bekkering et al., 2009 for some sport examples).

The preceding theoretical approaches underline the benefit of considering the body and the environment as potent candidates in shaping cognitive processes. Features of the body (embodiment) and of the environment (embedding) play an active role in cognitive processing. According to this view, cognition per se is based on interactions between the mind, body, and environment (see Rowlands, 2010). Embodiment can be described as cognitive processes resulting from bodily sensations located in an "inner space," while embedded cognition refers to interactions with an "outer space" beyond one's own body (Leman, 2008). It has already been suggested that cognitive processes may be better understood if certain environments, circumstances, tools, and technologies are considered as parts of an "extended cognitive system": a cognitive system that does not reside exclusively in one's head (extended cognition thesis, Clark, 2008; Clark & Chalmers, 1998). Certain body–environment relations can actually function as vehicles for cognition, by supporting purely cognitive operations (Clark, 2008).

## OVERVIEW OF EMPIRICAL RESEARCH

How can these theoretical approaches inform empirical research? Apart from acknowledging the fact that corporeal and environmental contributions can constitute parts of the cognitive system (an extended cognitive system), it is necessary to consider situations where (1) this is (optimally) possible and (2) this is (optimally) trainable. We hold that in order to investigate cognitive processes as they relate to the body and the environment, one has to investigate situations that involve all three components (cognition, body, and environment) where they unfold most prominently. Such situations are most prevalent in sports and the performing arts. Sports and the performing arts are ancient and culturally widespread activities, practiced on a daily basis, and often performed by individuals without much deliberate practice or expertise in them. They appear naturally in development through playful activities and spontaneous interactions. Still, through education and training, they can be cultivated to exceed established limits of the human potential. Practicing sports and performing arts, apart from performance outcomes that are aesthetically pleasurable to experience, brings about health and social-emotional benefits, both acute and long-term ones (MacDonald, Kreutz, & Mitchell, 2012; Williamson, 2004). Still, theoretical and empirical work combining frameworks such as the ones mentioned above is only recently emerging (e.g., D'Ausilio et al., 2015).

The assumption that cognition, body, and environment can be optimally investigated in applied motor performance contexts necessitates a methodological shift in the implementation of experimental designs for uncovering these relationships: research may benefit from moving beyond experimental tasks that stem from a cognitive psychology tradition to tasks that are commonplace to performing individuals (e.g., performing music, dancing, reacting to an opponent's movements, executing movement routines). This methodological shift can pose considerable challenges to the experimenter: learning and performance situations are complex and in continuous flux; they are often unique and difficult to reproduce. Examples of the challenges of studying motor behavior have been documented (Rosenbaum, 2005), and the potential benefits to the study of human behavior by investigating bodily movements have been outlined (de Gelder, 2009).

Innovative technologies that can capture the dynamics of performance at their nascence, such as motion capture, may be of assistance in investigating sports and performing arts contexts. Pioneering work by Johansson (1973) has shown that humans exhibit considerable sensitivity when observing movements, even if actions are depicted by just a few point-lights attached to an individual's body. In fact, human actions provide considerable information about an individual's characteristics (e.g., gender, identity) but also social-communicative signals (e.g., emotions, intentions) (for reviews see Blake & Shiffrar, 2007; Pavlova, 2012). A considerable benefit of capturing human motor performance is that experimental control can be implemented, thus, making the investigation of the dynamics of human (inter-) action possible. In this sense, capturing an individual's motor performance (e.g., by motion capture) and displaying it to observers to judge it (e.g., in point-light displays) is a useful tool to understand cognitive and social processes. Next, we outline some examples from empirical research in the domains of the performing arts (music and dance) and sports.

## Research in the Performing Arts

A main stream of research on the relations between perception and action deals with the influence of embodied motor knowledge on perceptual tasks. Because a great deal of actions in the realm of motor performance occurs in social situations (e.g., reacting to an opponent or a dance partner, playing in a musical ensemble), advanced perceptual skills are crucial for successful interactions and interpersonal synchronization. In this regard, a fundamental aspect of acting upon an environment is to be in control of one's actions, understand their properties, and be aware of their consequences to others in interactive situations. In a series of studies, the capacity of individuals to distinguish between self- or other-generated actions and between expressive and inexpressive movement properties were investigated (Sevdalis & Keller, 2009, 2010, 2011b, 2012). In these experiments, the motor performance of individuals was recorded by a motion capture system. The individuals performing in these experiments were instructed to execute certain actions in synchrony with music (i.e., dancing, walking, and clapping). Actions were executed in solo or dyadic conditions. After several months, the same individuals who performed the actions were invited back to the lab, to observe point-light displays of their performances. The movements depicted in these displays were also varying in terms of their properties, with manipulations applied to their spatial-temporal parameters (e.g., duration of the movement, number of point-light markers available). The task for the participants was to identify whether the depicted point-light figure was themselves or another individual, and whether the movements of the depicted individual were expressive or inexpressive. Results indicated that discrimination performance in these tasks was fairly robust (i.e., even under conditions of occlusion), and possible even in non-experts. Self-reported empathy indices were also associated positively with discrimination accuracy. These results suggest that perceiving the actions of oneself and others and their expressive qualities relies on the dynamic properties of actions, especially their spatial and temporal characteristics. More complex movements provide richer information on agency and expression intensity and are, thus, easier to distinguish when observed as point-light displays (cf. Daprati, Wriessnegger, & Lacquaniti, 2007; Loula, Prasad, Harber, & Shiffrar, 2005).

If non-experts may recognize their own spontaneous movements in highly individual and information-rich activities, then experts with deliberately trained motor skills should show comparable performance characteristics even for more constrained movements. Furthermore, because action competencies in a specific domain aid in perception, deliberately trained movements should resonate in the observers' motor system to a higher degree compared to everyday-life movements. For answering these questions, a study investigated highly trained movements of orchestral conductors in comparison with their gait (Wöllner, 2012). Each orchestral conductor was matched with two further conductors according to age, gender, and expertise in a forced-choice self-other paradigm. Conductors were more successful in identifying their own movements when they were presented with point-light displays of skilled conducting gestures as compared to point-light displays of walking. In addition, conductors perceived the quality of their own gestures to be higher than those of others, independently of individual awareness of agency. These results indicate that perceptual response mechanisms resonate more with one's own skilled actions, even if performers are not fully aware of their agency. Individual perception of high-quality actions, thus, matches internal models of movements better in one's domain of expertise.

Characteristics of motor expertise, especially domain-specific action competencies, are mirrored in perceptual accuracy and also in behavioral tasks when reacting to other's movements in social interactions (Wöllner & Cañal-Bruland, 2010). String musicians were able to synchronize more accurately and more consistently with the entry movements of a first violinist shown in several video clips. They benefited from their motor expertise in playing string instruments themselves, as compared to highly trained observers (musicians of other instruments who were visually experienced in watching a first violinist) or individuals without such experiences. Perceptual consistency was, thus, related to motor consistency, which has been described as a key characteristic of expert performance in the domain of music and beyond (cf. Konczak, van der Velden, & Jaeger, 2009; Schoonderwaldt & Altenmüller, 2014).

Further research suggests that motor training may even affect the perception of person-related cues in observers. In common movements such as gait, cues of the walking person's gender are perceivable in point-light displays (e.g., Pollick, Kay, Heim, & Stringer, 2005). In a study by Wöllner and Deconinck (2013), musically trained participants watched point-light displays of male and female orchestral conductors with various degrees of expertise. Multimodal experimental conditions (in which sound and vision was provided alone or in combination) and two further control conditions (walking and static images) were implemented. Observers were able to indicate the conductors' gender in point-light displays of walking, replicating previous research, but not consistently for the skilled movements of conducting; while gestures of novice conductors provided some cues of their gender, experienced conductors' gestures did not afford such cues. Differences between experienced and less-experienced conductors for both males and females were present in their motor behavior as recorded with a motion capture system. Consistency in vertical acceleration patterns—a prominent cue for synchronizing with conducting gestures—was higher in experts. Taken together, these lines of research provide hints for the close relationship between perceptual and motor processes in deliberately trained (conducting) and naturally occurring spontaneous movements (walking). In fact, it remains a challenging task to investigate how facets of motor performance are intertwined: some aspects are explicitly learned and deliberately trained, while others are more implicitly acquired as for many daily movements. These ways of learning have an impact not only on the perception of independent observers, but also on the internal models of the individuals carrying out certain actions, or on the interaction between individuals in tasks with temporal or spatial constraints.

In a recent study, such characteristics were put into experimental test in a learning experiment on musical ensemble coordination (Ragert, Schröder, & Keller, 2013). Expert pianists were instructed to learn musical pieces and perform them in a dyadic situation. Auditory information and body movement were recorded (the latter with a motion capture system) and compared. The results suggest that knowledge about the musical structure of a co-performer's part is important for the interpersonal alignment of ancillary body movements (i.e., head motion and body sway) linked to the music's phrasal and metric structure at long time scales. In other words, interpersonal alignment of body movements may be a natural feature of ensemble performance when co-performers are familiar with each other's parts. This tendency was also shown in previous research where even unintentional congruency between the duet musicians' body sway (Keller & Appel, 2010) or ancillary movements (unrelated to the sound production) such as head nods (Goebl & Palmer, 2009) were shown to contribute to the ensemble's optimal coordination.

## Research in Sports

The relationships between perception and performance have been additionally investigated in the domain of sports. Similarly to the domain of performing arts, research commonly focuses on the coupling between perceptual accuracy and performance practices, often in relation to motor expertise. In one study by Hohmann, Troje, Olmos, and Munzert (2011), professional basketball players and novice college students were asked to identify the type of basketball dribbles (e.g., between the legs, behind the back) when observing them as point-light displays. The results showed that experts performed better than novices in terms of identification accuracy and reaction times, but there was no perceptual advantage when observing actions generated by oneself, a teammate, or an unknown expert player. In another experiment (Hohmann et al., 2011), expert basketball players were asked to identify the dribbling actors (e.g., oneself, a teammate). Although identification of one's own actions was slightly better than identification of teammates, this difference did not reach significance. However, actor recognition was better for complex actions, such as dribbling, in comparison to a control condition displaying more stereotypical walking movements. Taken together, these results suggest that both information about an action and previous related training can affect perceptual performance.

Another interesting sport situation where the relations between training and performance become evident is to deliberately intend to deceive an opponent by one's actions. This is a challenging task for both expert and novice players, who need to realize the opponent's true intention and respond appropriately (Aglioti, Cesari, Romani, & Urgesi, 2008; Cañal-Bruland & Schmidt, 2009). Does expertise in performing an action enhance detection of deceptive intentions when observing movements? This question was examined by Sebanz and Shiffrar (2009), who asked experienced and novice basketball players to observe basketball passes and distinguish between fake or true ones. Results indicated that while experts' and novices' perceptual performance was similar when postural cues were displayed (i.e., static movie frames), experts outperformed novices when only kinematic information (i.e., point-light displays) was observed. In another study by Mori and Shimada (2013), anticipation of direction change in a running opponent was investigated in experienced and novice rugby players. Again, experts were better than novices in anticipating deceptive actions, irrespectively of whether actions were depicted as filmed sequences or as point-light displays. These findings highlight the importance of the impact of motor expertise in judging action outcomes.

Apart from cues detected in full body movements, action outcomes can be identified in simpler situations that include, for instance, anticipation of motor information in racket sports. In a series of experiments, prediction accuracy of badminton stroke direction and stroke depth was examined by recruiting expert and non-expert badminton players (Abernethy & Zawi, 2007; Abernethy, Zawi, & Jackson, 2008). The participants observed stroke movements in film and point-light conditions that introduced spatial and temporal occlusions to the kinematic information. The results showed that experts were better than non-experts in perceptual accuracy, especially when kinematic information was only depicted as point-light displays. Interestingly, experts could give accurate predictions by being more sensitive to early kinematic information (before the racquet-shuttle contact) and from kinematics that were functionally unrelated to the stroke (e.g., lower body movements). In contrast, non-expert participants were not attuned to such information and needed more stroke-specific information (e.g., from the racquet and the body movements concurrently).

The use of kinematic information was also investigated in another series of experiments on anticipation accuracy of tennis strokes (ground stroke vs lob stroke) by expert and novice tennis players (Shim, Carlton, & Kwon, 2006; Shim, Chow, Carlton, & Chae, 2005). In these experiments, anticipation accuracy was better than chance for both experts and novices, but non-experts' anticipation accuracy was not degraded by the use of point-light displays. However, with occlusion of the racket and forearm, the players' ability to determine the type of stroke reduced. Taken together, the findings from anticipation accuracy assessment in racket sports suggest that the relations between expertise and the dynamics of kinematic information between body parts and tools (i.e., racquet) are crucial for effective performance.

The deliberate use of critical kinematic information can be of assistance in learning situations, especially during the acquisition phase of a skill, when someone is learning by observation of prototype movements executed by a model. Breslin, Hodges, Williams, Kremer, and Curran (2006) investigated the importance of the relative motions of specific body parts (e.g., intra-limb vs inter-limb) for learning. Participants without previous experience of cricket bowling observed different versions of a point-light cricket bowler model performing an action, and they were asked to reproduce it. Participants were allocated to different learning groups that observed different amounts of kinematic information (e.g., full-body point-light display, intra-limb relative motions of the right bowling arm, inter-limp relative motions of the right and left wrists). The results showed that the groups that observed full-body and intra-limb displays were more accurate in the acquisition of the skill than the inter-limp observing group, suggesting that in early stages of skill acquisition, participants are more perceptually sensitive to the movements of action-specific body parts.

## CONCLUSION

In this chapter, we provided an overview of theoretical approaches and empirical studies on how embodied practices situated in performance environments such as sports and the performing arts (i.e., music and dance) can support cognitive processes, especially in perceptual and learning tasks. The emphasis was put on investigations that considered the interactions between cognitive and bodily aspects of performance in ecologically valid settings. Variants of embodied experience became evident in action-specific bodily expertise, self-generated actions, effector-specific training, and bodily extensions by the use of tools. It is remarkable that across diverse performance domains such as sports, music, and dance, similar experimental and methodological paradigms have been implemented. For instance, agency (i.e., experiencing one's self as being in control of one's actions and their effects) has been investigated in performers as distinct as dancers, orchestral conductors, and basketball players. Action intentions have been a topic of inquiry for expressive body movements and for deceptive body movements. Action anticipation has been studied in interactive contexts, including the movements of an opponent rugby player or the movements of a fellow piano player. These studies introduced manipulations that tested the effects of sensorimotor experience both in terms of overt motor behavior (e.g., spatial-temporal occlusions) and long-term deliberate practice (e.g., expert–novice comparisons). These similarities may provide the ground for essential interdisciplinary dialogue in future investigations, especially with regard to the transfer of motor expertise across performance domains (see also Sevdalis & Raab, 2014).

Despite the diversity in domain-specific motor performance requirements and practices, examples such as the previous can be considered as facets of the same fundamental ability of social cognition. Being able to perform tasks such as perceiving agents, intentions, and anticipation tendencies informs individuals about themselves and others. The processing of social signals available in movements and their deliberate use in interactions is serving as a mutual action organization or "orchestration." It requires awareness of oneself and awareness of the effects of one's signals on others; it also relies on the ability to take into account another individual's point of view (Goldman & de Vignemont, 2009; Frith & Frith, 2012). Consequently, being in control of bodily information and responding appropriately to it has obvious benefits for an individual or a team in interactive contexts: it can lead to a more accurate representation of the dynamics of the situation and a better organization of one's actions toward shared or competitive goals. It can promote efficient action execution while playing, individually or in unison, by more accurately perceiving and predicting one's own and others' movements. It can contribute in building rapport and empathy between interacting individuals.

Performance enhancement also poses considerable challenges to the individual. Action execution and perception are error prone and require time to reach certain standards (see also Gray, 2014). For example, a player's intention can be misunderstood; a musician's instrumental sounds may lag behind those of others. As the empirical research overview highlighted, there are often considerable differences between experts and novices in aspects of motor performance such as timing consistency, movement variability, and use of kinematic information.

Implementing motion capture technology-based training programs that target sensorimotor skills can be a promising path for the future. The transition from novice to expert performance can be supported by exploiting embodied motor knowledge for learning purposes. Learning from or about others in performance contexts requires intense engagement of one's sensorimotor skills and guidance from experienced instructors. It is still an open empirical issue to define the degree of influence of embodied experience on cognitive functioning in a long-term perspective and in teaching situations. Therefore, longitudinal studies are necessary that investigate how emerging sensory-motor skills are mutually affecting each other during development. Comparisons between older and younger individuals are also required, which control for expertise and age effects: these characteristics may additionally influence cognitive fluency and command of motor precision because there are critical time points in development when sensorimotor skills begin to fade. More studies that involve interacting individuals can also be beneficial because they will inform about the development of action–perception links in real time. To conclude, sports, music, and dance provide a wealth of possibilities for examining perception–action links and can be at the forefront of scientific inquiries in performance psychology.

## REFERENCES

Abernethy, B., & Zawi, K. (2007). Pickup of essential kinematics underpins expert perception of movement patterns. *Journal of Motor Behavior, 39*, 353–367.

Abernethy, B., Zawi, K., & Jackson, R. C. (2008). Expertise and attunement to kinematic constraints. *Perception, 37*, 931–948.

Aglioti, S. M., Cesari, P., Romani, M., & Urgesi, C. (2008). Action anticipation and motor resonance in elite basketball players. *Nature Neuroscience, 11*, 1109–1116.

Bekkering, H., De Bruijn, E., Cuipers, R., Newman-Norlund, R., Van Schie, H., & Muelenbroek, R. (2009). Joint action: neurocognitive mechanisms supporting human interaction. *Topics in Cognitive Science, 1*, 340–352.

Blake, R., & Shiffrar, M. (2007). Perception of human motion. *Annual Review of Psychology, 58*, 47–73.

Breslin, G., Hodges, N. J., Williams, A. M., Kremer, J., & Curran, W. (2006). A comparison of intra- and inter-limb relative motion information in modelling a novel motor skill. *Human Movement Science, 25*, 753–766.

Cañal-Bruland, R., & Schmidt, M. (2009). Response bias in judging deceptive movements. *Acta Psychologica, 130*, 235–240.

Clark, A. (2008). *Supersizing the mind: Embodiment, action, and cognitive extension*. Oxford and New York: Oxford University Press.

Clark, A., & Chalmers, D. J. (1998). The extended mind. *Analysis, 58*, 7–19.

Daprati, E., Wriessnegger, S., & Lacquaniti, F. (2007). Kinematic cues and recognition of self-generated actions. *Experimental Brain Research, 177*, 31–44.

D'Ausilio, A., Novembre, G., Fadiga, L., & Keller, P. E. (2015). What can music tell us about social interaction? *Trends in Cognitive Sciences, 19*, 111–114.

Frith, C. D., & Frith, U. (2012). Mechanisms of social cognition. *Annual Review of Psychology, 63*, 287–313.

de Gelder, B. (2009). Why bodies? Twelve reasons for including bodily expressions in affective neuroscience. *Philosophical Transactions of the Royal Society B: Biological Sciences, 364*, 3475–3484.

Gibson, J. J. (1979). *The ecological approach to visual perception*. Boston, MA: Houghton Mifflin.
Goebl, W., & Palmer, C. (2009). Synchronization of timing and motion among performing musicians. *Music Perception, 26*, 427–438.
Goldman, A., & de Vignemont, F. (2009). Is social cognition embodied? *Trends in Cognitive Science, 13*, 154–159.
Grafton, S. T. (2009). Embodied cognition and the simulation of action to understand others. *Annals of the New York Academy of Sciences, 1156*, 97–117.
Gray, R. (2014). Embodied perception in sport. *International Review of Sport and Exercise Psychology, 7*, 72–86.
Herwig, A., Beisert, M., & Prinz, W. (2013). Action science emerging: introduction and leitmotifs. In W. Prinz, M. Beisert, & A. Herwig (Eds.), *Action science: Foundations of an emerging discipline* (pp. 1–33). Cambridge, MA: MIT Press.
Hohmann, T., Troje, N. F., Olmos, A., & Munzert, J. (2011). The influence of motor expertise and motor experience on action and actor recognition. *Journal of Cognitive Psychology, 23*, 403–415.
Jeannerod, M. (2006). *Motor cognition: What actions tell the self*. New York: Oxford University Press.
Johansson, G. (1973). Visual perception of biological motion and a model for its analysis. *Perception & Psychophysics, 14*, 201–211.
Keller, P. E., & Appel, M. (2010). Individual differences, auditory imagery, and the coordination of body movements and sounds in musical ensembles. *Music Perception, 28*, 27–46.
Koch, I., Keller, P. E., & Prinz, W. (2004). The ideomotor approach to action control: implications for skilled performance. *International Journal of Sport and Exercise Psychology, 2*, 362–375.
Konczak, J., van der Velden, H., & Jaeger, L. (2009). Learning to play the violin: motor control by freezing, not freeing degrees of freedom. *Journal of Motor Behavior, 41*, 243–252.
Leman, M. (2008). *Embodied music cognition and mediation technology*. Cambridge, MA: MIT Press.
Lipps, T. (1883). *Grundtatsachen des Seelenlebens* [*Foundation principles of the mind*]. Bonn: Max Cohen & Sohn.
Lipps, T. (1903). *Grundlegung der Ästhetik* [*Foundation of aesthetics*]. Hamburg and Leipzig: Leopold Voss.
Lotze, R. H. (1852). *Medizinische Psychologie oder Physiologie der Seele* [*Medical psychology or physiology of the mind*]. Leipzig: Weidmannsche Buchhandlung.
Loula, F., Prasad, S., Harber, K., & Shiffrar, M. (2005). Recognizing people from their movement. *Journal of Experimental Psychology: Human Perception and Performance, 31*, 210–220.
MacDonald, R., Kreutz, G., & Mitchell, L. (Eds.). (2012). *Music, health and wellbeing*. New York: Oxford University Press.
Marsh, K. L., Richardson, M. J., Baron, R. M., & Schmidt, R. C. (2006). Contrasting approaches to perceiving and acting with others. *Ecological Psychology, 18*, 1–37.
McArthur, L. Z., & Baron, R. M. (1983). Toward an ecological theory of social perception. *Psychological Review, 90*, 215–238.
Mori, S., & Shimada, T. (2013). Expert anticipation from deceptive action. *Attention, Perception, and Psychophysics, 75*, 751–770.
Novembre, G., & Keller, P. E. (2014). A conceptual review on action-perception coupling in the musicians' brain: what is it good for? *Frontiers in Human Neuroscience, 8*, 603.
Pavlova, M. A. (2012). Biological motion processing as a hallmark of social cognition. *Cerebral Cortex, 22*, 981–995.
Pollick, F. E., Kay, J., Heim, K., & Stringer, R. (2005). Gender recognition from point-light walkers. *Journal of Experimental Psychology: Human Perception and Performance, 31*, 1247–1265.
Prinz, W. (1990). A common coding approach to perception and action. In O. Neumann & W. Prinz (Eds.), *Relationships between perception and action: Current approaches* (pp. 167–201). Berlin: Springer.

Ragert, M., Schröder, T., & Keller, P. E. (2013). Knowing too little or too much: the effects of familiarity with a co-performer's part on interpersonal coordination in musical ensembles. *Frontiers in Psychology, 4,* 368.

Rosenbaum, D. A. (2005). The Cinderella of psychology: the neglect of motor control in the science of mental life and behavior. *American Psychologist, 60,* 308–317.

Rowlands, M. (2010). *The new science of the mind: From extended mind to embodied phenomenology.* Cambridge, MA: MIT Press.

Schoonderwaldt, E., & Altenmüller, E. (2014). Coordination in fast repetitive violin-bowing patterns. *PLoS One, 9*(9), e106615.

Schubert, T. W., & Semin, G. R. (2009). Embodiment as a unifying perspective for psychology. *European Journal of Social Psychology, 39,* 1135–1141.

Sebanz, N., & Shiffrar, M. (2009). Detecting deception in a bluffing body: The role of expertise. *Psychonomic Bullettin & Review, 16,* 170–175.

Sevdalis, V., & Keller, P. E. (2009). Self-recognition in the perception of actions performed in synchrony with music. *Annals of the New York Academy of Sciences, 1169,* 499–502.

Sevdalis, V., & Keller, P. E. (2010). Cues for self-recognition in point-light displays of actions performed in synchrony with music. *Consciousness and Cognition, 19,* 617–626.

Sevdalis, V., & Keller, P. E. (2011a). Captured by motion: dance, action understanding, and social cognition. *Brain and Cognition, 77,* 231–236.

Sevdalis, V., & Keller, P. E. (2011b). Perceiving performer identity and intended expression intensity in point-light displays of dance. *Psychological Research, 75,* 423–434.

Sevdalis, V., & Keller, P. E. (2012). Perceiving bodies in motion: expression intensity, empathy, and experience. *Experimental Brain Research, 222,* 447–453.

Sevdalis, V., & Keller, P. E. (2014). Know thy sound: perceiving self and others in musical contexts. *Acta Psychologica, 152,* 67–74.

Sevdalis, V., & Raab, M. (2014). Empathy in sports, exercise, and the performing arts. *Psychology of Sport and Exercise, 15,* 173–179.

Shim, J., Carlton, L. G., & Kwon, Y. H. (2006). Perception of kinematic characteristics of tennis strokes for anticipating stroke type and direction. *Research Quarterly for Exercise and Sport, 77,* 326–339.

Shim, J., Chow, J. W., Carlton, L. G., & Chae, W. S. (2005). The use of anticipatory visual cues by highly skilled tennis players. *Journal of Motor Behavior, 37,* 164–175.

Williamon, A. (Ed.). (2004). *Musical excellence: Strategies and techniques to enhance performance.* New York: Oxford University Press.

Wilson, M. (2002). Six views of embodied cognition. *Psychonomic Bulletin & Review, 9,* 625–636.

Wilson, A. D., & Golonka, S. (2013). Embodied cognition is not what you think it is. *Frontiers in Psychology, 4,* 58.

Wöllner, C. (2012). Self-recognition of highly skilled actions: a study of orchestral conductors. *Consciousness and Cognition, 21,* 1311–1321.

Wöllner, C., & Cañal-Bruland, R. (2010). Keeping an eye on the violinist: motor experts show superior timing consistency in a visual perception task. *Psychological Research, 74,* 579–585.

Wöllner, C., & Deconinck, F. J. A. (2013). Gender recognition depends on type of movement and motor skill. Analysing and perceiving biological motion in musical and nonmusical tasks. *Acta Psychologica, 143,* 79–87.

# Chapter 15

# Auditory Action Perception

Christian Kennel[1], Alexandra Pizzera[2]
[1]*Department of Performance Psychology, German Sport University Cologne, Cologne, Germany;*
[2]*Institute of Sports and Sports Sciences, Heidelberg University, Heidelberg, Germany*

*Walking produces sounds through an individual's footsteps. If an individual hears the sound of a foot dragging on the ground, this may cause the individual to lift up the foot a little bit higher during the next step. Furthermore, the individual may use this experience in perceiving footstep sounds for subsequent steps. Consider athletes who, on hearing their own footstep sounds, make adjustments to stride while running over hurdles or sprinting toward the takeoff board for the long jump.*

Every action produces sound, which influences every subsequent action, leading to a strong interplay between the senses and motor activity. In this chapter, we elaborate on this action–auditory perception interplay. We begin with a brief description of auditory perception and then discuss how this perception is processed when focusing on action. We then present theoretical advances in this field and how these can explain the bidirectional link between perception and action from an auditory perspective. To round off this chapter, we discuss empirical studies that have explored the effects of natural as well as artificial movement sounds on the perception, learning, and control of bodily movements.

## AUDITORY PERCEPTION

Hearing is the ability to perceive sounds by detecting changes in the air pressure (or other medium) which are caused by vibrations. It is one of the remote senses, that is, it is used to collect information from a distance. These vibrations (sound waves), captured by the ear, are transduced into nerve impulses in the cochlea. Afterward, multiple brain areas such as the auditory cortex of the temporal lobe perceive the nerve impulses. Humans are able (with intra- and extra-individual variation) to hear sounds with a frequency between 20 Hz and 16 kHz and sound pressure between 0 and 120 dB (Faller & Schünke, 2012). This ability to detect and translate sound waves is highly sensitive, even to the tiniest changes in the intensity or type of sound wave. However, this is not the only important characteristic of the auditory system. In addition, humans are able to segregate an auditory percept into different streams (Bregman, Liao, & Levitan, 1990). This segregation happens hierarchically to obtain a meaningful percept. For example, one can follow what someone is saying at a loud party or focus on a specific instrument in an orchestra. People can separate sounds into meaningful streams because of the cognitive structures of the brain that enable them to perceive a relationship between several tones and then, for instance, as a result of pattern recognition, perceive music, which goes beyond the pure perception of individual sound waves. Auditory signals are processed in both a temporal and a spectral way. The temporal analysis in the inner ear and the spectral encoding (in fundamentals and harmonics) lead to a specific "neural card" (Lopez-Poveda, Palmer, & Meddis, 2010). This card is unconsciously classified and compared to familiar auditory files in the auditory cortex. Specific perceptual codes are responsible for the corresponding sound experience.

## AUDITORY ACTION PERCEPTION

An action, whether it is a target-oriented, fine-motor grasping action, an aimless gross-motor walking activity, or an instinctive dance movement in time to a favorite piece of music, generates an auditory product. Movement and noise are, thus, rather directly connected; even the physical explanation for noise is a moving sound wave. The variation of sounds caused by movement is inexhaustible. If one imagines, for instance, different sports disciplines (e.g., hurdling, triple jump, or basketball) and the sounds that typically accompany them, this variation becomes obvious. The potential usefulness of auditory information caused by action might be that it allows an evaluation of one's own or an opponent's movement. Thus, it is possible to assign specific auditory percepts to physical characteristics caused by particular movements. In general, humans perceive loudness, pitch, and timbre of an auditory stimulus. The perceived loudness of a stimulus depends on the amplitude of a sound wave. This amplitude is mainly connected to the momentum of a movement (e.g., a tennis serve or the takeoff in long jump). The perceived pitch depends on the frequency of an auditory stimulus. This pitch is, for instance, a function of the ground characteristics an athlete is running on (e.g., mechanical oscillations of different Tartan tracks). The perceived timbre of a stimulus is the quality of a sound. Humans can distinguish two sounds with identical loudness and pitch based on timbre. Timbre is in the domain of action sounds that are not of decisive importance, because natural movement variation causes a disparity in the first two physical characteristics (i.e., loudness and pitch). We outline this interdependency between movement and perception in the following section.

## BIDIRECTIONAL ACTION PERCEPTION COUPLING

Descartes postulated in the seventeenth century that action and perception are independent and need a translator to connect the so-called "sensory codes" to the "motor codes." To act within an environment, visual perception, for instance, creates a retinal code, which has to be translated into motor reactions in specific body coordinates. This view changed in the nineteenth century with the work of Lotze (1852) and James (1890). They sought to explain that there is no elementary difference between perception and action. The ideomotor principle describes that "every mental representation of a movement awakens to some degree the actual movement which is its object" (James, 1890). Lotze (1852) postulated that "perceived or even imagined actions can affect the execution of corresponding movements." Both described a connection between action and perception, but the directionality and the manner have, so far, not been explored.

How perceptual representations and action representations are linked and contribute to each other can be described by a number of theories that have been introduced by Pizzera (Chapter 13) and will be readdressed in this section, with specific application for the auditory sense (Hommel, Müsseler, Aschersleben, & Prinz, 2001; Prinz, 1997; see also Pizzera, Chapter 13). For instance, the common coding theory claims that there is a shared representation for action and perception. Hearing someone running and performing this action activate common computational codes. These "event codes" and "action codes" (i.e., the sensory codes and motor codes mentioned previously) are represented in a common medium and need no translator, because they overlap. The common coding theory is useful to depict the interdependency between action and perception, but it does not explain the structure or characteristics of the codes. The theory of event coding emphasizes the anticipated or intended effect of actions. Feature codes are proposed as a representational medium for perceptual contents and action plans. These feature codes are not specific to a certain action but spread the information as a bundle of codes (event) to the motor system. It remains unclear how observers process the perception of externally generated action (watching or hearing someone doing something).

Simulation theory (Gallese & Goldman, 1998) assumes that one's own motor system is involved when people predict actions. The motor system is thought to be part of a simulation network that provides information about an observed action by reactivating previous self-generated actions. Wilson and Knoblich (2005) proposed that various brain areas translate a perceived movement into a motor program, which acts as an *emulator* (Grush, 2004). This emulator simulates the perceived movement internally. When perceiving a movement (in the form of auditory, visual, sensorimotor, or multisensory stimuli), people compare the present stimuli to representations stored in memory. These representations are available as forward models (Wolpert & Flanagan, 2001) running offline (Schubotz, 2007). Thus, when a movement sound (or a video) is presented to a participant in a perceptual experiment, the participant compares the stimulus (delayed reafference) to predicted feedback (corollary discharge). The higher the match between predicted and actual feedback, the better is the perceptual performance. Additionally, the agent of a perceived stimulus can be identified by noting the sensory discrepancy between predicted and actual feedback (Blakemore, Wolpert, & Frith, 2000): A slight sensory discrepancy leads to the attribution as being "self-generated." A high sensory discrepancy leads to an attribution as being "caused by an external agent." This predictive account of the motor system can be generalized from action to event perception (Schubotz, 2007).

## INTERNAL MODEL AND REAFFERENCES

Traditional feed-forward models (Keele, 1981) assumed that a necessary motor command is already present at the beginning of a movement. Corrections or adaptations are only possible in late phases of a movement. Current models, by contrast, emphasize the role of sensory feedback (Aschersleben, Gehrke, & Prinz, 2001) for motor control. These internal models (Wolpert & Flanagan, 2001) postulate that there is always a predicted effect of the action, which is compared to the real action to adapt the movement. Two categories of internal models can be distinguished: the inverse models and the forward models (Kawato, 1999).

The forward models suggest that, using only the motor command, it is possible to predict the output. A forward model of an arm movement, for example, predicts the position and velocity (output) on the basis of the actual position and the motor command. An efference copy represents the output. Afterward, the efference copy and actual feedback are compared. If there is little or no discrepancy, the movement is classified as self-generated. A discrepancy leads to a classification as being other-generated (Blakemore et al., 2000). Theoretically, it is a simple and direct model, but given the complexity of movements with additional time constraints (especially in sports situations) and the interaction with the environment, among other issues, this simplicity is essential. The forward model is not a settled structure. It needs to be trained through practice (Effenberg & Mechling, 1999).

Inverse models orient on the output. They calculate the necessary motor command to manage the task, acting as a controller. It is an inverse model, for example, that calculates the motor command when a person wants to move his or her arm with a specific velocity and direction. The aim of inverse models is to produce the most efficient movement, as shown by the principle of cost control (more detailed in Wolpert & Ghahramani, 2000). Forward models and inverse models are not independent. Complex tasks contain a combination of both. An inverse model acts as a controller, and a forward model as a predictor. That means the efference copy from an inverse model, which is related to the environment, can be used to predict the related output with the help of a forward model.

Von Holst and Mittelstaedt described this prediction of sensory consequences in the principle of reafferences in 1950. They postulated that every (motor) command contains a copy (efference copy) that remains in the central nervous system. This copy represents a reference value. The actual feedback (reafferences) from the periphery is then compared to the efference copy. If they are in accordance with each other, the copy is deleted and the movement is considered as executed. If the reafference and the efference copy are not coincident, the execute command remains, a so-called exafference. This happens when the environment is not as expected and gives the organism the possibility to make an appropriate change. Von Holst and Mittelstaedt illustrated this principle with the help of the visual system of a fly (*eristalis*). In their study, they rotated the head,

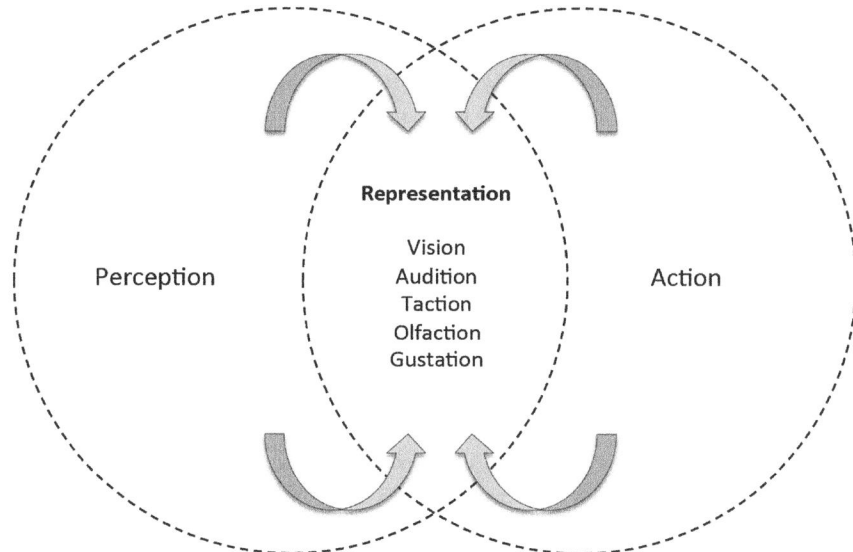

**FIGURE 1** The multisensory, bidirectional dependency of action and perception.

and therewith the visual system of the fly, by 180°. The optokinetic response showed compensation movement based on the incongruity of the efference copy and the reafference. Other studies showed comparable results for other senses. Bell applied the principle of reafferences to electric fish (*Gnathonemus petersii*). He showed reafferences in neuronal reactions in their electrosensory lateral line lobe to counteract a self-produced electric field (Bell, Myers, & Russell, 1974). Blakemore et al. (2000) demonstrated the principle with the fact that it is not possible to tickle yourself. Poulet and Hedwig (2002) carried out an experiment with crickets (*Gryllus bimaculatus*). An efference copy inhibits an interneuron while they are singing. Due to this inhibition, the auditory reafferences do not activate the interneuron. Frith (2005) suggested that these abilities to control self-generated movements, sounds, and touch lead to a more general understanding of the principle. In summary, different theoretical approaches, focused more on either perception or action, claim an interdependency of action and perception on various levels. Figure 1 highlights this general understanding.

## EMPIRICAL EVIDENCE

Several investigations dealing with motor experience, motor intention, motor learning, and motor competence have supported the idea that perception and action interact bidirectionally (for an overview, see Schütz-Bosbach & Prinz, 2007). One common way to investigate this interaction is to evaluate performance by presenting self-generated and other-generated past (recorded in a previous experiment) movements in a perceptual experiment. The assumption is that participants possess motor experience (and therewith, internal representations) for self-generated movements, which should lead to a higher perceptual quality of self-generated actions. Much of the empirical evidence has been derived from studies in the visual domain. These studies have shown that a higher quality of action (expertise or experience) leads to a higher perceptual quality when watching self-related videos (videos of oneself) of dart throwing (Knoblich & Flach, 2001), point-light displays of self-generated basketball dribbling (Hohmann, Troje, Olmos, & Munzert, 2011), or variations of one's own full-body movements (Loula, Prasad, Harber, & Shiffrar, 2005).

There is also a line of research stressing the importance of action sounds. For instance, participants were able to recognize their own clapping sounds (Flach, Knoblich, & Prinz, 2004), even when only the temporal pattern of the original claps was presented by uniform tones. Another study indicated that pianists can better recognize and synchronize with their self-related recordings than those of others (Keller, Knoblich, & Repp, 2007). Another line of research, not directly in the field of motor control but worth mentioning because it stresses the influence of auditory stimuli during the movement, focused on the other direction of the action–perception coupling, namely, the influence of perception on action. Examining the effect of auditory stimuli on the performance of a movement task, studies revealed that, for instance, sprint performance (Simpson & Karageorghis, 2006) or running endurance (Karageorghis et al., 2009) were significantly better when participants heard motivational music compared to no music. It remains unclear, however, if this increased performance is the result of rhythmic information embedded in the music or a positive effect on the mood of the athlete. Another study on running performance showed that auditory feedback, in the form of auditory commands during the movement, enhanced performance (Eriksson, Halvorsen, & Gullstrand, 2011). To sum up, the perception of auditory stimuli seems to have a great impact on action execution.

To understand what happens during the auditory perception of action in the brain and to determine if the mechanisms are similar to those observed during visual perception of action, neurophysiological studies have been conducted. Most of the studies in this domain use musicians because they create sound patterns by using hand movements and, therefore, show a tight coupling between sound and movement. The majority of these studies have suggested that there is some form of "action-listening" or "hearing-doing" mechanism in individuals (Bangert et al., 2006; Baumann et al., 2007; Haslinger et al., 2005; Lotze, Scheler, Tan, Braun, & Birbaumer, 2003). For instance, in a functional magnetic resonance imaging (fMRI) study, Lahav, Saltzman, and Schlaug (2007) trained nonmusicians to play a piece of music by ear (without music sheets). Subsequently, they monitored brain activity while the same participants listened to the newly acquired piece. Although participants did not perform any actions while listening, activation was found bilaterally in the frontoparietal motor-related network, coinciding with the neural circuits related to action observations. If the notes were presented in a different order or in an equally familiar but motorically unknown order, this network was much less activated, if at all. The behavioral data in this study revealed that participants increased their pitch recognition from 24% to 77% as a by-product of learning to play by ear. It would have been interesting, however, to investigate differences not only in the learning process, but also in the pitch recognition between participants learning to play the piece by ear and participants learning from notes. Nevertheless, similar results were reported in an fMRI study involving tapping actions accompanying musical rhythms (Chen, Penhune, & Zatorre, 2008). Furthermore, in studies using more complex whole-body movements, researchers investigated the neural substrate of biological motion perception while participants listened to footsteps (Bidet-Caulet, Voisin, Bertrand, & Fonlupt, 2005). Besides an activation in the auditory cortex, an auditory attentional network was shown to share frontal and parietal areas previously found in visual attention paradigms. In addition, there was activation in a posterior superior temporal sulcus (STSp) region, overlapping the same temporal human motion area found with visual input. The authors, therefore, proposed that independent of the sensory modality input, an area of the STSp region might be "supramodal." Similarly, other studies also found that the STSp responds to visual, auditory, and tactile stimulation (Barraclough, Xiao, Baker, Oram, & Perrett, 2005; Bruce, Desimone, & Gross, 1981). These and similar findings on the multisensory-based approach were summarized and reviewed by Làdavas (2008). She concluded that brain regions that are similar and/or neighboring the regions used in unimodal perception process the integration of different sensory modality inputs. She further pointed out that adaptive advantages should be considered with regard to neurorehabilitation, for instance, for the recovery of unisensory deficits.

One function of such multisensory-integrative networks, in which the same brain areas are activated during action perception and different modalities, might be the facilitation of movement patterns. Because basic human movement itself already produces a variety of auditory information, in the next section, we focus on everyday actions such as walking as well as naturally occurring sounds in sports.

## Auditory Action Perception with Natural Movement Sounds

When individuals walk, they usually make sounds, resulting from the contact phase of the feet meeting the ground. Such footstep sounds can contain valuable information about the person walking (e.g., sex or mood) but also about the surface the person is walking on (Ekimov & Sabatier, 2006; Saarela & Hari, 2008). Studies examining the interaction of acoustic and visual information during walking have come to similar conclusions. People are quite accurate (average of 96%) in perceiving the motion direction of a walker moving across the auditory scene when listening to his or her footsteps (Bidet-Caulet et al., 2005). They are also very accurate in visually identifying and detecting walking humans (for a review, see Blake & Shiffrar, 2007). The interaction between modalities, however, becomes obvious when individuals make use of auditory cues during the visual analysis of human motion. Specifically, individuals showed increased sensitivity to visual displays when auditory cues (footstep sounds) were present (Thomas & Shiffrar, 2010). This was not the case if visual displays were presented with simple tones, which leads to the conclusion that the use of auditory cues might apply only to meaningful sounds or even natural movement sounds. In addition, a study using tap dancing as sounds demonstrated that these movement sounds need to be in synchrony (temporally coincident) with the visual display, showing the functional combination of audiovisual information (Arrighi, Marini, & Burr, 2009).

Studies examining complex whole-body movements found similar results. Golfers were able to discriminate sound recordings of their self-generated movement from recordings of others' movements, except when the other person's sounds had nearly the same temporal factors (Murgia, Hohmann, Galmonte, Raab, & Agostini, 2012). Agostini, Righi, Galmonte, and Bruno (2004) showed that hammer throwers optimized their performance via training with auditory feedback. The sounds were generated by the movement of the hammer flying through the air. Recent research confirmed that naturally occurring movement sounds in the form of auditory reafferences contain useful information in many respects. The information delivered by movement sounds plays a role in both action and perception. In a large research project,[1] hurdling was used as a representative of a complex whole-body movement that offers a rhythmical step structure and amplitude range that is audible during performance. The experiments demonstrated, on the one hand, the interdependency of motor experience and perceptual ability and, on the other hand, the influence of auditory reafferences on motor performance during a complex movement task (Kennel, Hohmann, & Raab, 2014). More specifically, people were better in recognizing their own past movement sounds than movement sounds from others. This means with increasing quality of action and related representations, auditory perception increases. Furthermore, this perception is not based on specific auditory features. Manipulations of temporal structure or amplitude range did not decrease perceptual performance (Kennel et al., 2014). Auditory perception has a specific influence on action (motor control) as well. The quality of action in a complex movement task deteriorated significantly with a manipulation of the auditory feedback (Kennel et al., 2015). Hence, simple sounds that occur as a by-product of human movement can help reveal the interdependency of action and perception.

The above-described studies showed that individuals use auditory reafferences to perceive and control bodily actions. A mismatch between the predicted sensory consequences and the actual auditory feedback can lead to adaptive changes and performance declines if feedback is delayed.

---

1. This large research project was supported by the German Research Foundation (DFG) – ID RA 940/13-1.

## Auditory Action Perception with Artificial Movement Sounds (Sonification)

Most of the research described above was conducted with natural movement sounds. The question remains what, exactly, is perceived with regard to the movement sounds. As the hurdle study by Kennel et al. (2014) showed, rather than specific features of the sound being relevant, the whole combination of sound features appeared to contribute, leading to the idea of a Gestalt sound. In addition, it remains unclear whether this is true only for natural movement sounds or could also apply to artificial sounds. In the following sections, we discuss this aspect in more detail. We describe the processing of artificial movement sounds (i.e., sonification), reporting on the development of this research field and several studies.

Sonification is described as synthetically generated acoustic information (Kramer, 1994; Schaffert, Mattes, & Effenberg, 2010). By displaying kinematic parameters acoustically, numerical data are made audible. Due to its measurement system, sonification can display acoustic information online (concurrent to action execution), enabling athletes to sensitize to the rhythm of the movement. With regard to its implementation, usually acceleration measures of material (e.g., boat during rowing) or body parts (e.g., arms during breaststroke) are recorded. These are then mapped to specific tones on the musical tone scale and related to pitch. For this so-called sonification method of parameter mapping (Hermann, 2008; Kramer et al., 1998), the tone scale refers to the Musical Instrument Digital Interface (MIDI) standard. For instance, using the example of on-water rowing, the point of zero boat acceleration is represented by the middle C on the Western musical tone scale. Therefore, the tone pitch increases with increasing boat acceleration and decreases with decreasing boat acceleration (for more detailed information, see Schaffert, Mattes, & Effenberg, 2011). Microelectromechanical acceleration sensors are used to measure acceleration. The data are transformed with software such as Pure Data, enabling real-time graphical programming techniques (Puckette, 2007). A systematic review of mapping strategies for sonification concluded that pitch is the most used auditory dimension in sonification applications, mostly employed to "sonify" kinematic quantities (Dubus & Bresin, 2013).

In this context, researchers developed a real-time sonification framework for all common MIDI environments based on temporal (acceleration) and spatial (orientation) data from inertial sensors (Brock, Schmitz, Baumann, & Effenberg, 2012). Brock et al. proposed a five-step process for preparing effective auditory stimuli to be used as feedback to support motor learning:

Step 1. Motion performance
Step 2. Motion data stream (inertial motion capturing)
Step 3. Motion data representations and kinematic parameter selection
Step 4. MIDI control message (transforming motion data into sound)
Step 5. Auditory feedback

For more detailed information on the different steps, please see Brock et al.

Studies examining the link between action and perception using auditory information through sonification have typically distinguished effects on motor perception, motor control, and motor learning. In the following section, we review and discuss research for each of these areas.

*Motor Perception*

Studies have shown that individuals are able to correctly perceive and identify movement patterns through sonification (Effenberg, 2005; Schaffert et al., 2011). A study combining behavioral and neurophysiological data examined the effect of sonification of a breaststroke movement on action perception (Schmitz et al., 2013). Healthy, young nonathletes were asked to judge differences in the movement velocity of a human avatar performing the breaststroke in swimming. In the congruent condition, the visual display was accompanied by movement sonification, matching two kinematic parameters of the visual model. In the incongruent condition, two different chords were linked to the visual model that showed similar timbre and pitch ranges as in the congruent condition, but with no correspondence between chord switching and movement kinematics. The results revealed that participants were better able to judge differences in movement velocity in the congruent condition. These results were supported by neurophysiological data in that fMRI showed enhanced activity in areas known to be important multisensory sites (superior and medial posterior temporal regions) and the insula bilaterally and the precentral gyrus on the right side. These results were in accordance with a study using a similar design but a different movement, namely, the counter movement jump (Scheef et al., 2009). Enhanced activation was found in area V5/MT bilaterally, if visual movement displays of the counter movement jump were accompanied concordantly by sonified movements (multimodal condition), compared to the unimodal purely visual or purely auditory conditions. Even with artificial stimuli, this brain region seems to play a role in multimodal motion integration, also providing some explanation of the facilitatory effects of motor perception in behavioral studies.

A study with everyday upper limb actions examined how, exactly, sonification should be configured to gain information-rich sound sequences (Vinken et al., 2013). According to the authors, it is still unclear what movement feature should be selected, what kinematic–acoustic mapping pattern is effective, and how many dimensions should be considered. In a repeated-measures design, participants were asked to discriminate six everyday upper limb actions, which were sonified using seven kinds of kinematic–acoustic mappings. Participants showed comprehensibility of the artificial movement sounds as well as short-term learning effects. There were no differences between different dimensional encodings, indicating a high efficiency for intermodal pattern discrimination. Taken together, these two studies show that individuals are able to correctly perceive sonified movement, even human gross motor movements. In the next section, we explore if this can be used in motor control and learning as well.

## Motor Control and Learning

In a study with rowers, athletes received acoustic information on the boat's forward motion (acceleration) during on-water training sessions using the sonification method (Schaffert et al., 2011). Six junior national rowing athletes received online-generated sound that characterized the rhythm of the rowing cycle related to tone pitch. The results revealed significant boat velocity increases for the conditions in which acoustic feedback was given. Interestingly, the participants preferred transmission via loudspeakers, so that they could still perceive natural rowing sounds in addition. This goes along nicely with the results presented in the first section of this chapter. The method of sonification was also applied in dance, using auditory information in the learning process of two different dance moves (Grosshauser, Bläsing, Spieth, & Hermann, 2011). Dancers received real-time sonified information on selected biomechanical parameters such as foot pressure, knee angles, and foot acceleration. A self-developed sensor kit attached to the body of the dancer (sensors fixed on the foot, insole soft pads, and force-sensitive resistor inserted into the plastic foam of the shoe, and a goniometer attached to the knee joint) served as an apparatus for data acquisition. The sensor data showed differences between correct and incorrect executions that were not clearly visible from pure visual observation, as is usually provided by a coach. This enabled feedback on errors that are often not clearly observable, making error correction possible from trial to trial. Self-reports by dancers and coaches also showed that the dancers improved their trained moves with sonification and also gave positive feedback on its usability. However, a limitation of this study is the lack of empirical proof. In gymnastics, sonification was used to support the learning process of the circles performance on a pommel horse. Performance increased significantly with the use of this feedback system during training, which provided trajectory information in the form of an auditory signal (Baudry, Leroy, Thouvarecq, & Chollet, 2006).

Possible explanations for such perception–action links in the auditory domain derive from studies on sensorimotor adaptation. Typically, participants are asked to point at targets while visual feedback of the hand is distorted. Due to this distortion, pointing accuracy decreases and then gradually recovers. Aftereffects then occur as soon as the distortion is removed, resulting in information on the recalibration of the sensorimotor system. Studies have shown that such aftereffects apply to both the visual and auditory modality (Craske, 1966; Kagerer & Contreras-Vidal, 2009). Specifically, independent of whether individuals received visual feedback with regard to visual targets or auditory feedback (continuous real-time feedback approximating the noise of an approaching and detonating grenade while the hand approaches the target) with regard to auditory targets, the magnitude of adaptive recalibration was similar (Schmitz & Bock, 2014).

Another study with a similar focus—although with slightly different stimulus generation—was conducted in weight lifting (Murgia, Sors et al., 2012). In a within-subject design, participants performed three bench lifts, each in an experimental condition and a control condition (no stimulation). In the experimental condition, participants received an auditory stimulus that consisted of an initial countdown, a low-intensity sound corresponding to the down phase of the movement, and a high-intensity sound associated with the pressing phase. The participants showed higher exerted power during lifting in the experimental condition, taking the average of the three experimental trials. However, limitations of this study need to be considered. For instance, there was no variation in sound characteristics, and measures other than the maximal power could have provided more information on the specific effects of this auditory information (i.e., physiological parameters). In addition, stimuli were not linked to the participants' performance in the form of continuous real-time feedback, as is usually found during online sonification.

## Outlook in Applied Fields

Auditory information generated by actions has received little attention, but recent research (Kennel et al., 2014, 2015) suggests there are several practical implications. Applications are not limited to the sports domain but are also valid for rehabilitation or prevention. On the one hand, auditory feedback allows coaches, therapists, or individuals themselves to evaluate the execution of movements quite well—for in rhythmic, fast, or cyclic/continuous movements, this auditory feedback is very effective. On the other hand, auditory feedback also influences movement execution. Within the sports domain, the area of application is obvious. Coaches have already reported using auditory feedback to assess movement, but for many of them, this means giving athletes verbal feedback on movement quality or specific phases of the movement. In contrast, natural auditory feedback is not yet widely used. Here, we refer to intrinsic auditory feedback, that is, the auditory information that is produced through movement itself (natural movement sounds) or through sonification methods (kinematic movement parameters derived from the movement itself).

So far, auditory training has been based on artificial and information-poor sounds to train rhythmical movement. Auditory reafferences, however, contain more dimensional information, and their application may result in qualitatively higher training success. Future research should investigate the influence of auditory stimuli in areas such as rehabilitation or specific movement relearning. In this context, the consideration of auditory reafferences might help to unravel hidden movement concepts in an additional sense. In short, the practical implications of auditory reafferences seem to be manifold but so far still mostly unexplored, and many questions remain.

For instance, even if positive effects of acoustic feedback were evident in different studies, what should be the frequency of exposure? How often should athletes receive such information? What about retention? If acoustic feedback influences action perception and execution, how long do these effects last? For how long should feedback be given to achieve certain effects, and for how long will these effects last? Are recommendations similar to those in the field of visual feedback on motor learning and control?

In sports, coaches might now ask, after receiving information about using auditory feedback, what they should use in the training process, audio or visual information. Currently, visualization plays the dominant role in technique analysis. However, there are limitations of visual observation (of time-sensitive events) not shared by auditory perception. For instance, especially for very fast movements, auditory information can be represented with a clearer time resolution, due to the higher processing speed of auditory versus visual perception (Shams, Kamitani, & Shimojo, 2000). In addition, auditory information offers ideal feedback of temporal structure and rhythm (MacPherson, Collins, & Obhi, 2009). The use of auditory feedback might be advantageous especially since the eyes are already busy during movement control. Moreover, if cross-modal effects are considered, the two modalities apparently interact for optimal information processing. For instance, six experiments showed that perceptual organization in the auditory modality can influence perceptibility in the visual modality (Vroomen & de Gelder, 2000). Participants received a high tone embedded in a sequence of low tones during a task to detect a visual target in a changing sequence of visual distractions. Controlling for other explanations by slightly changing the methodological procedures in a series of experiments, the authors concluded that the auditory information improved the detection of the visual target. Though some might argue that the development of such auditory information for training is time-consuming and not worth the effort, the advancement of new technology might open the field to new opportunities and ideas for training optimization.

## CONCLUSION AND FUTURE RESEARCH

Research on the perception–action link, so far, has shown an obvious bias toward the visual and sensorimotor domains, neglecting the potential contribution of auditory processing to action perception and motor control. However, simple and clear examples are numerous when considering the possible performances of blind athletes or the variety of sounds perceivable in a well-attended sports hall. The studies presented in this chapter have shown that both natural and artificial movement sounds support the perception, learning, and control of human movements. The fact that some features of movement sounds can be represented in a similar way becomes obvious when rhythmic structure is considered. Although natural movement sounds seem closest to action execution, synthetically generated acoustic information can contain the same acoustic structure. What is not yet known is if the mechanisms found for natural movement sounds are similar to those found in sonification studies. What feature of the sound is actually perceived or should be sonified for enhanced action perception and execution? Is it the combination of all features in a sound, as suggested by studies from Kennel et al. (2014) for natural movement sounds, or are specific features such as amplitude or pitch (representing certain angles or accelerations such as in sonification) of more importance? Is this different depending on whether natural or artificial movement sounds are presented?

Future studies could explore these questions from a behavioral as well as a neurophysiological point of view. A major difference that becomes obvious just by asking these questions is that sonification enables access to otherwise mute phases of a movement. When recording movement sounds, biomechanical parameters such as angles cannot be perceived from sounds. This is where sonification offers clear advantages.

To sum up, actions are represented similarly, independent of whether they are performed, seen, or heard. The current chapter has shown that cross-modal effects are apparent when it comes to perceiving, learning, and controlling human movements and that these cross-modal effects expand to the interaction of perception and action, especially in the domain of sports. Therefore, when we hear or see someone running along a track, a slight dragging sound of the runner's foot on the ground might not only cause the runner to adjust his or her running technique, but might also provoke some change in our own next footstep.

## REFERENCES

Agostini, T., Righi, G., Galmonte, A., & Bruno, P. (2004). The relevance of auditory information in optimizing hammer throwers performance. In P. B. Pascolo (Ed.), *Biomechanics and sports* (pp. 67–74). Vienna, Austria: Springer.

Arrighi, R., Marini, F., & Burr, D. (2009). Meaningful auditory information enhances perception of visual biological motion. *Journal of Vision, 9*(25), 1–7.

Aschersleben, G., Gehrke, J., & Prinz, W. (2001). Tapping with peripheral nerve block. *Experimental Brain Research, 136*(3), 331–339.

Bangert, M., Peschel, T., Schlaug, G., Rotte, M., Drescher, D., Hinrichs, H., et al. (2006). Shared networks for auditory and motor processing in professional pianists: evidence from fMRI conjunction. *Neuroimage, 30*, 917–926.

Barraclough, N. E., Xiao, D., Baker, C. I., Oram, M. W., & Perrett, D. I. (2005). Integration of visual and auditory information by superior temporal sulcus neurons responsive to the sight of actions. *Journal of Cognitive Neuroscience, 17*, 377–391.

Baudry, L., Leroy, D., Thouvarecq, R., & Chollet, D. (2006). Auditory concurrent feedback benefits on the circle performed in gymnastics. *Journal of Sports Sciences, 24*, 149–156.

Baumann, S., Koeneke, S., Schmidt, C. F., Meyer, M., Lutz, K., & Jancke, L. (2007). A network for audio-motor coordination in skilled pianists and non-musicians. *Brain Research, 1161*, 65–78.

Bell, C. C., Myers, J. P., & Russell, C. J. (1974). Electric organ discharge patterns during dominance related behavioral displays in *Gnathonemus petersii* (Mormyridae). *Journal of Comparative Physiology, 92*, 201–228.

Bidet-Caulet, A., Voisin, J., Bertrand, O., & Fonlupt, P. (2005). Listening to a walking human activates the temporal biological motion area. *Neuroimage, 28*, 132–139.

Blake, R., & Shiffrar, M. (2007). Perception of human motion. *Annual Review of Psychology, 58*, 47–73.

Blakemore, S. J., Wolpert, D., & Frith, C. (2000). Why can't you tickle yourself? *Neuroreport, 11*, 11–16.

Bregman, A. S., Liao, C., & Levitan, R. (1990). Auditory grouping based on fundamental frequency and formant peak frequency. *Canadian Journal of Psychology, 44*, 400–413.

Brock, H., Schmitz, G., Baumann, J., & Effenberg, A. O. (2012). If motion sounds: movement sonification based on inertial sensor data. *Procedia Engineering, 34*, 556–561.

Bruce, C., Desimone, R., & Gross, C. G. (1981). Visual properties of neurons in a polysensory area in superior temporal sulcus of the macaque. *Journal of Neurophysiology, 46*, 369–384.

Chen, J. L., Penhune, V. B., & Zatorre, R. J. (2008). Listening to musical rhythms recruits motor regions of the brain. *Cerebral Cortex, 18*, 2844–2854.

Craske, B. (1966). Intermodal transfer of adaptation to displacement. *Nature, 210*, 765.

Dubus, G., & Bresin, R. (2013). A systematic review of mapping strategies for the sonification of physical quantities. *PLoS One, 8*, 1–28.

Effenberg, A. O. (2005). Movement sonification: effects on perception and action. *IEEE Multimedia, 12*(2), 53–59.

Effenberg, A. O., & Mechling, H. (1999). The contribution of motor-acoustic information to motor learning. In P. Parisi, F. Pigozzi, & G. Prinzi (Eds.), *Sport science '99 in Europe. Proceedings of the 4th annual congress of the ECSS* (p. 780). Rome, Italy.

Ekimov, A., & Sabatier, J. M. (2006). Vibration and sound signatures of human footsteps in buildings. *The Journal of the Acoustical Society of America, 120*, 762.

Eriksson, M., Halvorsen, K. A., & Gullstrand, L. (2011). Immediate effect of visual and auditory feedback to control the running mechanics of well-trained athletes. *Journal of Sports Sciences, 29*, 253–262.

Faller, A., & Schünke, M. (2012). *Der Körper des Menschen: Einführung in Bau und Funktion.* Georg Thieme Verlag.

Flach, R., Knoblich, G., & Prinz, W. (2004). Recognizing one's own clapping: the role of temporal cues. *Psychological Research, 69*, 147–156.

Frith, C. (2005). The self in action: lessons from delusions of control. *Consciousness and Cognition, 14*, 752–770.

Gallese, V., & Goldman, A. (1998). Mirror neurons and the simulation theory of mind-reading. *Trends in Cognitive Sciences, 2*, 493–501.

Grosshauser, T., Bläsing, B., Spieth, C., & Hermann, T. (2011). Wearable sensor based real-time sonification of motion and foot pressure in dance teaching and training. *Journal of the Audio Engineering Society, 1*, 1–10.

Grush, R. (2004). The emulation theory of representation: motor control, imagery, and perception. *Behavioral and Brain Sciences, 27*, 377–442.

Haslinger, B., Erhard, P., Altenmüller, E., Schroeder, U., Boecker, H., & Ceballos-Baumann, A. O. (2005). Transmodal sensorimotor networks during action observation in professional pianists. *Journal of Cognitive Neuroscience, 17*, 282–293.

Hermann, T. (2008). Taxonomy and definitions for sonification and auditory display. In *Proc. 14th int. conference on auditory display (ICAD)*. Paris.

Hohmann, T., Troje, N. F., Olmos, A., & Munzert, J. (2011). The influence of motor expertise and motor experience on action and actor recognition. *Journal of Cognitive Psychology, 23*, 403–415.

Hommel, B., Müsseler, J., Aschersleben, G., & Prinz, W. (2001). The theory of event coding (TEC). A framework for perception and action planning. *Behavioral and Brain Sciences, 24*, 849–878.

James, W. (1890). *The principles of psychology.* New York: Dover Publications.

Kagerer, F. A., & Contreras-Vidal, J. L. (2009). Adaptation of sound localization induced by rotated visual feedback in reaching movements. *Experimental Brain Research, 193*, 315–321.

Karageorghis, C. I., Mouzourides, D. A., Priest, D. L., Sasso, T. A., Morrish, D. J., & Walley, C. L. (2009). Psychophysical and ergogenic effects of synchronous music during treadmill walking. *Journal of Sport & Exercise Psychology, 31*, 18–36.

Kawato, M. (1999). Internal models for motor control and trajectory planning. *Current Opinion in Neurobiology, 9*, 718–727.

Keele, S. W. (1981). Behavioral analysis of motor control. In V. Brooks (Ed.), *Handbook of physiology: Motor control*. Washington, DC: American Physiological Society.

Keller, P. E., Knoblich, G., & Repp, B. H. (2007). Pianists' duet better when they play with themselves: on the possible role of action simulation in synchronization. *Consciousness and Cognition*, *16*, 102–111.

Kennel, C., Hohmann, T., & Raab, M. (2014). Action perception via auditory information: agent identification and discrimination with complex movement sounds. *Journal of Cognitive Psychology*, *26*, 157–165.

Kennel, C., Pizzera, A., Hohmann, T., Schubotz, R. I., Murgia, M., Agostini, T., et al. (2014). The perception of natural and modulated movement sounds. *Perception*, *43*, 796–804.

Kennel, C., Streese, L., Pizzera, A., Justen, C., Hohmann, T., & Raab, M. (2015). Auditory reafferences: the influence of real-time feedback on movement control. *Frontiers in Psychology*, *6*, 69–75.

Knoblich, G., & Flach, R. (2001). Predicting the effects of actions: interactions of perception and action. *Psychological Science*, *12*, 467–472.

Kramer, G. (Ed.). (1994). *Auditory display: Sonification, audification, and auditory interfaces. Santa Fe Institute studies in the sciences of complexity. Proceedings, volume XVIII*. Reading: Addison-Wesley.

Kramer, G., Walker, B., Bonebright, T., Cook, P., Flowers, J., Miner, N., et al. (1998). *Sonification report: Status of the field and research agenda*. NSF Sonification White Paper– Master.

Lahav, A., Saltzman, E., & Schlaug, G. (2007). Action representation of sound. *Journal of Neuroscience*, *27*, 308–314.

Làdavas, E. (2008). Multisensory-based approach to the recovery of unisensory deficit. *Annals of the New York Academy of Sciences*, *1124*, 98–110.

Lopez-Poveda, E. A., Palmer, A. R., & Meddis, R. (Eds.). (2010). *The neurophysiological bases of auditory perception*. Springer.

Lotze, H. (1852). *Medizinische Psychologie: oder Physiologie der Seele*. Weidmann.

Lotze, M., Scheler, G., Tan, H. R. M., Braun, C., & Birbaumer, N. (2003). The musician's brain: functional imaging of amateurs and professionals during performance and imagery. *Neuroimage*, *20*, 1817–1829.

Loula, F., Prasad, S., Harber, K., & Shiffrar, M. (2005). Recognizing people from their movement. *Journal of Experimental Psychology: Human Perception and Performance*, *31*, 210–220.

MacPherson, A. C., Collins, D., & Obhi, S. S. (2009). The importance of temporal structure and rhythm for the optimum performance of motor skills: a new focus for practitioners of sport psychology. *Journal of Applied Sport Psychology*, *21*, 48–61.

Murgia, M., Hohmann, T., Galmonte, A., Raab, M., & Agostini, T. (2012). Recognising one's own motor actions through sound: the role of temporal factors. *Perception*, *41*, 976–987.

Murgia, M., Sors, F., Vono, R., Franco Muroni, A., Delitala, L., Di Corrado, D., et al. (2012). Using auditory stimulation to enhance athletes' strength: an experimental study in weightlifting. *Review of Psychology*, *19*, 13–16.

Poulet, J. F., & Hedwig, B. (2002). A corollary discharge maintains auditory sensitivity during sound production. *Nature*, *418*, 872–876.

Prinz, W. (1997). Perception and action planning. *European Journal of Cognitive Psychology*, *9*, 129–154.

Puckette, M. (2007). *The theory and technique of electronic music*. World Scientific Publishing Co. Pte. Ltd.

Saarela, M. V., & Hari, R. (2008). Listening to humans walking together activates the social brain circuitry. *Social Neuroscience*, *3*, 401–409.

Schaffert, N., Mattes, K., & Effenberg, A. O. (2010). A sound design for acoustic feedback in elite sports. In S. Ystad, M. Aramaki, R. Kronland-Martinet, & K. Jensen (Eds.), *Auditory display. 6th international symposium, CMMR/ICAD 2009*. (pp.143–165). Berlin, Germany: Springer.

Schaffert, N., Mattes, K., & Effenberg, A. O. (2011). An investigation of online acoustic information for elite rowers in on-water training conditions. *Journal of Human Sport & Exercise, 6*(2), 392–405.

Scheef, L., Boecker, H., Daamen, M., Fehse, U., Landsberg, M. W., Granath, D., et al. (2009). Multimodal motion processing in area V5/MT: evidence from an artificial class of audio-visual events. *Brain Research, 1252*, 94–104.

Schmitz, G., Bahram, M., Hammer, A., Heldmann, M., Samii, A., Münte, T. F., et al. (2013). Observation of sonified movements engages a basal ganglia frontocortical network. *BMC Neuroscience, 14*(32), 1–11.

Schmitz, G., & Bock, O. (2014). A comparison of sensorimotor adaptation in the visual and in the auditory modality. *PLoS One, 9*, e107834.

Schubotz, R. I. (2007). Prediction of external events with our motor system: towards a new framework. *Trends in Cognitive Sciences, 11*, 211–218.

Schütz-Bosbach, S., & Prinz, W. (2007). Perceptual resonance: action-induced modulation of perception. *Trends in Cognitive Sciences, 11*, 349–355.

Shams, L., Kamitani, Y., & Shimojo, S. (2000). Illusions: what you see is what you hear. *Nature, 408*, 788.

Simpson, S. D., & Karageorghis, C. I. (2006). The effects of synchronous music on 400-m sprint performance. *Journal of Sports Sciences, 24*, 1095–1102.

Thomas, J. P., & Shiffrar, M. (2010). I can see you better if I can hear you coming: action-consistent sounds facilitate the visual detection of human gait. *Journal of Vision, 10*, 1–11.

Vinken, P. M., Kröger, D., Fehse, U., Schmitz, G., Brock, H., & Effenberg, A. O. (2013). Auditory coding of human movement kinematics. *Multisensory Research, 26*, 533–552.

Von Holst, E., & Mittelstaedt, H. (1950). The principle of reafferences. *Naturwissenschaften, 37*, 464–476.

Vroomen, J., & de Gelder, B. (2000). Sound enhances visual perception: cross-modal effects of auditory organization on vision. *Journal of Experimental Psychology: Human Perception and Performance, 26*, 1583–1590.

Wilson, M., & Knoblich, G. (2005). The case of motor involvement in perceiving conspecifics. *Psychological Bulletin, 131*, 460–473.

Wolpert, D. M., & Flanagan, J. R. (2001). Motor prediction. *Current Biology, 11*, 729–732.

Wolpert, D. M., & Ghahramani, Z. (2000). Computational principles of movement neuroscience. *Nature Neuroscience, 3*, 1212–1217.

Chapter 16

# Visual Perception in Expert Action

Rita de Oliveira
*School of Applied Sciences, London South Bank University, London, UK*

*Vision is one of the six sensory systems that we use to know and interact with our environment but has been singled out as the most important form of exteroception for motor control. The reason for this implicit upgrade is probably that many human actions are directed at objects or targets beyond our immediate physical contact. The only link between these objects and us is the pattern of light reflected from their surfaces, and yet, we act upon them with great ease. No doubt, humans make significant strides in establishing appropriate relations between perceptions and actions at early stages of their development. When my nephew Rodrigo was three months old, it took him considerable perseverance and a lot of jerky movements to finally grasp the toy my mother was patiently holding and rambling. But once the relations between perceptions and actions are better established, humans can be incredibly skillful at interacting with distant objects, even when the constraints imposed on the interaction are severe and a high degree of precision is required. Like many other sportive tasks, basketball shooting is characterized by tight temporal constraints, limited spatial variation, and high accuracy demands. How basketball players manage to consistently throw a ball through the basket, even if severely challenged by their opponents, is a remarkable feat!*

The problem of visual perception and action has occupied scientists for years (Gibson, 1979/1986), and the present chapter is but another step in understanding the intricate relations between action and perception in such a context where few errors are allowed and few are made. Much of the research reviewed in the present chapter was conducted to uncover the visual basis of basketball shooting but is generalizable to other skills and different populations. Expert basketball shooting is a relevant skill because to succeed in it, a player needs to have developed a very close link between perception and action (Fajen, Riley, & Turvey, 2008). Therefore, understanding how this link works in detail will inform the mechanisms underlying other complex skills and the development of those skills. Basketball shooting consists of throwing a ball on a parabolic flight that passes through a metal rim twice the size of the ball at about 3-m height. Common shooting types are the free throw and the jump shot. Free throws are taken in less than 10s from the 4.6-m line without opposition. Jump shots can be taken from anywhere in the field, usually in the presence of opponents, and imply that the ball is released while the player is airborne. Conventional knowledge stipulates that *players must see the basket before they shoot*, but this can be incorrect in two ways. First, it is not granted that vision is required before the shot, as opposed to during the shot. While vision gathered before the movement may be useful, it may also be insufficient or unnecessary for accurate shooting. This temporal aspect is relevant because it gives insight into the timely interaction between visual perception and action. Second, it is not certain that the player must actually see the basket, as opposed to merely looking at it. The location of the target may be perceived through various information sources, not necessarily retinal ones. This spatial aspect is relevant because it gives insight into the optical basis of goal-directed movement. In what follows, I describe in more detail these temporal and spatial aspects of visual perception and action, backed up with the relevant literature. Next, I briefly review the available literature on the visual perception of basketball shooting and introduce six experiments in which the temporal and spatial aspects of basketball shooting are investigated.

## STATE OF RESEARCH

### Temporal Aspects of Visual Perception and Action

One of the most intensive debates in the study of the visual control of action is whether visual information is used primarily for the planning of goal-directed movements or for the online guidance of such movements. At the base of this controversy are important philosophical and methodological differences (Abernethy & Sparrow, 1992; Gibson, 1979/1986; Williams, Davids, & Williams, 1999). Nevertheless, the theoretical arguments and experimental results that are put forward in support of either view are often appropriate, even though they are sometimes applicable only to a subset of actions. I start by presenting the arguments for the offline use of visual information, followed by the arguments for the online use of visual information. Next, I discuss the timing of visual information pick-up, or in other words, whether there is a moment relative to a given action when picking up visual information is preferred, sufficient, and optimal.

### *Offline Use of Visual Information for the Control of Action*

Several authors have argued that many movements are controlled on the basis of visual information gathered before movement initiation. Broadly speaking, three arguments have been advanced for the offline use of visual information in the control of action. The first argument is that visuomotor delays sometimes exceed the duration of the movement, which would render online corrections of the trajectory impossible. The visuomotor delay is the duration it takes for visual information to be used in motor control. Although it is a physiological latency period in the sense that stimulation must travel from the sensory system to the musculoskeletal system, it has mostly been measured in terms of changes to the movement kinematics in near aiming movements. Over time, improved techniques and accumulated research led to estimations of visuomotor delays as short as 80 ms, with the precise value depending on several factors including the type of task, the type of error correction, and the experience of the participant (Carlton, 1992). For example, an elbow extension executed at maximal velocity over 100 degrees on the horizontal plane would last approximately 130 ms (Kistemaker, van Soest, & Bobbert, 2006). Supposedly, in fast movements such as this, visual feedback plays a minimal role, implying that the movement is largely guided by visual information gathered before its initiation. Recent theoretical developments that qualify this argument (Desmurget & Grafton, 2000) will be discussed later.

The second argument is that in order to plan the parameters of movement execution, long fixations on the target would be required before movement initiation. This is a consistent finding in the literature. Longer target fixations are found in more complex tasks, and have been associated with higher levels of expertise and more accurate performance (Janelle et al., 2000; Vickers, 1996; Williams, Singer, & Frehlich, 2002), although badminton seems to be an exception (Abernethy & Russell, 1987). Invariably, these authors surmise that the amount of time spent in looking at an object or target reflects the duration and accuracy of movement parameterization. If the movement is to be planned in part or in total before its initiation, then it seems crucial to have enough time available to pick up advance visual information. Unfortunately, no attempts have been made of investigating the mechanism underlying the beneficial effects of longer fixations under dynamic task conditions (cf., Klostermann, Kredel, & Hossner, 2014).

The third argument is that movement production and performance are largely similar when full vision is available during execution compared to when vision is eliminated. A movement executed in total darkness will exhibit the same characteristic kinematic features as in good lighting conditions such that it can be identified as a kick, a punch, or a throw. Also, its endpoint will likely be in the vicinity of the target; thus, in many tasks, visual information pickup during execution has been considered an advantage rather than a necessity (Williams et al., 1999). In this connection, authors have argued that through practice, performers may become less dependent on the use of online visual information (but, see Proteau, 1992), either because the need to make corrections during movement execution is reduced (Schmidt, 1988), or because reliance on other sources of sensory information is increased (Bennet & Davids, 1995; Davids, Palmer, & Savelsbergh, 1989; Fleishman & Rich, 1963).

## *Online Use of Visual Information for the Control of Action*

Another group of authors have argued that movements are controlled on the basis of visual information picked up during movement execution. In general, four arguments have been used for the online use of visual information in the regulation of action. The first argument is that in the absence of online vision, performance always deteriorates to some extent. Visual occlusion during movement execution has been demonstrated to have detrimental effects in a broad variety of actions. One reason is that, under such circumstances, corrective adjustments to the ongoing trajectory cannot be effectuated on the basis of updated visual information. The consequence is increased endpoint variability and error in both near- and far-aiming tasks (Khan et al., 2006; Oudejans, van de Langenberg, & Hutter, 2002; Westwood, Heath, & Roy, 2003). Another reason for the decline of performance due to the removal of online visual information is the resulting increase in sway reflecting worse postural control and balance (Blanchard et al., 2007; Herpin et al., 2010; Robertson, Tremblay, Anson, & Elliott, 2002; Sullivan & Hooper, 2005).

The second argument is that movement itself generates information, which amounts to saying that perception and action are functionally coupled in the attainment of a particular goal. Constancies in the relation between movement and perception, for instance the fact that the eyes converge more when looking at a closer object, are reliable information sources that can guide movement (Gibson, 1979/1986). In learning to move in a particular setting, children engage in exploratory behavior revealing persisting and changing characteristics of the environment and objects therein (Adolphe, 1995; Mark, Jiang, King, & Paasche, 1999; de Oliveira & Wann, 2011). Later, performers may exploit the learned links between perception and action to successfully perform challenging tasks like crossing a street or running to catch a fly ball (Michaels & Oudejans, 1992; Oudejans, Michaels, Bakker, & Davids, 1999; Oudejans, Michaels, van Dort, & Frissen, 1996). In tasks where perception and action are functionally coupled in real time, online vision may be the only means to attain the goal because, in part, the relevant visual information is brought about through movement (e.g., Bootsma & van Wieringen, 1990; Lee, Lishman, & Thomson, 1982; de Oliveira et al., 2009a, 2009c).

A related argument is that in more dynamic tasks and situations events occur in real time; thus, relevant information appears on the fly. Also, the relations between performer and object may change unpredictably, implying that actions need to be constantly adjusted to comply with novel or more precise information. For instance, in driving or sailing, new and more detailed visual information is constantly appearing and needs to be acted upon, and the same holds for running over rough terrain or when negotiating obstacles while walking (de Oliveira & Wann, 2011; Patla & Greig, 2006; Reynolds & Day, 2005). Likewise, in shooting a basketball while being airborne, updated visual information is necessary for the guidance of actions in much the same way as it is necessary to overcome a defender, because the relative positions between performer and target are changing continuously (Esteves, de Oliveira, & Araújo, 2011; Oudejans et al., 2002; Raab, de Oliveira, & Heinen, 2009).

The fourth argument is that better spatial and temporal accuracy is achieved toward the end of goal-directed actions (e.g., Lee et al., 1982; Lee, Young, & Rewt, 1992). As we will see next, this may have several reasons. Some information sources, for instance binocular disparity, become more precise as the target object and the observer approach each other (Cutting & Vishton, 1995). In addition, toward the end of the movement, performers may resort to different sources of visual information than those that are used in performing the initial portion of the movement (Caljouw, van der Kamp, & Savelsbergh, 2004). Thus, later, rather than earlier, information sources may be more specific to the temporal and spatial demands of the task (Bootsma & van Wieringen, 1990; Caljouw et al., 2004). Also, because information is conveyed over time, it is likely that accuracy improves as the performer takes the time to perceive the event, aspects of the object or layout (Gibson, 1979/1986). For these reasons, visual information toward the end of the movement may have increased informational value for the accuracy of goal-directed actions, especially in a dynamic environment. In conclusion, it seems that both offline and online visual information are used in the control of complex, goal-directed movements, but their relative contribution may depend on characteristics of both task and performer. While visual information gathered before movement initiation might be used to direct actions with relative accuracy, online visual information permits corrections or adaptations in more dynamic environments with marked consequences for endpoint accuracy.

## Timing of Optical Information Pick-Up

Many studies of the temporal aspects of visual information pick-up have focused on either of two aspects. The first is the duration of visuomotor delays that we referred to earlier. The second is the minimal duration of information pick-up, or in other words, the amount of time necessary for object and environmental features to be perceived. Results on these topics are variable because both durations depend on characteristics of the performer as well as the task, but visuomotor delays of 80 ms and minimal pick-up durations of 100 ms have been reported in literature (Carlton, 1992). A related temporal aspect of optical information pick-up that has received far less research attention is the timing during an action when visual information would be most useful. For instance, in catching a ball, it appears that certain portions of the ball flight are more informative than others. Thus, if one would be allowed to view only 200 ms of its flight, catching performance would depend on its timing relative to the catch. Viewing the ball very early in the ball's flight would require the prediction of the remaining trajectory, whereas viewing only the very end of the flight would mean that part of that information could no longer be used for movement control because of the visuomotor delay.

Previous research has shown that visual information about the ball picked up closer to the moment of catching results in improved accuracy (Whiting, Gill, & Stephenson, 1970). However, in this research, the timing and the amount of viewing time varied together making it impossible to draw conclusions regarding each in isolation. In table tennis, Ripoll and Fleurance (1988) found that expert players stabilize head and eyes on an intermediate location between the point of bounce and the point of contact with the ball, that is, relatively late during the hitting movement. In cricket batting, participants follow the ball after it bounces almost until the moment of contact, that is, well into the batting movement (Land & McLeod, 2000). A late timing of optical information pick-up also seems to be advantageous in the control of the lower limbs. For instance, seeing the final portion of the ball flight has been found to be useful in controlling a soccer ball (Williams & Weigelt, 2002), looking only two steps ahead is sufficient to land the feet on targets along the travel path (Patla & Vickers, 2003), and late optical information is used in regulating the footfalls to a given demarcation point such as the take-off board in the long jump (Lee et al., 1982). During the approach phase, the variability of footfalls relative to the take-off board reduces drastically during the last four steps prior to actually hitting the board. This well-established finding, which holds for jumpers of all skill levels, demonstrates that the visual control in goal-directed locomotion emerges late and continues until the end of the action (Berg, Wade, & Greer, 1994; Hay, 1988; Scott, Li, & Davids, 1997). An interesting discussion point raised by Lee et al. (1982) is what visual information guides the regulation of gait for a successful positioning of the jumping foot relative to the take-off board. The same type of question can be asked with respect to many other tasks because, evidently, different sources of visual information are used in the control of different actions.

## Spatial Aspects of Visual Perception and Action

Let us start with a brief dialog. How do you catch a ball? I look at it and see its speed and distance. No, you cannot do that. One of the reasons so few people ask about the spatial basis of visual perception and action is that few people realize what type of stimulation can be detected by humans. Humans simply do not have sensory cells that react to velocity or distance! But, we do have sensory cells reacting to many other stimuli. The question of what visual information sources underlie the control of actions has been strongly inspired by the work of Gibson (1979/1986). His radical theory proposed that the traditional information-processing theory was not good enough to explain the majority of behavior. Gibson was the first researcher to understand and advise on how to land airplanes. He found that the optic flow field specified the actions of the pilot/plane and could, therefore, be used to navigate. The optic flow field is ubiquitous and a direct consequence of our own movement. Think of a spaceship traveling through space, and you will immediately recall the distant stars looming in and passed you; that is an example of the optic flow field. Gibson's theory of direct perception draws heavily on the notion that the requisite information is available in the ambient optic array, and it is detailed enough to guide action (Gibson, 1979/1986; Warren & Fajen, 2004). Several information sources have been identified pertaining to object properties, the layout of the environment, and observer characteristics.

To perceive the layout of the environment and the depth and size of objects therein, we can use sources of visual information like eye convergence (take your finger to your nose and you will feel the strain in eye muscles), accommodation (a reflex to focus on objects by changing lens shape and pupil size), binocular disparity (the difference in the image at the right and left eye, the principle is also used in 3D movies and astronomy technology), motion perspective (the visual field in the distance moves slower), relative size (objects in the distance appear smaller), relative density (the visual field in the distance is denser with more contours), height in the visual field (objects in the distance appear higher in the visual field), and occlusion (objects in the distance are occluded by objects in the foreground; Cutting & Vishton, 1995; Ono & Wade, 2005). Some of these sources of visual information vary together and are only useful within a limited range of distances. Many are used in 2D paintings depicting 3D scenes. Observers use different sources of information for objects at different distances and may resort to alternative sources depending on the prevailing constraints. The richness of information and the flexibility in its use is a great asset for observers, as it allows them to obtain the required information to perform a particular perceptual or perceptuomotor task. However, for researchers of perception, it poses formidable methodological challenges because the wealth of information sources available for judging, perceiving, and acting upon objects demands meticulous control of the sources available. Nevertheless, important strides have been made to examine the contribution of many sources.

For instance, in reviewing the role of binocular eye movements for depth perception, Collewijn and Erkelens (1990) concluded that eye vergence is informative only for objects within reach, while disparity, which controls vergence, can be used to perceive objects within 3 m. In testing the relative contribution of changing size and changing disparity in the perception of object motion in depth, it was found that task characteristics influence their relative effectiveness. For example, slow objects viewed for brief periods of time seemed to be judged on the basis of their changing size, whereas for fast objects viewed for a longer duration, stereopsis prevailed (Regan, Beverley, & Cynader, 1979). (Stereopsis comes about because people with two eyes see the world simultaneously from two different positions which are about 7.6 cm apart. This principle is applied to 3D movies and intergalactic exploration!) Also noteworthy is that these effects were accompanied by large interpersonal differences. The use of binocular parallax and motion parallax have been investigated in walking, yielding poor results at distances between 1 and 5 m when another information variable, angular elevation, was available (see later, Philbeck & Loomis, 1997).

Generalizations over different tasks and observers need to be made with caution because the environment is so rich in information sources and because observers are so resourceful in using alternative ones. When an observer looks at an approaching object in the optic flow field, the image of the object expands continuously until the moment of contact; thus, the optical looming pattern can inform about the time it will take for contact to occur between the object and the observer. Time-to-contact has been defined as the inverse of the expansion rate of the retinal image (or a closed optical contour), and it is known as the optical variable tau (Lee, 1976). Though tau has been investigated predominantly in the control of catching (Caljouw, 2004; Tresilian, 1993), variants thereof have been examined in the control of braking, the timing of landing a somersault, and the regulation of the approach for the long jump (Goodman & Liebermann, 1992; Lee, 1976; Lee et al., 1982, 1992). Time-to-contact is one of the few attempts to identify and formalize mathematically an optical variable that can be used in the control of action, and perhaps therefore, it has triggered a wealth of research. Other optical variables and information sources have been identified, although their corresponding mathematical laws of control have not been identified. As an observer moves about the environment, the optic flow field, which is the pattern of motion visible at the eye, also informs about motion and immobility, direction of heading, and steering (Gibson, 1979/1986).

Following Gibson, several researchers have tested the use of optic flow in the guidance of heading, steering, walking, postural control, and the perception of layout (Domini & Caudek, 2003; Wann & Land, 2000; Wann & Swapp, 2000; Warren, Kay, & Yilmaz, 1996; Warren, Kay, Zosh, Duchon, & Sahuc, 2001; Wilkie & Wann, 2002, 2005; Wu, He, & Ooi, 2005). Another optical variable that has been brought to the fore is the optical acceleration of the tangent of a projectile's elevation angle. Variables related to the optical acceleration of a fly ball have been shown to guide timely locomotion to the place of interception (Chapman, 1968; McBeath, Shaffer, & Kaiser, 1995). Importantly, this source of information can be picked up through visual kinesthesis as well as proprioception as the observer directs his or her gaze and visually tracks the ball (Oudejans et al., 1999).

One premise that underlies the identification of viable information sources (as well as the corresponding laws of control) is the specificity of visual information in relation to perception and action. It maintains that a single optical variable exists that directly relates to the to-be-perceived property and the to-be-performed action (Michaels & Carello, 1981). As explained earlier, the optical variable tau specifies time-to-contact, optic flow specifies the direction of motion, and optic acceleration specifies the location of interception with a fly ball. However, the premise of a one-to-one mapping between an informational variable and action has received much empirical opposition by authors proposing the use of more than one variable to guide a given action. It has been shown that when two information sources specify the same property, accuracy is not improved by the availability of both. However, when they are incongruent with each other, perceptual accuracy diminishes in proportion to their incongruence (Kim & Grocki, 2006).

Information sources can also be used when they do not specify the appropriate action, provided there is some constancy in the environment that can be taken into account. An obvious environmental constancy is gravity, which may be exploited in perceiving limb orientation (Cohen & Welch, 1992; van de Langenberg,

Kingma, & Beek, 2008). Other environmental constancies can be taken into account in particular tasks. For instance, as the observer directs her or his eyes to a target on the floor, the viewing angle informs about the direction of the target. It is the assumption that the target lays on the floor that allows angular declination to be informative about distance (Ooi, Wu, & He, 2001; Philbeck & Loomis, 1997; see also de Oliveira, Oudejans & Beek, 2009). So, although we do not have sensory cells for distance, looking in the right direction does relate directly to distance if and only if there is an even floor. Perceptions and actions are also calibrated to observer characteristics like body size or eye height that inform about which actions are possible and which are not (Esteves et al., 2011; Mark, 1987; Rybarczyk, Coelho, Cardoso, & de Oliveira, 2014; Warren & Whang, 1987). In conclusion, the spatial aspects of the environment and objects can be perceived and acted upon through the use of adequate visual information sources. Such variables may be used independently, combined with other sources of information, or calibrated to stable properties of the environment and observer.

## APPLIED SCIENCE: THE VISUAL CONTROL OF BASKETBALL SHOOTING

After having reviewed a selected part of the vast literature on the temporal and spatial aspects of visual perception and action, I now briefly review the available studies on the visual control of basketball shooting. Earlier research on the visual control of basketball shooting suggests that eye and head stabilization relative to the target are crucial for successful performance in the free throw and jump shot alike (Ripoll, Bard, & Paillard, 1986). In addition, although visual acuity does not seem to be a necessary asset for performance (Applegate & Applegate, 1992), expert shooters fixate relatively long on the target before initiating the free throw (Vickers, 1996). This long fixation has been interpreted as evidence for movement programming, thus the free throw was considered to be mostly pre-programmed and subsequently executed in an open-loop fashion. However, critics charged that this interpretation would be invalid in more dynamic instances, namely in jump shooting or when the shooting kinematics allow vision of the target during movement execution (Oudejans et al., 2002). In support of this assertion, it was demonstrated that seeing the target only during the final shooting movements provided enough information for accurate jump shooting. (As an aside, the kinematics of the arms in basketball shooting determine whether or not the basket is visible during the last elbow extension. If the propulsion hand remains below the line of sight, the target is occluded by hands and ball during the last elbow extension. This is a low (hand) shooting style used by participants in the study of Vickers. If the propulsion hand rises above the line of sight, the target is visible during the last elbow extension. This is a high (hand) shooting style used by participants in the study of Oudejans et al.) Although both the studies of Vickers and Oudejans et al. report interesting results, the authors offered divergent interpretations as to how the visual and motor systems would interact.

Following this research, the temporal aspects of visual perception in basketball shooting were investigated in a series of three experiments. The first study examined the preferred timing of optical information pick-up by imposing intermittent viewing on expert basketball shooters. Players used LCD goggles that became clear and opaque at pre-defined intermittent cycles. While wearing these goggles, players were asked to take a step and a jump shot. de Oliveira, Oudejans, and Beek (2006) found that players preferred to see the basket as late as possible during the shooting movements. The second study used the same goggles, but they were not pre-programmed (de Oliveira, Huys, Oudejans, van de Langenberg, & Beek, 2007). Instead, there were four visual conditions where there was either full vision, or vision was occluded, and the movement was delayed by 0, 1, or 2 s. Although most shots still landed in the vicinity of the basket, end-point accuracy was significantly better under full-vision, and not significantly different between the delay conditions. Also inter-join coordination strength decreased with the severity of the visual conditions, indicating poor coordination. This result corroborated the preference for late viewing. It also denied support to the notion of a pre-program and instead reinforced the notion of online visual control as a fundamental tenet of this complex skill. The third study examined gaze behavior during the basketball jump shot and free throw (de Oliveira, Oudejans, & Beek, 2008). The gaze behavior showed that players using different styles and executing different techniques abode by the same principle of looking as late as possible as permitted by the style and technique. This program of research employed three different techniques to examine the same question and systematically found that basketball players show a marked preference for controlling their actions online during execution. In a direct application of these principles of online control, Oudejans (2012) devised a training intervention to improve three-point basketball shooting performance. The experimental group wore liquid crystal goggles that were made opaque during a dribble and a step and made transparent in the final 500–800 m before ball release. The results showed a statistically significant improvement of 8% in the experimental group but not in the control groups (cf., Oudejans, Koedijker, Bleijendaal, & Bakker, 2005).

Another study was conducted that examined the spatial aspects of basketball shooting. In three experiments, we examined what information sources might be used in guiding the online execution of the basketball shoot (de Oliveira, Oudejans & Beek, 2009). First, we identified the vector that describes the relative position between the player's eye and the basket. This vector can be expressed as an angle or direction (which would be available from the pattern of contraction and extension of the neck and eye muscles) and by a distance (which would be available from convergence or binocular disparity). We also took into account the two constants that include the height of the basket and the horizontality of the floor. Then, we systematically eliminated information sources and measured the outcomes of performance. At the end of the three experiments, we concluded that players use the angle of elevation to guide their shooting actions. The angle of elevation that coincides with the direction of the basket is given by the pattern of muscle contraction at the neck and eyes such that if players are closer to the basket they will increase the angle until they are looking straight up, and if they are further from the basket they will decrease the angle until they are almost looking straight ahead (cf., Warren, 1998). Interestingly, the angle of elevation can only be a useful information source when the height of the basket is taken into account. Therefore, it is important that the players calibrate their visual perception and action to relevant constants in the environment.

This program of research investigated the temporal and spatial aspects of the visual perception for basketball shooting. The temporal aspects were the preferred timing of optical information pick-up, the issue of online versus offline control, and the gaze behavior during shooting preparation and execution. Together, the results of these studies underscore the importance of late visual information picked up for the online control of basketball shooting. We also investigated the spatial aspects by identifying the information sources that players use to determine the spatial location of the basket. In this study, the angle of elevation subtended by the line of sight as the player looks at the basket was singled out as the primary information source for perceiving the distance between player and basket. These results have important implications for the interpretation of previous studies, as well as theoretical implications that provide guidelines for future research.

## Reinterpretation of Previous Studies

These results call for a reinterpretation of previous studies on the visual guidance of basketball shooting. Only two studies have directly addressed the issue of online versus offline use of visual information during basketball shooting, and their interpretations were contradictory. Vickers (1996), who recorded gaze behavior of low-style expert shooters, found that they looked at the target for about 1 s before movement initiation and claimed that they needed this time to pre-program their movement. Oudejans et al. (2002), who occluded the vision of high-style expert shooters in selected moments of the shot, found that they required vision of the target only during the last 350 ms before ball release and claimed that they used visual information online to control their movements. Overall, the findings of the basketball studies (de Oliveira et al., 2006, 2007, 2008) are in accordance with previous results, though not with previous interpretations. The results of de Oliveira and collaborators (2006, 2007) are in agreement with the conclusion of Oudejans et al. (2002) that basketball players use visual information online to control movement execution but extend those results to low-style shooters as well. The results of de Oliveira and collaborators (2008) replicate the finding of Vickers (1996) that low-style shooters look at the target for 1 s, and they provide evidence for the suggestion that high-style shooters look at the target while airborne (Oudejans et al., 2002), but they extend those findings by mapping the gaze behavior to different shooting styles and different shooting techniques.

The results of de Oliveira and collaborators (2009) indicate that expert basketball players use the angle of elevation to guide their movements. This is a particularly interesting finding in itself, but also in relation to previous findings, which it may help explain in retrospect. In 1986, Ripoll, Bard, and Paillard found that stabilization of both head and eyes on the target is a determinant of success and a mark of expertise. They speculated that such stabilization served a postural function in correctly orientating the trunk toward the basket, but it now seems that anchoring head and eyes on the target also plays an instrumental role in picking up angle of elevation information (cf., Herpin, et al., 2010). This interpretation is also consistent with the earlier finding of Applegate and Applegate (1992; Mann, Ho, de Souza, Watson & Taylor, 2007) that visual acuity is not crucial for accurate performance. Even under the blurriest condition in their study, participants still shot accurately (cf., Ryu, Abernethy, Mann, & Poolton, 2015). In that condition, so the authors contended, the basket and backboard could still be discerned, and I argue that players could still fixate a point on the target, thereby having angle of elevation information. Finally, the visual fixation of the basket, as found by Vickers (1996), may be necessary for accurate shooting because it establishes a solid link between the player and the target that allows the player to reliably pick up the elevation angle continuously as the movement unfolds.

## Methodological Considerations

In the experimental work reported about basketball shooters, there were several choices made with regard to participants, experimental paradigms, and performance measures that warrant more justification and discussion. The participants in all six experiments were experienced top-level basketball players, as well as the best shooters of their respective teams. Expertise in sports results from thousands of hours of practice involving countless repetitions of specific actions such as taking free throws or jump shooting. Expertise is mostly defined on the basis of experience and level of competition as well as nontrivial performance measures like the percentage of hits. The prevailing assumption in studies of expertise is that, through practice, the links between perceptual and motor systems become well attuned, making performance more efficient. The performer becomes sensitive to the most useful information sources, and the information used is better integrated into motor control (e.g., de Oliveira, Lobinger, & Raab, 2014). Under this assumption, changes in expert performance in experimental settings are attributable to experimental manipulation and not to random motor fluctuations or other factors. This is one reason why expert performance offers a privileged setting for examining the visual basis of complex perceptual-motor skills.

Besides the benefits of studying expert performance, the task we studied is also advantageous for three other reasons. First, it can be recreated in the laboratory (i.e., provided the laboratory is large) permitting the use of equipment such as movement registration systems in a controlled environment. Second, the goal of the task is maintained, namely that of scoring a point by having the ball pass through the rim. Third, the task we studied is part of the training routines of the participants. Basketball shooting is also a complex perceptual-motor skill that involves a whole-body movement, and thus, the present findings are likely to have parallels in other relevant activities and actions such as walking, aiming, and catching (cf., de Oliveira et al., 2009b).

In most studies, we used a visual occlusion paradigm to probe the visual basis of basketball shooting. This technique is particularly useful in investigating the pick-up and use of visual information. Temporal occlusion can inform about the contribution of picking up visual information in a particular period of the movement. For instance, if visual occlusion during elbow extension deteriorates performance, then this period of vision is probably necessary to performance. Spatial occlusion can inform about the relative importance of information sources for performing the task at hand. For instance, if visual occlusion of the ground has no effects on performance, its contribution is probably minimal. In one study, we examined the gaze behavior in basketball shooting, which allowed us to verify our assumption that if players can look at the basket during shooting, they will. Gaze tracking is probably the most used technique in investigating vision in sports because it allows the experimenter to assess what the participant is looking at with minimal performance disruption. However, great caution should be exerted in interpreting fixation location and duration as indicating, respectively, an area of interest and the amount of information processed (Williams & Ericsson, 2005). Also, gaze tracking does not disclose whether such aspects of behavior are necessary or contiguous to performance. However, as illustrated in de Oliveira et al. (2008), a detailed, time-continuous analysis of gaze behavior may reveal systematic differences between experimental conditions that can be given a meaningful interpretation in combination with other findings.

Clearly, the most direct performance measure in basketball shooting is the percentage of hits. However, because the actual distance as well as the direction of the error can inform about the quality of the visual information available, it was necessary to devise a means to assess shooting errors with greater discriminative power. To this effect, it was important to use two synchronized cameras that permitted the recovery of the three-dimensional trajectory of the ball and consequent estimation of its landing position on the plane of the rim. Besides providing a very accurate estimation of ball position along both axes on the plane of the rim, this procedure permits precise estimations when the ball trajectory is interrupted by the backboard or rim. This is an innovative technique that can be used in other settings, for instance, in recovering the trajectory of a fly ball or a frisbee. Overall, it is important to consider that these methods provide insight into the visual basis of action on a strictly behavioral level. An inherent limitation of this type of research is that internal processes cannot be uncovered but only inferred from the results of controlled manipulations.

## FUTURE RESEARCH

The previous studies show that when visual information is reduced or occluded during movement execution, along with all sources of visual information, expert shooters still manage to land the ball in the target vicinity. This suggests that visual information gathered before movement initiation is in use and that the errors observed are a consequence of the deterioration of the visual information previously gathered. However, there is another possibility that can explain this result. In the absence of visual information, the player may still exploit online kinesthetic information about body configuration and perform accordingly. In other words, it is possible that the visual information gathered before the trial is used primarily to orient the player to the target, to obtain an estimated anchoring point, and that kinaesthetic information is still used to guide the movements online given that anchor point. In this light, it would be interesting to train players to perform in the absence of vision during movement execution and evaluate the accuracy of their spatial representations as well as their reliance on kinaesthetic information. These are interesting topics for further research as they could have important implications for the understanding of whether and how perceptual and representational information sources are used in the guidance of complex motor skills.

At this point, it is unclear whether it is necessary to visually fixate one particular location in the environment to perform a shooting action accurately. Clearly, players do look at the target (de Oliveira et al., 2008; Vickers, 1996), and we think this is instrumental for picking up the angle of elevation, but our results do not rule out the use of other information sources that are available to the player. For example, the horizontal distance to the basket, which relates geometrically to the location of the basket, is a candidate variable, and its contribution could be tested by having a single dot glowing at eye level underneath the basket. The disadvantages of this candidate information source is that if the player looks at the dot during the actual shooting action, it is likely that inaccuracies result from the changes in kinematics, after all, looking at the basket is part and parcel of shooting. If they were free to move head and eyes about, we would expect the participants to direct them to the estimated position of the basket and to shoot in accordance with that guessed position. This would imply that participants still make use of angular information but can use an estimated anchoring point on the basket based on other information sources.

Some personal and task characteristics may influence the usefulness of different information sources. The height of the observation point, or eye level, is one of them. The angular elevation or declination is geometrically dependent on the eye level of the observer, and therefore, its discriminative power is delimited by this characteristic. For instance, if the basket is at the eye level of a player, angular information is useless simply because there is little or no angle to be detected. As the difference between eye level and the height of the basket becomes larger, the discriminative power of angular information increases accordingly. Conversely, the discrimination between targets placed on the floor is also dependent on eye level. At greater distances, target discrimination is facilitated with increased eye level. Beyond basketball shooting and walking to targets, further research on the use of angular elevation in other tasks could prove interesting. For instance, when driving vehicles with great inertia, like trucks and buses, it is of paramount importance to perceive and react timely to objects on the road for which a heightened position of the driver is functional if the driver uses angular information. Differences in the type of traffic, obstacles, and maneuvers should all be taken into account when designing vehicles not only to guarantee the best performance of the car, but also to optimize the perceptual-motor responses of the driver. Through systematic research into exciting topics like these, more insight can be gained into the visual basis of complex perceptual motor skills. By the way, *Players must look at the basket during their shot*!

## ACKNOWLEDGMENT

The author would like to acknowledge the contribution of Prof. Peter Beek to an earlier version of this chapter, which appeared in the PhD thesis Visual Perception for Basketball Shooting.

## REFERENCES

Abernethy, B., & Russell, D. G. (1987). The relationship between expertise and visual search strategy in a racquet sport. *Human Movement Science, 6*, 283.

Abernethy, B., & Sparrow, W. A. (1992). The rise and fall of dominant paradigms in motor behaviour research. In J. J. Summers (Ed.), *Approaches to the study of motor control and learning* (pp. 3–45). Amsterdam: Elsevier Science Publishers.

Adolphe, K. (1995). Psychophysical assessment of toddlers' ability to cope with slopes. *Journal of Experimental Psychology: Human Perception and Performance, 21*, 734–750.

Applegate, R. A. (1992). Set shot shooting performance and visual acuity in basketball. *Optometry and Vision Science, 69*, 765–768.

Bennet, S., & Davids, K. (1995). The manipulation of vision during the powerlift squat: exploring the boundaries of the specificity of learning hypothesis. *Research Quarterly for Exercise and Sport, 66*, 210–218.

Berg, W., Wade, M., & Greer, N. (1994). Visual regulation of gait in bipedal locomotion: revisiting Lee, Lishman, and Thomson (1982). *Journal of Experimental Psychology: Human Perception and Performance, 20*, 854–863.

Blanchard, Y., McVeigh, R., Graham, M., Cadet, M., Mwilambwe, K., & Scott, C. (2007). The influence of ambient lighting levels on postural sway in healthy children. *Gait & Posture, 26*, 442–445.

Bootsma, R. J., & van Wieringen, P. C. (1990). Timing an attacking forehand drive in table tennis. *Journal of Experimental Psychology: Human Perception and Performance, 16*, 21–29.

Caljouw, S. R. (2004). Timing of goal-directed hitting: impact requirements change the information-movement coupling. *Experimental and Brain Research, 155*, 135–144.

Caljouw, S. R., van der Kamp, J., & Savelsbergh, G. J. P. (2004). Catching optical information for the regulation of timing. *Experimental Brain Research, 155*, 427–438.

Carlton, L. G. (1992). Visual processing time and the control of movement. In L. Proteau & B. C. Elliott (Eds.), *Vision and motor control* (pp. 3–31). Amsterdam: Elsevier Science.

Chapman, S. (1968). Catching a baseball. *American Journal of Physics, 36,* 368–370.
Cohen, M. M., & Welch, R. B. (1992). Visual-motor control in altered gravity. In L. Proteau & B. C. Elliott (Eds.), *Vision and control* (pp. 153–175). Amsterdam: Elsevier Science.
Collewijn, H., & Erkelens, C. J. (1990). Binocular eye movements and the perception of depth. In E. Kowler (Ed.), *Eye movements and their role in visual and cognitive processes* (pp. 213–261). Amsterdam: Elsevier.
Cutting, J., & Vishton, P. (1995). Perceiving layout and knowing distances: the integration, relative potency, and contextual use of different information about depth. In E. Epstein & S. Rogers (Eds.), *Perception of space and motion* (pp. 69–117). London: Academic press.
Davids, K., Palmer, D., & Savelsbergh, G. J. P. (1989). Skill level, peripheral vision and tennis volleying performance. *Journal of Human Movement Studies, 16,* 191–202.
de Oliveira, R. F. (2007). *Visual perception for basketball shooting*. (Doctorate thesis) VU University Amsterdam. Ipskamp: Amsterdam. ISBN: 978-90-9022139-7.
de Oliveira, R. F., Damisch, L., Hossner, E.-J., Oudejans, R. R. D., Raab, M., Volz, K., et al. (2009b). The bidirectional links between decision-making, perception and action. *Progress in Brain Research, 174,* 85–93.
de Oliveira, R. F., Huys, R., Oudejans, R. R. D., van de Langenberg, R., & Beek, P. J. (2007). Basketball jump shooting is controlled online by vision. *Experimental Psychology, 54*(3), 180–183.
de Oliveira, R. F., Lobinger, B. H., & Raab, M. (2014). An adaptive toolbox approach to the route to expertise in sport. *Frontiers in Psychology, 5*(709), 1–4.
de Oliveira, R. F., Oudejans, R. R. D., & Beek, P. J. (2006). Late information pick-up is preferred in basketball shooting. *Journal of Sports Sciences, 24*(9), 933–940.
de Oliveira, R. F., Oudejans, R. R. D., & Beek, P. J. (2008). Gaze behaviour in basketball shooting: further evidence for online control. *Research Quarterly for Exercise and Sport, 79*(3), 399–404.
de Oliveira, R. F., Oudejans, R. R. D., & Beek, P. J. (2009a). Experts appear to use angle of elevation information for basketball shooting. *Journal of Experimental Psychology: Human Perception and Performance, 35*(3), 750–761.
de Oliveira, R. F., Oudejans, R. R. D., & Beek, P. J. (2009c). Experts appear to use angle of elevation information for basketball shooting. *Journal of Experimental Psychology: Human Perception and Performance, 35*(3), 750–761.
de Oliveira, R. F., & Wann, J. P. (2011). Driving skills of young adults with developmental coordination disorder: regulating speed and coping with distraction. *Research in Developmental Disabilities, 32,* 1301–1308.
Desmurget, M., & Grafton, S. (2000). Forward modeling allows feedback control for fast reaching movements. *Trends in Cognitive Sciences, 4,* 423.
Domini, F., & Caudek, C. (2003). 3-D structure perceived from dynamic information: a new theory. *Trends in Cognitive Sciences, 7,* 444–449.
Esteves, P. T., de Oliveira, & Araújo, D. (2011). Posture-related affordances guide attack in basketball. *Psychology of Sport and Exercise, 12*(6), 639–644.
Fajen, B. R., Riley, M. A., & Turvey, M. T. (2008). Information, affordances, and the control of action in sport. *International Journal of Sport Psychology, 40,* 79–107.
Fleishman, E., & Rich, S. (1963). Role of kinesthetic and spatial-visual abilities in perceptual-motor learning. *Journal of Experimental Psychology, 66,* 6–11.
Gibson, J. J. (1979/1986). *The ecological approach to visual perception*. Hillsdale, NJ: Lawrence Erlbaum Associates.
Goodman, D., & Liebermann, D. G. (1992). Time to contact as a determiner of action: vision and motor control. In L. Proteau & B. C. Elliott (Eds.), *Vision and motor control* (pp. 335–349). Amsterdam: Elsevier Science.

Hay, J. (1988). Approach strategies in the long jump. *International Journal of Sport Biomechanics, 4*, 114–129.

Herpin, G., Gauchard, G. C., Lion, A., Collet, P., Keller, D., & Perrin, P. P. (2010). Sensorimotor specificities in balance control of expert fencers and pistol shooters. *Journal of Electromyography and Kinesiology, 20*, 162–169.

Janelle, C., Hillman, C., Apparies, R., Murray, N., Meili, L., Fallon, E., et al. (2000). Expertise differences in cortical activation and gaze behavior during rifle shooting. *Journal of Sport and Exercise Psychology, 22*, 167–182.

Khan, M., Franks, I., Elliott, D., Lawrence, G., Chua, R., Bernier, P., et al. (2006). Inferring online and offline processing of visual feedback in target-directed movements from kinematic data. *Neuroscience and Biobehavioral Reviews, 30*, 1106–1121.

Kim, N.-G., & Grocki, M. J. (2006). Multiple sources of information and time-to-contact judgments. *Vision Research, 46*, 1946.

Kistemaker, D., van Soest, A., & Bobbert, M. (2006). Is equilibrium point control feasible for fast goal-directed single-joint movements? *Journal of Neurophysiology, 95*, 2898–2912.

Klostermann, A., Kredel, R., & Hossner, E.-J. (2014). The quiet-eye without a target: the primacy of visual information processing. *Journal of Experimental Psychology: Human Perception and Performance, 40*(6), 2167–2178.

Land, M. F., & McLeod, P. (2000). From eye movements to actions: how batsmen hit the ball. *Nature Neuroscience, 3*, 1340–1346.

Lee, D. N. (1976). A theory of visual control of braking based on information about time-to-collision. *Perception, 5*, 437–459.

Lee, D. N., Lishman, J. R., & Thomson, J. A. (1982). Regulation of gait in long jumping. *Journal of Experimental Psychology: Human Perception and Performance, 8*, 448–459.

Lee, D. N., Young, D., & Rewt, D. (1992). How do somersaulters land on their feet? *Journal of Experimental Psychology: Human Perception and Performance, 18*, 1195–1202.

Mann, D. L., Ho, N. Y., De Souza, N. J., Watson, D. R., & Taylor, S. J. (2007). Is optimal vision required for the successful execution of an interceptive task? *Human Movement Science, 26*(3), 343–356.

Mark, L. S. (1987). Eyeheight-scaled information about affordances: a study of sitting and stair climbing. *Journal of Experimental Psychology: Human Perception and Performance, 13*, 361–370.

Mark, L., Jiang, Y., King, S., & Paasche, J. (1999). The impact of visual exploration of judgments of whether a gap is crossable. *Journal of Experimental Psychology: Human Perception and Performance, 25*, 287–295.

McBeath, M. K., Shaffer, D. M., & Kaiser, M. K. (1995). How baseball outfielders determine where to run to catch fly balls. *Science, 268*, 569–573.

Michaels, C. F., & Carello, C. (1981). *Direct perception*. Englewood Cliffs: NJ: Prentice-Hall.

Michaels, C. F., & Oudejans, R. R. D. (1992). The optics and actions of catching fly balls: zeroing out optical acceleration. *Ecological Psychology, 4*, 199–222.

Ono, H., & Wade, N. (2005). Depth and motion in historical descriptions of motion parallax. *Perception, 34*, 1263–1273.

Ooi, T., Wu, B., & He, Z. J. (2001). Distance determined by the angular declination below the horizon. *Nature, 414*, 197–200.

Oudejans, R. R. D. (2012). Effects of visual control training on the shooting performance of elite female basketball players. *International Journal of Sports Science and Coaching, 7*(3), 469–480.

Oudejans, R. R. D., Koedijker, J., Bleijendaal, I., & Bakker, F. C. (2005). The education of attention in aiming at a far target: training visual control in basketball jump shooting. *International Journal of Sport and Exercise Psychology, 3*, 197–221.

Oudejans, R. R. D., Michaels, C. F., Bakker, F. C., & Davids, K. (1999). Shedding some light on catching in the dark: perceptual mechanisms for catching fly balls. *Journal of Experimental Psychology: Human Perception and Performance, 25*, 531–542.

Oudejans, R., Michaels, C., van Dort, B., & Frissen, E. (1996). To cross or not to cross: the effect of locomotion on street-crossing behavior. *Ecological Psychology, 8*, 259–267.

Oudejans, R. R. D., van de Langenberg, R. W., & Hutter, R. I. (2002). Aiming at a far target under different viewing conditions: Visual control in basketball jump shooting. *Human Movement Science, 21*, 457–480.

Patla, A. E., & Greig, M. (2006). Any way you look at it, successful obstacle negotiation needs visually guided on-line foot placement regulation during the approach phase. *Neuroscience Letters, 397*, 110–114.

Patla, A. E., & Vickers, J. N. (2003). How far ahead do we look when required to step on specific locations in the travel path during locomotion? *Experimental Brain Research, 148*, 133–138.

Philbeck, J. W. & Loomis, J. M. (1997). Comparison of two indicators of perceived egocentric distance under full-cue and reduced-cue conditions. *Journal of Experimental Psychology: Human Perception and Performance, 23*, 72–85.

Proteau, L. (1992). On the specificity of learning and the role of visual information for movement control. In L. Proteau & B. C. Elliott (Eds.), *Vision and motor control* (pp. 67–103). Amsterdam: Elsevier Science.

Raab, M., de Oliveira, R. F., & Heinen, T. (2009). How do people perceive and generate options? *Progress in Brain Research, 174*, 49–59.

Regan, D., Beverley, K., & Cynader, M. (1979). The visual perception of motion in depth. *Scientific American, 241*, 136–151.

Reynolds, R. F., & Day, B. L. (2005). Visual guidance of the human foot during a step. *Journal of Physiology, 569*, 677–684.

Ripoll, H., Bard, C., & Paillard, J. (1986). Stabilization of head and eyes on target as a factor in successful basketball shooting. *Human Movement Science, 5*, 47–58.

Ripoll, H., & Fleurance, P. (1988). What does keeping one's eye on the ball mean? *Ergonomics, 31*, 1647–1654.

Robertson, S., Tremblay, L., Anson, J., & Elliott, D. (2002). Learning to cross a balance beam: implications for teachers, coaches and therapists. In K. Davids, G. Savelsbergh, S. J. Bennett, & J. V. D. Kamp (Eds.), *Interceptive actions in sport: Information and movement* (pp. 109–125). London: Routledge.

Rybarczyk, Y., Coelho, T., Cardoso, T., & de Oliveira, R. F. (2014). Effect of avatars and viewpoints on performance in virtual world: efficiency versus telepresence. *EAI Endorsed Transactions on Creative Technologies, 1* (1). http://dx.doi.org/10.4108/ct.1.1.e4.

Ryu, D., Abernethy, B., Mann, D. L., & Poolton, J. M. (2015). The contributions of central and peripheral vision to expertise in basketball: how blur helps to provide a clearer picture. *Journal of Experimental Psychology: Human Perception and Performance, 41*(1), 167–185.

Schmidt, R. A. (1988). *Motor control and learning: A behavioural emphasis* (2nd ed.). Champaign, IL: Human Kinetics.

Scott, M. A., Li, F. X., & Davids, K. (1997). Expertise and the regulation of gait in the approach phase of the long jump. *Journal of Sports Sciences, 15*, 597–605.

Sullivan, E., & Hooper, S. (2005). Effects of visual occlusion and fatigue on motor performance in water. *Perceptual and Motor Skills, 100*, 681–688.

Tresilian, J. R. (1993). Four questions of time to contact: a critical examination of research on interceptive timing. *Perception, 22*, 653–680.

van de Langenberg, R., Kingma, I., & Beek, P. J. (2008). The perception of limb orientation depends on the center of mass. *Journal of Experimental Psychology: Human Perception and Performance, 34*(3), 624–639.

Vickers, J. N. (1996). Visual control when aiming at a far target. *Journal of Experimental Psychology: Human Perception and Performance, 22*, 342–354.

Wann, J., & Land, M. (2000). Steering with or without the flow: is the retrieval of heading necessary? *Trends in Cognitive Sciences, 4*, 319.

Wann, J. P., & Swapp, D. K. (2000). Why you should look where you are going. *Nature Neuroscience, 3*, 647–648.

Warren, W. H. (1998). Perception of heading is a brain in the neck. *Nature Neuroscience, 1*, 647–649.

Warren, W. H., & Fajen, B. R. (2004). From optic flow to laws of control. In L. M. Vaina, S. A. Beardsley, & S. K. Rushton (Eds.), *Optic flow and beyond* (pp. 307–337). Norwell, MA: Kuwer.

Warren, W. H. J., Kay, B., & Yilmaz, E. (1996). Visual control of posture during walking: functional specificity. *Journal of Experimental Psychology: Human Perception and Performance, 22*, 818–838.

Warren, W. H. J., Kay, B. A., Zosh, W. D., Duchon, A. P., & Sahuc, S. (2001). Optic flow is used to control human walking. *Nature Neuroscience, 4*, 213–216.

Warren, W. H. J., & Whang, S. (1987). Visual guidance of walking through apertures: body-scaled information for affordances. *Journal of Experimental Psychology: Human Perception and Performance, 13*, 371–383.

Westwood, D. A., Heath, M., & Roy, E. A. (2003). No evidence for accurate visuomotor memory: systematic and variable error in memory-guided reaching. *Journal of Motor Behavior, 35*, 127–133.

Whiting, H. T. A., Gill, E. B., & Stephenson, J. M. (1970). Critical time intervals for taking in flight information in a ball-catching task. *Ergonomics, 13*, 265–272.

Wilkie, R. M., & Wann, J. P. (2002). Driving as night falls: the contribution of retinal flow and visual direction to the control of steering. *Current Biology, 12*, 2014–2017.

Wilkie, R. M., & Wann, J. P. (2005). The role of visual and nonvisual information in the control of locomotion. *Journal of Experimental Psychology: Human Perception and Performance, 31*, 901.

Williams, A. M., Davids, K., & Williams, J. G. (1999). *Visual perception and action in sport*. London: E & FNSpon Routledge.

Williams, A. M., & Ericsson, K. A. (2005). Perceptual-cognitive expertise in sport: some considerations when applying the expert performance approach. *Human Movement Science, 24*, 283.

Williams, A. M., Singer, R. N., & Frehlich, S. G. (2002). Quiet eye duration, expertise, and task complexity in a near and far aiming task. *Journal of Motor Behavior, 34*, 197–207.

Williams, A., & Weigelt, C. (2002). Vision and proprioception in interceptive actions. In K. Davids, G. Savelsbergh, B. SJ, & J. V. D. Kamp (Eds.), *Interceptive actions in sports: Information and movement* (pp. 90–108). London: Routledge.

Wu, J., He, Z. J., & Ooi, T. L. (2005). Visually perceived eye level and horizontal midline of the body trunk influenced by optic flow. *Perception, 34*, 1045–1060.

# Section E

# Performance under Pressure of Individuals or Teams and Other Performance Phenomena of Emotion–Cognition Interactions

17. Bridging the Gap between Emotion and Cognition: An Overview (Sylvain Laborde)    275
18. Performing under Pressure: Influence of Personality-Trait–Like Individual Differences (Emma Mosley and Sylvain Laborde)    291
19. The Influence of Hormonal Stress on Performance (Franziska Lautenbach and Sylvain Laborde)    315
20. Performing under Pressure: High-Level Cognition in High-Pressure Environments (K. Werner)    329

## Overview

This section aims to address how emotions and cognition are affected by pressure and how they influence performance under pressure (Chapter 17). Chapter 18 starts by delineating the influence of stable and enduring characteristics of individuals, referred to as personality trait–like individual differences, on aspects related to how individuals deal with pressure (e.g., decision-making, coping). Pressure will induce physiological and hormonal changes in individuals, and getting a grip on those changes through specific interventions should help to improve performance under pressure, as introduced in Chapter 19. The way high-level cognition processes function under pressure, such as creativity and problem solving, will be addressed in Chapter 20.

Chapter 17

# Bridging the Gap between Emotion and Cognition: An Overview

**Sylvain Laborde[1,2]**
[1]*Department of Performance Psychology, Institute of Psychology, German Sport University, Cologne, Germany;* [2]*UFR STAPS, EA 4260, University of Caen, Caen, France*

Why do some people struggle to find the words during a public presentation, though the speech was flowing while rehearsing? Why do musicians who practice their instrument flawlessly all day long still experience stage fright and make mistakes when giving a concert? Why do students, who studied extremely hard for their final examination, get very anxious when taking the test and just cannot structure their thoughts? How can it happen that James Lebron, one of the best NBA players, regularly misses crucial free throws during competitions, though this is a closed skill that one can extensively train?

When there is something at stake, the pressure to perform comes immediately into play. Pressure represents "any factor or combination of factors that increases the importance of performing well on a particular occasion" (Baumeister, 1984, p. 610). Pressure usually triggers stress (e.g., Laborde, Raab, & Kinrade, 2014) and emotions, in particular anxiety (e.g., Laborde, Lautenbach, Allen, Herbert, & Achtzehn, 2014). Hence, pressure has the potential to influence cognitive processes, for example, evidenced in a recent special issue on emotion and decision-making (Laborde, Dosseville, & Raab, 2013). Complementary to this account, neuroscientists made clear that cognition and emotion are embedded (Damasio, 1994), in the sense that effective cognition can only happen with a working functional connection to emotions, which ultimately motivates the investigation of the emotion–cognition dyad in Section E. This introductory chapter to Section E provides the reader with the basics to understand the relevant theories attempting to bridge the gap between emotion and cognition in performance contexts. In it, we will point out the current challenges of the field and offer a critical view on the theories. Finally, we will show how theory development and the pertinent combination of different research methods can help both basic research and the applied field with the development of specific interventions.

## EMOTIONS AND OTHER AFFECTIVE PHENOMENA

For the longest time, stress has been the focus in research on affective phenomena and cognition (Lazarus, 2000). This is no longer the case: in the last few years, there has been a shift toward emotions. **Stress** is considered as a unidimensional concept representing the degrees of external pressure or disturbed reactions and can be indexed in terms of arousal or activation (Lazarus, 2000). Emotions, in contrast, offer a richer, multidimensional view.

Given the fact that within the realm of affective states these concepts are often used interchangeably, I first want to define them shortly to get a better grip on them. **Affects** are acknowledged to represent the whole diversity of the phenomena experienced by an individual. They are associated with a hedonic tone and include preferences, attitudes, feelings, moods, and emotions. They can be considered as encompassing all these affective terms (Scherer, 2005). **Mood** and **emotions** differ in that emotions are shorter in duration, lasting from a few seconds to a couple of minutes, and they are triggered by a specific event. Mood can last longer (one or several days) and are not related, in particular, to one event or situation, such as the recurrent experience of negative moods during depression (Feldman–Barrett, 1998; Lench, Flores, & Bench, 2011; Scherer, 2005).

The focus here is on emotions because of their adaptive role, which will be decisive to achieve peak performance. The changes provoked by an emotion are supposedly facilitative to an adequate response to the environmental changes that triggered that emotion in the first place (Lench et al., 2011). According to Scherer (2005), emotions have five main components: the cognitive component (appraisal), the neurophysiological component (bodily symptoms), the motivational component (action tendencies), the motor expression component (facial and vocal expression), and the subjective feeling component (emotional experience). *Two main perspectives have been taken to conceptualize emotions, the discrete approach and the dimensional approach.* The **discrete approach** describes each emotion as having a specific cognitive content as well as specific appraisal properties (Scherer, 2005). This is, for example, the case for the cognitive-motivational-relational theory of Lazarus (2000). It identifies each emotion through specific cognitive content and appraisal properties referred to as "core relational theme." The **dimensional approach** considers that any emotion can be defined along a continuum of two main dimensions: valence and arousal. Valence or hedonic tone refers to experiences ranging from negative/unpleasant to positive/pleasant. Arousal, also called activation or intensity, refers to a sense of mobilization and ranges from low arousal to high arousal (Feldman–Barrett, 1998). Both valence and arousal play an important evolutionary role. First, valence indicates whether things are going well or badly regarding the existence of the individual. Second, arousal, which is close to the idea of action readiness, illustrates that emotions help us to respond to environmental stimulation or reach a goal (Frijda, 1986). High levels of arousal may, therefore, motivate actions more than low levels of arousal.

Both approaches, discrete and dimensional, are relevant for performance. We can potentially derive similar emotion regulation strategies from both approaches; however, the underlying reasons might differ. For the discrete approach, the focus would be on identifying those specific emotions that can either help or hinder the individual to perform in a specific setting. Subsequently, we can select the strategies to act on emotion-specific appraisal to elicit/control the targeted emotions. For the dimensional approach, the goal would be to identify the appropriate levels of valence and intensity required for the task and use specific strategies to reach them.

Albeit this chapter is entitled emotion and cognition, we might refer below to research based on affect or moods, because, as we have already mentioned, terms are still used interchangeably in the literature, and our section could benefit from integrating those works to get a more complete overview on the topic.

## COGNITION: A NECESSARY DISTINCTION BETWEEN EXECUTIVE AND NON-EXECUTIVE FUNCTIONS

Cognition (see also Raab, Chapter 1) has been divided between executive and non-executive functions. Such differences are necessary to specify the expected influence of emotions. Executive functions represent "a collection of top-down control processes used when going on automatic or relying on instinct or intuition would be ill-advised, insufficient or impossible" (Diamond, 2012, p. 135). Executive functions, for example, make it possible to "mentally play with ideas; taking the time to think before acting; meeting novel, unanticipated challenges; resisting temptations and staying focused" (Diamond, 2012, p. 135). Researchers now agree on three core executive functions: inhibition, working memory, and cognitive flexibility (Diamond, 2012; Lehto, Juujärvi, Kooistra, & Pulkkinen, 2003; Miyake & Friedman, 2012). Higher order cognitive functions are built on those core functions, such as reasoning, problem-solving, decision-making, and planning (Collins & Koechlin, 2012).

In contrast, **non-executive functions** are based on processes driven automatically or reflexively by stimulation (Thayer, Hansen, Saus–Rose, & Johnsen, 2009). This can occur when the person is instructed to be passive toward an event. Non-executive tasks, for example, are simple reaction time and choice reaction time tasks. These kinds of tasks do not require short-term memory in addition to controlled and focused attention, or manipulation of new information (Cowan, 1988). Finally, the main aspect that differentiates executive from non-executive tasks is that the former requires active attention and the latter one passive attention toward an event (Posner & Raichle, 1999).

Both executive and non-executive functions are relevant for performance, according to task requirements. In contrast to what was thought for a long time, human beings do not work like computers. They experience affective states, which influence their cognitive processes, and not only in a negative way, as it was long believed (Damasio, 1996).

## PERFORMANCE-ORIENTED THEORIES BRIDGING THE GAP BETWEEN EMOTION AND COGNITION

From an evolutionary perspective, emotions have been at the core of our struggle to survive and, more prominently in today's world, to perform. There is, thus, a consensus to acknowledge the strong impact emotions have on human performance (Eccles et al., 2011). This acknowledgment gave birth to different theories, which aimed to explain the influence of emotions on performance in general. We will first review these theories related to performance in general for their potential to inform specific theory specification/development, before focusing on theories addressing cognition specifically. Finally, we provide a critical overview of all these theories.

### General Emotion–Performance Theories

We identified four main theories concerned with the emotion–performance relationship: Lazarus's (2000) cognitive-motivational-relational theory, Hanin's (2000) individual zone of optimal functioning, the biopsychosocial model of challenge and threat model (Blascovich, Seery, Mugridge, Norris, & Weisbuch, 2004), and the theory of challenge and threat states in athletes (Jones, Meijen, McCarthy, & Sheffield, 2009).

#### The Cognitive-Motivational-Relational Theory

The cognitive-motivational-relational theory (Lazarus, 2000) postulates that individuals continuously appraise their ongoing relationship with the environment. There is a core relational theme for each emotion, and when the appraisal corresponds to a core relational theme, an emotion will arise. The core relational theme for each emotion is linked to a specific action tendency. According to Lazarus (2000), the function of emotion is to facilitate adaptation and, by extension, performance. The cognitive-motivational-relational theory postulates that the influence of emotion on performance will depend on the match between the action tendencies derived from the core relational theme and the task demands (Lazarus, 2000), like anger facilitating gross muscular peak force performance (Woodman et al., 2009).

#### Individual Zone of Optimal Functioning

Issued from the sport domain, the individual zone of optimal functioning theory (Hanin, 2000) assumes that "two constructs related to energizing and organizing aspects of emotion may account for the impact of emotions upon performance process: energy mobilization (demobilization) and energy utilization (misuse)" (Hanin, 2000, p. 84). As a general rule, the interaction of specific emotional content (anxiety, anger, etc.) with specific emotional intensity (high, moderate, or low) will produce specific optimal or dysfunctional effects on athletic performance (Hanin, 2007). For example, a high intensity of anxiety might be detrimental to performance with a tremendous increase in worrisome thoughts as well as with an elevated physiological arousal hindering efficient movement coordination, whereas a moderate level of anxiety could provide the athlete with exactly the right amount of mobilization required to perform well. In addition, the functionality of emotion is expected to be determined according to a resource-matching hypothesis, that is, an evaluation of the individual resources at hand in association with the demands of the task (Hanin, 2007). For example, anxiety, a negatively toned emotion, could still be perceived as optimal by an athlete if he is convinced that he has the resources to face the demands of the competition, and that the rise of heart rate accompanying anxiety should not be considered as a threat signal but rather as an indicator that the organism is getting ready for the competition.

In the cognitive-motivational-relational theory and the individual zone of optimal functioning, emphasis is placed on appraisal, of the emotion core relational themes for Lazarus (2000) and of emotion functionality with the resource-matching hypothesis for Hanin (2007). Other theories, the biopsychosocial model (Blascovich et al., 2004) and the theory of challenge and threat states in athletes (Jones et al., 2009), put the emphasis on different appraisal components: the challenge and threat appraisals.

### The Biopsychosocial Model of Challenge and Threat

The biopsychosocial model (Blascovich et al., 2004) is not concerned with emotions directly but with the evaluation of a situation in terms of challenge and threat. It postulates that prior to engaging in a task, individuals evaluate both the demands of the task and the resources they possess to deal with these demands. It is important to notice that this evaluation occurs only in motivated performance situations and when the individual is actively engaged in the task. A challenge state is expected to occur when an individual thinks he/she has enough resources to cope with the demands of the task, whereas a threat state occurs when the person thinks he/she does not possess the resources to meet the demands of the task. Demand and resource appraisals can occur consciously, unconsciously (i.e., automatically), or both (Blascovich, 2008). Given the fact that appraisal occurs most of the time without the individual being aware of it, a critical aspect of the biopsychosocial model is that challenge and threat states are better indexed not with self-report measures, but rather objectively via different patterns of neuroendocrine and cardiovascular responses (Blascovich, 2008). For both, challenge and threat states, an increase of the sympathetic-adrenomedullary activation causing the release of catecholamines is expected. In addition, in the threat state, there will also be an elevated pituitary-adrenocortical activation provoking a release of cortisol (Seery, 2011). As a consequence, a challenge state is marked by a relatively higher cardiac output and lower total peripheral resistance compared to a threat state. This higher cardiac output associated with a lower resistance in the blood vessels allows more efficient delivery of oxygen to the working muscles and to the brain.

The biopsychosocial model offers a very nice avenue because it uses objective emotion components to build its predictions, going further in comparison to theories purely relying on subjective appraisals assessed by self-reports. In the sport domain, the biopsychosocial model was adapted to give birth to the theory of challenge and threat states in athletes (Jones et al., 2009).

### Theory of Challenge and Threat States in Athletes

Based partially on the biopsychosocial model and specified in the sport domain, the theory of challenge and threat states in athletes (Jones et al., 2009) assumes that challenge and threat appraisals will have different motivational, emotional, and physiological consequences for performance. The challenge or threat state in response to competition is determined by self-efficacy, control perception, and achievement goals (Jones et al., 2009). A challenge state can be accompanied by either positive (e.g., hope) or negative emotions (e.g., anxiety) and is expected to be helpful to performance. A threat state is only accompanied by negative emotions and is thought to harm performance. This is different from the individual zone of optimal functioning, where positive (in terms of hedonic valence) emotions are thought to have the potential to harm performance, as well. In addition, challenge and threat states are thought to influence effort, attention, decision-making, physical functionality, and hence, athletic performance. Contrary to the biopsychosocial model, the theory of challenge and threat states in athletes does address cognition; however, the framework proposed appears to be very loose theoretically (i.e., to summarize, cognition is expected to be improved by a challenge state), and hence, the predictions that could result from this appear too unspecific to be tested, emphasizing the critical need to have theories specifically addressing cognition.

## Performance-Oriented Emotion–Cognition Theories

More specific than theories addressing performance in general, several theories focus on cognition, providing a higher level of specificity and explanatory power. Whereas some frameworks try to provide an overview of all variables playing a role in emotion–cognition interactions, other theories are more specific, providing either a cognitive account (theory of reinvestment and attentional control theory) or neurophysiological account (neurovisceral integration model) of emotion–cognition interactions. We review here four emotion–cognition theories that could explain behavior in a performance context.

### Boxes and Arrows Frameworks

In an attempt to capture the complexity of the emotion–cognition relationship, some researchers chose the approach of using global frameworks integrating all the parameters they thought to play a role in such relationship. For example, during the last decade, Tenenbaum and colleagues proposed several frameworks (Tenenbaum, 2003; Tenenbaum, Basevitch, Gershgoren, & Filho, 2013;

Tenenbaum et al., 2009) attempting to provide a comprehensive overview to understand the emotion–cognition relationship, or sometimes focused on one aspect of cognition, such as decision-making. The most recent framework (Tenenbaum et al., 2013) was presented as a conceptual framework linking appraisal, emotions, and cognitions under pressure, relying on the individual zone of optimal functioning (Hanin, 2000) for the emotion part. Tenenbaum et al. (2013) predict that emotions appraised as functional and pleasant are more likely to be associated with optimal decision-making. Although we acknowledge their effort in trying to conceptualize and bring together all variables that might influence the emotion–cognition interactions, such frameworks suffer from two main drawbacks. The first being the causality supposedly reflected by the direction of the arrows is often theoretically questionable. Second, the level of specificity should be clarified so that other researchers can test the model (e.g., what is understood exactly by "optimal decision-making"). As the purpose of a theory should be to provide testable hypotheses to explain phenomena, all in all, such descriptive models seem of little utility for both research and applied purposes. In conclusion, theoretical frameworks need to be sufficiently specified to allow for empirically testable hypotheses.

A similar though different approach would be, for example, to use path analysis to explain how emotions relate to (cognitive) performance (e.g., Laborde, Dosseville, Guillén, & Chávez, 2014; Nicholls, Polman, & Levy, 2012). With path analysis, the predictions of the proposed model grounded in theory have a sufficient level of specificity to be tested empirically; hence, the final model would be supported by empirical evidence.

Moving from general frameworks to more specific theories, we continue with two approaches that are often put forward to provide a cognitive account of performance breakdown under pressure: self-focus or distraction (Masters & Maxwell, 2008).

## Theory of Reinvestment

The theory of reinvestment (Masters & Maxwell, 2008) is based on the conscious processing hypothesis (Masters, 1992). It assumes that the anxiety raised by pressure will cause a decrease in performance, due to the fact that athletes are consciously controlling skills when facing stressful situations. This conscious control makes those skills more fragile and more susceptible to disruption. Another explanation is that the explicit processes used when reinvesting under pressure consume working memory, and the reduced function of working memory then debilitates automatic processing, causing skill breakdown under pressure (Masters & Maxwell, 2004). If the reinvestment theory was formerly conceived only in terms of motor performance, it has been refined with a cognitive component, that is, decision-making (Kinrade, Jackson, Ashford, & Bishop, 2010). This makes the theory of reinvestment a potential candidate to further explain the emotion–cognition relationship, despite its main focus on anxiety.

## Attentional Control Theory

In line with distraction theories, the attentional control theory (Eysenck, Derakshan, Santos, & Calvo, 2007) represents a development of the processing efficiency theory (Eysenck & Calvo, 1992). The processing efficiency theory predicts that cognitive anxiety, in the form of worry, reduces the processing and storage capacity of working memory, hence reducing the resources available for a given task. Attentional control theory further assumes that "anxiety impairs efficient functioning of the goal-directed attentional system and increases the extent to which processing is influenced by the stimulus-driven attentional system. In addition to decreasing attentional control, anxiety increases attention to threat-related stimuli." (Eysenck et al., 2007, p. 336). Two core executive functions are then expected to be affected by the adverse consequences of anxiety: inhibition and shifting (that we referred to earlier as cognitive flexibility). However, the authors precise that anxiety may not impair the quality of performance (i.e., performance effectiveness) when it leads to the use of compensatory strategies, such as enhanced effort or an increased use of processing resources (Eysenck et al., 2007). After the reinvestment theory and the attentional control theory that provide a *cognitive account* of the relationship between emotion (i.e., anxiety) and cognition, the next section with the neurovisceral integration model (Thayer et al., 2009) offers a *neurophysiological basis* to explain the emotion–cognition relationship, based on heart rate variability.

## Neurovisceral Integration Model

*Heart rate variability*, which represents the oscillation in the interval between consecutive heart beats (Malik, 1996), has been recognized as an indicator for stress (Porges, 1995) and several other emotions (Kreibig, 2010). A model based on heart rate variability seems, therefore, to have a place here to explain the emotion–cognition relationship. Heart rate variability allows identification of the branch of the autonomic nervous system that is mediating heart rate (i.e., sympathetic and parasympathetic, Malik, 1996). According to the neurovisceral integration model (Thayer et al., 2009), heart rate variability indexes through vagal activity the activity of the prefrontal cortex and, thereby, cognition. The heart and the brain are connected bidirectionally, with efferent messages from the brain affecting the heart and afferent messages from the heart affecting the brain. Heart rate variability would serve as an index of the degree to which the core integration system guided by the medial prefrontal cortex is integrated with the brainstem nuclei that directly regulate the heart, mainly through the vagus activity. *According to the neurovisceral integration model, higher cognitive effectiveness is linked to greater activity of the parasympathetic system* (Thayer et al., 2009). This relationship originates from the common structures and networks at stake for cardiac control regulation and cognitive regulation. Optimal functioning of the prefrontal cortex ensures that the flow of activity along neural pathways will establish adequate mappings between input, internal states, and outputs needed to perform a given task (Miller & Cohen, 2001), therefore enabling flexible responses to changing environments (Thayer et al., 2009). Such flexibility is ensured by the central executive, whose role is to control and to coordinate different inhibitory, attentional, and memory functions into higher order cognitive functions (Thayer et al., 2009).

In summary, the neurovisceral integration model offers testable predictions concerning the emotion–cognition relationship, based on an objective physiological marker (i.e., heart rate variability). Despite this explanatory appeal power, this model might not fit all performance contexts, in particular as soon as movement is involved, as heart rate variability is influenced by movement (Krantz, Kreutz, Ericson, & Theorell, 2010). In the following, we provide a critical analysis of the theories reviewed and suggest research avenues where the field should go.

## A Critical View of the Theories Reviewed

As a first comment, all theories we review in this section are nomothetic, assuming a similar global functioning for all individuals, except the individual zone of optimal functioning, which is ideographic and assumes a differential functioning for each individual. Second, a point already evidenced in the labeling of the subheadings is that the first category we reviewed, emotion–performance theories, does not integrate cognition. The only exception is the theory of challenge and threat states in athletes (Jones et al., 2009); however, the predictions remain very unspecific (i.e., challenge states associated with improved cognition), so it is potentially of little use. *Despite this drawback, emotion–performance theories could to some extent inform further theory building*, pointing core elements that could be potential good candidates, such as emotion appraisal, considering, for example, core emotion themes, resource matching, and challenge and threat appraisals. See Table 1 for a summary of the theories reviewed.

Regarding the theories that directly address cognition, all attempts to draw models based on boxes and arrows should reach a high level of specificity to allow prediction testing. The theory of reinvestment and the attentional control theory focus only on the negative consequences of emotion under pressure, potentially due to their focus on anxiety. As a consequence, theories should consider a larger diversity of emotions, or researchers have to make clear why we need specific theories for specific emotions. Concerning the neurovisceral integration model, it is one of the theories offering the highest level of specificity, assuming an exclusive relationship between heart rate variability and executive functions, while no relationship is expected with non-executive functions. However, even if this model is very appealing and has received experimental support in a performance domain that we discuss extensively here (i.e., sport), we have to acknowledge that the task used in these studies was a laboratory task where participants were seated and not moving (Laborde & Raab, 2013; Laborde, Raab, et al., 2014), and findings are

TABLE 1 Summary of the Theories Reviewed and Their Respective Predictions Regarding the Emotion–Cognition Relationship

| | Theory/Framework/Model | Emotion | Cognition | Predictions |
|---|---|---|---|---|
| General emotion–performance theories | Cognitive-motivational-relational theory | Discrete approach (nomothetic) | Not mentioned | No specific prediction |
| | Individual zone of optimal functioning | Discrete approach (ideographic) | Not mentioned | No specific prediction |
| | Biopsychosocial model of challenge and threat | Focus on challenge and threat appraisals | Not mentioned | No specific prediction |
| | Theory of challenge and threat states in athletes | Focus on challenge and threat appraisals | Cognitive functioning, in general, and, specifically, decision-making | Challenge state has a positive influence on cognitive functioning and specifically on decision-making |
| Performance-oriented emotion–cognition theories | Conceptual framework linking appraisal, emotions, and cognitions under pressure | Discrete approach (ideographic, based on the individual zone of optimal functioning) | Cognitive processes, such as decision-making | Emotions appraised as functional and pleasant are associated with optimal decision-making |
| | Theory of reinvestment | Anxiety | Working memory | Anxiety triggers self-focus |
| | Attentional control theory | Anxiety | Working memory, inhibition, and cognitive flexibility | Anxiety triggers distraction |
| | Neurovisceral integration model | Focus on physiological component—heart rate variability | Executive and non-executive functions | Higher parasympathetic activity linked with higher performance on executive tasks |

then hard to transfer to the ecological situation on the field. *In summary, the neurovisceral integration model does not integrate the fact that in many performance domains cognition is enacted into movement.* This drawback could be related, for example, to the distinction made by Raab, Masters, and Maxwell (2005) regarding the what and how aspects of decision-making, offering a perspective of how cognition is embedded into action. In other words, for performance, it is not enough to know which action individuals decide to perform (i.e., action selection), but also how they plan to perform it (i.e., action execution). Moreover, it is likely that the neurovisceral integration model is perhaps not adapted to this conceptualization of cognition embedded into action in the sense that motor performance requires action readiness. Action readiness is certainly not best explained by a high parasympathetic activity, but perhaps more by specific cardiovascular parameters such as the ones present in a challenge state, like predicted by the biopsychosocial model (Blascovich et al., 2004). Consequently, further endeavors should either test both theories competitively or try to develop a new one that would allow precise predictions according to the (motor) demands of the situations (e.g., high or low level of arousal required).

This critical review of the existing theories is crucial, and not only at the basic research level. A careful focus on theory testing and development would ensure that the interventions designed based on these theories would be efficient in optimizing the emotion–cognition relationship. For example, if the theory postulates that emotion appraisal influences cognition, then interventions have to focus on the way individuals appraise the situation. If the theory postulates that heart rate variability influences cognition, then interventions should focus on ways to influence heart rate variability.

In addition to theory, focus could be made as well on methods. Only a few theories go beyond the subjective experience of the individual, thus ignoring the other emotion components as defined by Scherer (2005). For example, so far concerning the neurophysiological level, the theories we reviewed were mainly based on cardiovascular variables (i.e., the neurovisceral integration model and the biopsychosocial model). No theories relied, for example, on brain methods, such as electroencephalography or functional magnetic resonance imaging (fMRI). Using such methods could be important to first understand and then intervene effectively with noninvasive brain stimulation methods such as transcranial magnetic stimulation (TMS) and transcranial direct current stimulation (tDCS) (for measurement considerations, see also Tenenbaum & Fihlo, Chapter 3). In addition to the methodology concerned with emotion, we should ensure an appropriate testing of cognition considering the ecological validity of the different performance domains. A recent review (Marasso, Laborde, Bardaglio, & Raab, 2014) provides an insightful perspective on how to solve the challenges raised by the constraints between being ecological and isolating experimental factors when studying cognition (i.e., in this particular case, decision-making).

## CONTENT OF SECTION E

The content of the section is outlined below. First, emotions were considered in this introductory chapter mainly at the state level, but it is also important to consider them at the trait level (Laborde, Raab, & Dosseville, 2013; Lazarus, 2000). The trait level corresponds to the stable aspects related to our emotional climate, which will be reflected into personality trait–like individual differences (Laborde, Breuer–Weissborn, & Dosseville, 2013). This will be addressed by Mosley and Laborde (Chapter 18).

Second, considering the physiological parameters linked to emotions, according to Charmandari, Tsigos, and Chrousos (2005), there are two major neuroendocrine systems that have been shown to adapt the organism to stress situations: the HPA-axis (with cortisol as a biomarker) and the autonomic nervous system. This autonomic nervous system is then divided into two branches—the sympathetic nervous system and the parasympathetic nervous system—with, notably, heart rate variability as a marker. If we already considered heart rate variability in this section, what is happening at the hormonal level, and specifically regarding cortisol, still requires our attention, and this will be addressed by Lautenbach and Laborde (Chapter 19).

Third, if the core executive functions—inhibition, working memory, and cognitive flexibility—were already examined under pressure, this is not that much the case for the higher cognitive functions, such as creativity, reasoning, problem solving, and decision-making, as they were identified by Diamond (2012). Werner (Chapter 20) will show how these advanced cognitive functions work under pressure.

## CONCLUSION

What can public speakers, musicians, students, athletes, and any other individuals performing under pressure learn from this introductory section? If emotions can be detrimental to cognition under pressure, they can also be a precious help for those knowing how to master them. In particular, specific interventions (e.g., targeting appraisal, heart rate variability) based on the theories we reviewed could help providing the individual with the necessary resources to cope with the demands of the task being faced, according to the evolutionary role emotions play regarding behavior adjustment to constantly changing environmental stimuli. However, what we get from this section is that sound theories addressing the emotion–cognition relationship linked to performance are still lacking or are incomplete. Hence, based on this review, we encourage future researchers not only to test existing theories, but also to consider theory development (Hardcastle, 1996), following the recommendations established in this chapter to better address the challenges raised by the investigation of the emotion–cognition relationship in performance domains.

## REFERENCES

Baumeister, R. F. (1984). Choking under pressure: self-consciousness and paradoxical effects of incentives on skillful performance. *Journal of Personality and Social Psychology, 46*, 610–620.
Blascovich, J. (2008). Challenge and threat. In A. J. Elliot (Ed.), *Handbook of approach and avoidance motivation* (pp. 431–445). New York: Psychology Press.
Blascovich, J., Seery, M. D., Mugridge, C. A., Norris, R. K., & Weisbuch, M. (2004). Predicting athletic performance from cardiovascular indexes of challenge and threat. *Journal of Experimental Social Psychology, 40*, 683–688. http://dx.doi.org/10.1016/j.jesp.2003.10.007.
Charmandari, E., Tsigos, C., & Chrousos, G. (2005). Endocrinology of the stress response. *Annual Review of Physiology, 67*, 259–284. http://dx.doi.org/10.1146/annurev.physiol.67.040403.120816.
Collins, A., & Koechlin, E. (2012). Reasoning, learning, and creativity: frontal lobe function and human decision-making. *PLoS Biology, 10*(3), e1001293. http://dx.doi.org/10.1371/journal.pbio.1001293.
Cowan, N. (1988). Evolving conceptions of memory storage, selective attention, and their mutual constraints within the human information–processing system. *Psychological Bulletin, 104*(2), 163–191.
Damasio, A. (1994). *Descartes' error: Emotion, reason, and the human brain*. New York, NY: Grosset/Putnam.
Damasio, A. (1996). The somatic marker hypothesis and the possible functions of the prefrontal cortex. *Philosophical Transactions of the Royal Society of London. Series B, Biological sciences, 351*, 1413–1420. http://dx.doi.org/10.1098/rstb.1996.0125.
Diamond, A. (2012). Executive functions. *Annual Review of Psychology, 64*, 135–168. http://dx.doi.org/10.1146/annurev-psych-113011-143750.
Eccles, D. W., Ward, P., Woodman, T., Janelle, C. M., Le Scanff, C., Ehrlinger, J., et al. (2011). Where's the emotion? how sport psychology can inform research on emotion in human factors. *Human Factors, 53*, 180–202. http://dx.doi.org/10.1177/0018720811403731.
Eysenck, M. W., & Calvo, M. G. (1992). Anxiety and performance: the processing efficiency theory. *Cognition & Emotion, 6*, 409–434. http://dx.doi.org/10.1080/02699939208409696.
Eysenck, M. W., Derakshan, N., Santos, R., & Calvo, M. G. (2007). Anxiety and cognitive performance: attentional control theory. *Emotion, 7*, 336–353.
Feldman-Barrett, L. (1998). Discrete emotions or dimensions? the role of valence focus and arousal focus. *Cognition & Emotion, 12*, 579–599. http://dx.doi.org/10.1080/026999398379574.
Frijda, N. H. (1986). *The emotions*. Cambridge: University Press.
Hanin, Y. (2000). *Emotions in sport*. Champaign, IL: Human Kinetics.
Hanin, Y. (2007). Emotions in sport: current issues and perspectives. In G. Tenenbaum, & R. Eklund (Eds.), *Handbook of sport psychology* (3rd ed.) (pp. 31–58). New York, NY: Wiley.
Hardcastle, V. G. (1996). *How to build a theory in cognitive science*. Albany, NY: State University of New York Press.
Jones, M. V., Meijen, C., McCarthy, P. J., & Sheffield, D. (2009). A theory of challenge and threat states in athletes. *International Review of Sport and Exercise Psychology, 2*, 161–180. http://dx.doi.org/10.1080/17509840902829331.
Krantz, G., Kreutz, G., Ericson, M., & Theorell, T. P. (2010). Bodily movements influence heart rate variability (HRV) responses to isolated melodic intervals. *Music and Medicine, 3*(2), 108–113. http://dx.doi.org/10.1177/1943862110387612.
Kinrade, N. P., Jackson, R. C., Ashford, K. J., & Bishop, D. T. (2010). Development and validation of the decision–specific reinvestment scale. *Journal of Sports Sciences, 28*, 1127–1135. http://dx.doi.org/10.1080/02640414.2010.499439.

Krantz, G., Kreutz, G., Ericson, M., & Theorell, T. P. (2010). Bodily movements influence heart rate variability (HRV) responses to isolated melodic intervals. *Music and Medicine*. http://dx.doi.org/10.1177/1943862110387612.

Kreibig, S. D. (2010). Autonomic nervous system activity in emotion: a review. *Biological Psychology, 84*, 394–421. http://dx.doi.org/10.1016/j.biopsycho.2010.03.010.

Laborde, S., Breuer–Weissborn, J., & Dosseville, F. (2013). Personality-trait-like individual differences in athletes. In *Advances in the psychology of sports and exercise* (pp. 25–60). New York, NY: Nova.

Laborde, S., Dosseville, F., Guillén, F., & Chávez, E. (2014). Validity of the trait emotional intelligence questionnaire in sports and its links with performance satisfaction. *Psychology of Sport and Exercise, 15*, 481–490. http://dx.doi.org/10.1016/j.psychsport.2014.05.001.

Laborde, S., Dosseville, F., & Raab, M. (2013). Introduction, comprehensive approach, and vision for the future. [Special issue] Emotions and decision making in sports. *International Journal of Sport & Exercise Psychology, 11*, 143–150. http://dx.doi.org/10.1080/1612197X.2013.773686.

Laborde, S., Lautenbach, F., Allen, M. S., Herbert, C., & Achtzehn, S. (2014). The role of trait emotional intelligence in emotion regulation and performance under pressure. *Personality and Individual Differences, 57*, 43–47. http://dx.doi.org/10.1016/J.Paid.2013.09.013.

Laborde, S., & Raab, M. (2013). The tale of hearts and reason: the influence of mood on decision making. *Journal of Sport & Exercise Psychology, 35*, 339–357.

Laborde, S., Raab, M., & Dosseville, F. (2013). Emotions and performance: valuable insights from the sports domain. In C. Mohiyeddini, M. Eysenck, & S. Bauer (Eds.), *Handbook of psychology of emotions: Recent theoretical perspectives and novel empirical findings* (Vol. 1) (pp. 325–358). New York, NY: Nova.

Laborde, S., Raab, M., & Kinrade, N. P. (2014). Is the ability to keep your mind sharp under pressure reflected in your heart? Evidence for the neurophysiological bases of decision reinvestment. *Biological Psychology, 100C*, 34–42. http://dx.doi.org/10.1016/j.biopsycho.2014.05.003.

Lazarus, R. S. (2000). How emotions influence performance in competitive sports. *The Sport Psychologist, 14*, 229–252.

Lehto, J. E., Juujärvi, P., Kooistra, L., & Pulkkinen, L. (2003). Dimensions of executive functioning: evidence from children. *British Journal of Developmental Psychology, 21*(1), 59–80. http://dx.doi.org/10.1348/026151003321164627.

Lench, H. C., Flores, S., & Bench, S. (2011). Discrete emotions predict changes in cognition, judgment, experience, behavior, and physiology: a meta–analysis of experimental emotion elicitations. *Psychological Bulletin, 137*, 834–855. http://dx.doi.org/10.1037/a0024244.

Malik, M. (1996). Heart rate variability. Standards of measurement, physiological interpretation, and clinical use. Task Force of the European Society of Cardiology and the North American Society of Pacing and Electrophysiology. *European Heart Journal, 17*, 354–381.

Marasso, D., Laborde, S., Bardaglio, G., & Raab, M. (2014). A developmental perspective on decision making in sports. *International Review of Sport and Exercise Psychology*, 1–23. http://dx.doi.org/10.1080/1750984x.2014.932424.

Masters, R. S. (1992). Knowledge, knerves and know–how: the role of explicit versus implicit knowledge in the breakdown of a complex motor skill under pressure. *British Journal of Psychology, 83*, 343–358.

Masters, R. S., & Maxwell, J. P. (2004). Implicit motor learning, reinvestment and movement disruption: what you don't know won't hurt you? In A. M. William, & N. J. Hodges (Eds.), *Skill acquisition in sport: Research, theory and practice* (pp. 207–228). London, UK: Routledge.

Masters, R. S., & Maxwell, J. P. (2008). The theory of reinvestment. *International Review of Sport and Exercise Psychology, 1*, 160–183. http://dx.doi.org/10.1080/17509840802287218.

Miller, E. K., & Cohen, J. D. (2001). An integrative theory of prefrontal cortex function. *Annual Review of Neuroscience, 24*, 167–202.

Miyake, A., & Friedman, N. P. (2012). The nature and organization of individual differences in executive functions: four general conclusions. *Current Directions in Psychological Science, 21*(1), 8–14. http://dx.doi.org/10.1177/0963721411429458.

Nicholls, A. R., Polman, R. C., & Levy, A. R. (2012). A path analysis of stress appraisals, emotions, coping, and performance satisfaction among athletes. *Psychology of Sport and Exercise, 13*, 263–270. http://dx.doi.org/10.1016/j.psychsport.2011.12.003.

Porges, S. W. (1995). Cardiac vagal tone: a physiological index of stress. *Neuroscience and Biobehavioral Reviews, 19*(2), 225–233.

Posner, M. I., & Raichle, M. I. (1999). *Images of mind*. New York: Scientific American Library.

Raab, M., Masters, R. S., & Maxwell, J. P. (2005). Improving the 'how' and 'what' decisions of elite table tennis players. *Human Movement Science, 24*, 326–344. http://dx.doi.org/10.1016/j.humov.2005.06.004.

Scherer, K. R. (2005). What are emotions? And how can they be measured? *Social Science Information, 44*, 695–729. http://dx.doi.org/10.1177/0539018405058216.

Seery, M. D. (2011). Challenge or threat? Cardiovascular indexes of resilience and vulnerability to potential stress in humans. *Neuroscience and Biobehavioral Reviews, 35*(7), 1603–1610. http://dx.doi.org/10.1016/j.neubiorev.2011.03.003.

Tenenbaum, G. (2003). Expert Athletes: an integrated approach to decision making. In *Expert performance in sports: Advances in research on sport expertise* (Vol. 61) (pp. 191–218).

Tenenbaum, G., Basevitch, I., Gershgoren, L., & Filho, E. (2013). Emotions–decision-making in sport: theoretical conceptualization and experimental evidence. *International Journal of Sport and Exercise Psychology, 11*(2), 151–168. http://dx.doi.org/10.1080/1612197x.2013.773687.

Tenenbaum, G., Hatfield, B. D., Eklund, R. C., Land, W. M., Calmeiro, L., Razon, S., et al. (2009). A conceptual framework for studying emotions–cognitions–performance linkage under conditions that vary in perceived pressure. In M. Raab, J. Johnson, & H. Heekeren (Eds.), *Progress in brain research* (Vol. 174) (pp. 159–178). Elsevier.

Thayer, J. F., Hansen, A. L., Saus-Rose, E., & Johnsen, B. H. (2009). Heart rate variability, prefrontal neural function, and cognitive performance: the neurovisceral integration perspective on self-regulation, adaptation, and health. *Annals of Behavioral Medicine, 37*, 141–153. http://dx.doi.org/10.1007/s12160-009-9101-z.

Woodman, T., Davis, P. A., Hardy, L., Callow, N., Glasscock, I., & Yuill-Proctor, J. (2009). Emotions and sport performance: an exploration of happiness, hope, and anger. *Journal of Sport & Exercise Psychology, 31*, 169–188.

Chapter 18

# Performing under Pressure: Influence of Personality-Trait-Like Individual Differences

**Emma Mosley[1], Sylvain Laborde[2,3]**
[1]*Centre for Event and Sport Research, Bournemouth University, UK;* [2]*Department of Performance Psychology, Institute of Psychology, German Sport University, Cologne, Germany;* [3]*UFR STAPS, EA 4260, University of Caen, Caen, France*

Will is an amateur boxer who is fighting in the heavy weight final against an unbeaten opponent whom he has wanted to beat since he started competing. He has suffered defeats from this athlete before but always picks himself back up and is determined he can win this time. He has put in hours of training because he wants to be perfect and uses his high standards as his motivation. The match begins, and he sees this opportunity as a challenge. During the match, he notices that his opponent is tired and reads his emotional language well. He flies in with a right hook and knocks the opponent to the ground and wins the final.

On the other hand, Sarah is an international junior tennis player about to compete in a qualifier for junior Wimbledon, which means the world to her. She is an outright perfectionist and everything has to be perfect, even dropping a single point just is not good enough. During the match, she loses a set, and this kick-starts her normal reaction to become anxious and very pessimistic about the remaining time in the match. She consciously tries to control her movements to perform to her perfect standards. Her performance unravels, and she loses the match; she slams her racket to the floor in frustration and storms off court.

Predicting performance under pressure can be a tricky business, with many theories providing different explanations and not one being able to provide a 100% prediction of this. If anything, a 100% prediction of performance is impossible; however, one area of interest within this domain is personality. The interest lies in the belief that, for the most part, the construct is stable, and therefore, an individual's personality is not likely to change, regardless of conditions (Boyle, 2008; Pervin & Cervone, 2010), as a constant personality has been suggested to have an underlying influence over behavior irrespective of situation (Aidman & Schofield, 2004). Therefore, if personality is stable in a range of situations (Boyle, 2008; Pervin, 2003) and influences behavior (Aidman & Schofield, 2004), this then advocates personality as a useful predictor of performance in pressurized environments. Specifically, this chapter will shift away from broad measures of personality such as the big five personality dimensions (i.e., Allen, Greenlees, & Jones, 2013; Garbarino, Chiorri, & Magnavita, 2014; Kaiseler, Polman, & Nicholls, 2012) and focus on the many other individual differences located at the trait level that provide support for understanding people's behavior under pressure. These individual differences were grouped together under the umbrella term of personality trait-like individual differences (PTLIDs) by Laborde, Breuer-Weissborn, and Dosseville (2013).

Within the current context, individual differences relate to the different traits that "make up" an individual and how this make-up has the potential to influence behavior within the performance environment. Laborde et al. (2013) defined them as follows:

*"Reflecting psychological individual differences not belonging to the main conceptualization of personality (i.e. big five), but which are considered as traits linked to personality"* (p. 26).

A recent review by Laborde et al. (2013) highlighted a range of PTLIDs, the influence they have on sporting performance, and their applications to the sporting environment. The work by Laborde et al. (2013) provides a basis for the current chapter, which has a specific emphasis on performance under pressure.

PTLIDs have demonstrated their relevance within pressurized environments that involve facing challenges and high levels of stress such as sport (e.g., Laborde, Lautenbach, Herbert, Allen, & Achtzehn, 2014), business (e.g., Luszczynska & Cieslak, 2005), and academia (e.g., Qualter, Gardner, Pope, Hutchinson, & Whiteley, 2012). They have also shown importance in situations that can be a matter of life or death such as fire-fighting (e.g., Maddi, Harvey, Resurreccion, Giatras, & Raganold, 2007), human surgery (e.g., Malhotra, Poolton, Wilson, Ngo, & Masters, 2012), and combat (e.g., Maddi, Matthews, Kelly, Villarreal, & White, 2012). The role of PTLIDs has been highlighted in these situations and has been suggested to potentially influence the stress-coping response (Bolger & Zuckerman, 1995) and moderate the stress process (Aidman & Schofield, 2004), meaning that personality influences facilitative or debilitative behaviors within pressurized situations, which consequently affects performance outcomes.

## PTLIDs AND THE INFLUENCE ON PERFORMANCE UNDER PRESSURE

Within this section, each trait is defined, contextualized, and its influence on performance under pressure is discussed. Figure 1 demonstrates the traits that will be addressed within the chapter. Their size denotes the amount of research present within the criteria of the particular trait and performance pressure. They will be addressed in alphabetical order.

**FIGURE 1** PTLIDs addressed in the following sections (Note: This wordle reflects the number of studies investigating the respective PTLID together with performance).

## Competitive Trait Anxiety

### Definition and Background

Competitive trait anxiety (CTA) is a behavioral predisposition to perceive competitive situations as a threat and then respond with state anxiety levels that are disproportionate to the levels of objective threat (Martens, Vealey, & Burton, 1990). It is important to note that CTA is a form of domain-specific anxiety and was developed to predict how athletes respond in competitive sporting situations (Martens et al., 1990). CTA helps to predict performance as it can affect the subjective competitive situation, which is how the individual views the environment through cognitive appraisal (Martens et al., 1990). For example, if a footballer taking a penalty high in CTA is predisposed to view the situation as threatening, then thoughts would direct to the shot, which could lead to a greater somatic (bodily) response, which could result in impaired performance (Weinberg & Gould, 2011).

### Influence on Performance under Pressure

Within the sporting environment, high levels of CTA are more likely to have a debilitating effect on performance in competitive situations (Smoll & Smith, 1990). For example, in a study of professional ballet dancers, those high in CTA displayed maladaptive coping strategies when facing competition (Barrell & Terry, 2003). Therefore, it is feasible to suggest that lower levels of CTA are beneficial for performance. Weinberg and Genuchi (1980), for example, conducted a field study of golfers in competitions, finding that those with lower levels of CTA had less state anxiety and performed better than those with moderate or high CTA levels. Lower levels of CTA have also been related to higher confidence on competition days (Zeng, Leung, & Wenhao, 2008), positive post competition affective states (Cerin & Barnett, 2011), and lower levels of state anxiety (Murray & Janelle, 2007; Weinberg & Genuchi, 1980). CTA has also been linked with perfectionism, and research has shown that athletes with higher levels of maladaptive perfectionism tend to have higher levels of CTA (Gotwals & Dunn, 2007; Gotwals, Dunn, Causgrove, Dunn, & Gamache, 2010). As lower levels of CTA appear to be desirable for performance, researchers have used psychological skills training to help reduce CTA, which has been found to have a positive effect on CTA levels, for example, within national shooting athletes (Ma & Kim, 2011). The theory of CTA has also been applied to other performance settings such as musicians, where over 80% of the musicians reported competitive anxiety symptoms (Miller & Chesky, 2004), suggesting that CTA research should not be confined to sport.

## Trait Emotional Intelligence

### Definition and Background

Trait emotional intelligence (EI) is considered a personality trait rather than a cognitive ability and involves self-perceptions, which embrace the subjective nature of emotion (Petrides, Pita, & Kokkinaki, 2007). Although there have been arguments within the literature surrounding the make-up of EI, trait EI has been considered a personality dimension (Petrides et al., 2007) as opposed to knowledge or an ability (Nelis, Quoidbach, Hansenne & Mikolajczak, 2009) and, therefore, will be discussed at the trait level within this chapter. Nelis et al. (2009) have stated that trait EI refers to dispositions that are emotionally related, thus causing tendencies to behave in a predetermined manner in emotional situations. It is suggested that individuals high in trait EI are able to effectively control and modify emotions through implementing strategies, a process known as emotion regulation (Gross & Thompson, 2007). Trait EI positively influences emotion regulation, which promotes beneficial effects including coping under stress (Laborde, Brull, Weber, & Anders, 2011).

### Influence on Performance under Pressure

It has been shown that higher trait EI produces superior performance under pressure in a range of performance settings such as academic exams (Qualter et al., 2012), experimental tasks (learning and decision-making; Laborde, Dosseville, & Scelles, 2010), and sport (Laborde, Raab, & Kinrade, 2014). Trait EI has shown beneficial effects in both long-term performance achievements (e.g., Qualter et al., 2012) and in short-term, pressurized performance (e.g., Laborde et al., 2010). Although long-term performance is important, pressure can often manifest within a particular situation or event such as a presentation. For example, students who had higher trait EI experienced less negative affect during an unfamiliar knowledge recall test (Laborde et al., 2010). Trait EI has also been linked to physiological responses during stressful situations, with individuals higher in trait EI displaying a better physiological resistance to stress (Laborde et al., 2011; Laborde, Lautenbach, & Allen, 2015) and successfully predicted cortisol secretion in pressurized tennis serving (Laborde, Lautenbach, et al., 2014). This demonstrates that higher levels of trait EI have an influence over the physiological stress response, which shows its protective role over the negative effects of stress, which in turn can positively affect performance in pressurized environments (Laborde, Lautenbach, et al., 2014). Furthermore, trait EI has shown predictive abilities over performance in pressure situations such as academic exams (Qualter et al., 2012).

## Hardiness

### Definition and Background

Hardiness is a personality style that helps a person cope, withstand (Gentry & Kobasa, 1984; Weinberg & Gould, 2011), and actively engage in transformational coping when faced with stressful events (Quick, Wright, Adkins, Nelson, & Quick, 2013). Transformational coping allows the person to reframe the stressful situation and perceive it as an opportunity rather than a threat (Nelson & Simmons, 2003). The trait is made up of three factors, which include the following: a sense of control over external events, commitment in daily life, and a challenge perspective if unexpected changes occur (Kobasa, 1979). The three counterparts of hardiness amalgamate, which results in the individual working harder to transform potentially stressful situations into opportunities (Maddi, 2004). Therefore, as this trait develops, it forms the pathway for resilience in stressful environments, which ultimately results in performance enhancement through active coping (Maddi, 2006).

### Influence on Performance under Pressure

Individuals high in hardiness have shown better performance under stress in a range of demanding environments and occupations such as the military (e.g., Maddi et al., 2012), academia (Maddi, Harvey, Khoshaba, Fazel, & Resurreccion, 2009), sport (Hanton, Neil, & Evans, 2013), fire-fighting (Maddi et al., 2007), and business (Luszczynska & Cieslak, 2005). For example, Hanton et al. (2013) examined the hardiness levels of 510 collegiate and club athletes who had competed to a county level or higher. They found that the athletes who rated higher in hardiness had lower levels of both cognitive (worry) and somatic (bodily symptoms) anxiety, higher levels of self-confidence, and better coping (Hanton et al., 2013). This suggests that when athletes high in hardiness are put in pressure situations, they respond in a facilitative way to the negative stressors in the environment. Hardiness also predicts better performance longitudinally and helps to buffer stress within a pressurized environment (De La Vega, Ruiz, Gomez, & Rivera, 2013; Maddi et al., 2012; Westman, 1990). This is demonstrated by Maddi et al. (2012) as they assessed hardiness in army cadets who trained within a pressurized environment to prepare them for their occupational duties. They found that hardiness successfully predicted performance in academic and physical tests, which suggests hardiness facilitates performance under pressure through an inclination to transformational learning within stressful environments (Maddi et al., 2012).

## Mental Toughness

### Definition and Background

Mental toughness can be defined as "a collection of experientially developed and inherent sport-specific and sport-general values, attitudes, emotions, and cognitions that influence the way in which an individual approaches, responds to, and appraises both negatively and positively construed pressure, and adversities to consistently achieve his or her goals" (Gucciardi, Gordon, & Dimmock, 2009, p. 67). The conceptualization and belonging of mental toughness has been argued in the domains of personality, "traitness," state of mind, and psychological characteristics (Crust, 2007; Gucciardi, Hanton, Gordon, Mallett, & Temby, 2015). It could be suggested that the construct lies within the realms of PTLIDs as it has been based within the theoretical dimensions of hardiness (Clough & Earle, 2002). It shares the three Cs of hardiness (control, commitment, and challenge) but has an addition of confidence; this allows the individual to be confident in their ability to overcome negative experiences (Clough & Earle, 2002). A further distinction of mental toughness, away from other concepts such as resilience and hardiness, is that it plays a role in positive challenging situations, for example, winning streaks in football (Gucciardi et al., 2009; Jones, Hanton, & Connaughton, 2007).

### Influence on Performance under Pressure

The majority of research within mental toughness is based around its origin within sport; however, the concept is branching out to other performance domains. Within these domains, it is suggested to facilitate thriving in challenging, adverse, and pressure situations (Bull, Shambrook, James, & Brooks, 2005; Crust, 2007). A recent study by Gucciardi et al. (2015), for example, examined mental toughness in a range of achievement environments including students, athletes, employees, and army candidates. The collective results showed that mental toughness has important effects on performance, goal processes, and the ability to thrive under stress; moreover, it also endures across situations and time (Gucciardi et al., 2015). Similarly, mental toughness was shown to predict how successfully athletes will cope with the stresses of competition (Nicholls, Levy, Polman, & Crust, 2011). Furthermore, enhancement of mental toughness at a younger age appears to be an important factor in development and future performance (Bell, Hardy, & Beattie, 2013; Gucciardi, 2011; Gucciardi & Jones, 2012). For example, Bell et al. (2013) carried out a longitudinal study in which elite cricketers received intervention and education around the area of mental toughness to help them perform in threatening conditions. Post training the cricketers had increased in the trait and showed significant improvements in performance indicators, such as indoor batting assessments, when compared to the control group (Bell et al., 2013). The evidence suggests that mental toughness provides an individual with resources to not only successfully manage performance under pressurized environments but to also approach these environments in a more facilitative manner (Gucciardi & Jones, 2012).

## Optimism and Pessimism

### Definition and Background

Carver and Scheier (2001) define optimists as those who expect good experiences in the future, and pessimists are those who expect bad experiences.

A further distinction between optimists and pessimists is related to the concept of dispositional optimism, which is an individual's generalized expectation of either positive or negative outcomes (Scheier & Carver, 1992). Linked to this, individuals also develop explanatory styles, which are methods of interpreting both positive and negative events (Abramson, Seligman, & Teasdale, 1978; Buchanan & Seligman, 1995). Peterson (2000) suggested when regarded as a stable trait, optimism may assist individuals in regulating their own behavior. For example, in challenging and threatening environments, optimists tend to assert more confidence, goal-directed behavior, and have belief that the adversity can be overcome (Carver & Scheier, 2001). Conversely, the behavior of pessimists in the same environment leads to having doubts, being more hesitant, disengaging effort, and anticipating catastrophe (Carver & Scheier, 2001). Therefore, it is suggested that those higher in dispositional optimism cope better in pressure situations because of greater psychological adjustment (Scheier & Carver, 1985). Similarly, those who have optimistic explanatory styles when facing adversity are more likely to view it as a challenge to be overcome (Peterson, 2000) and develop more confidence for future adversity (Seligman, 1990). Perhaps one negative aspect of optimism is the denial of reality that may suppress the instinctual nature of behavior (Peterson, 2000); although in a variety of demanding settings, optimism is associated with psychological well-being (Scheier, Carver, & Bridges, 2000).

## Influence on Performance under Pressure

There are themes that have been highlighted within research that demonstrate the role of optimism in performing under pressure that include coping style, dealing with failure, and superior performance (Laborde et al., 2013). Strutton and Lumpkin (1993) found that professional salespersons' job-related stress was mediated by levels of dispositional optimism due to the use of problem-focused coping; similar findings have been mirrored in athletes (Grove & Heard, 1997). When individuals use problem-focused coping, it encourages goal-directed behavior by changing or removing the source of stress (Folkman & Lazarus, 1985). Concerning coping with failure, Seligman, Nolen-Hoeksema, Thorton, and Thornton (1990) manipulated feedback to a group of elite swimmers, for example, 1.5 s was taken off a 100-m swimming performance. The results showed that after negative events (manipulated feedback), swimmers who possessed an optimistic explanatory style went on to swim the same or better in the second swim, whereas pessimistic swimmers' performance deteriorated (Seligman et al., 1990). This suggests that optimistic swimmers, on average, perform better under pressure and that optimism could be used as a performance predictor, especially when following defeat (Seligman et al., 1990). More recently, this was replicated within football and basketball, and findings are consistent with previous research (Gordon, 2008). Optimism is also linked with successful performance; for example, Chemers, Hu, and Garcia (2001) have found that optimism was strongly related to performance outcomes within first-year college students. Similarly, restaurant managers found to have higher levels of dispositional optimism reported lower levels of stress and job burnout, which ultimately led to higher job satisfaction (Hayes & Weathington, 2007). Furthermore, optimism can be developed with attributional style training, to help sustain performance under pressure, and this has been shown within the research context (i.e., Parkes & Mallett, 2011).

## Perfectionism

### Definition and Background

Perfectionism is a personality characteristic, which is defined as "striving for flawlessness and setting exceedingly high standards for performance, accompanied by tendencies for overly critical evaluations" (Stoeber, 2011, p. 128). When competing at the highest level, where optimal or near perfect performance is required to succeed, it is understandable that the majority of competitive athletes possess this trait (Dunn, Gotwals, & Dunn, 2005; Gould, Dieffenbach, & Moffett, 2002). There have been some arguments over perfectionism being a purely negative trait that promotes self-defeating outcomes and unhealthy behavior patterns (Flett & Hewitt, 2005) that may lead to detrimental performance effects. It has been recently refined by Stoeber (2011) that perfectionism has two main concepts: perfectionistic strivings and perfectionistic concerns. Perfectionistic strivings are associated with aiming to achieve high standards of performance (Stoeber, 2011), positive emotions (Kaye, Conroy, & Fifer, 2008), and motivation that is facilitative for performance (Stoeber & Becker, 2008). Conversely, perfectionistic concerns are associated with the following: evaluation from others, performance fear, meeting personal expectations (Stoeber, 2011), and fear of failure (Sagar & Stoeber, 2009).

### Influence on Performance under Pressure

Both positive and negative effects of perfectionism on performance are present in a range of pressure settings. It is suggested that perfectionistic striving promotes facilitative behaviors to help improving performance, for example, focusing on accuracy rather than speed (Stoeber & Kersting, 2007; Stoeber, Uphill, & Hotham, 2009) and having goal-directed behavior to increase motivation (Kaye et al., 2008; Stoeber & Eismann, 2007). In a study of triathletes, those who scored higher on perfectionistic strivings outperformed their low-scoring counterparts, thus demonstrating the effects of perfectionism on competitive performance outcomes (Stoeber et al., 2009). However, Altstötter-Gleich, Gerstenberg, and Brand (2012) used a stress-inducing concentration task and found that perfectionistic concerns successfully predicted better performance, although this was paired with negative affect, which may not be conducive to future performances (Sagar & Stoeber, 2009). Similarly, maladaptive perfectionism showed a relationship with trait anger under athletic competitive situations, which results in greater dispositional tendencies for anger within sport (Dunn, Gotwals, Dunn, & Syrotuik, 2006). A qualitative study by Gucciardi, Longbottom, Jackson, and Dimmock (2010) examined the experiences of golfers choking under pressure, and one theme that emerged was perfectionistic tendencies in performance. One golfer said, "Your high expectations of yourself can create an unwanted source of pressure" (Gucciardi et al., 2010, p. 69): the quote demonstrates how the golfers set high standards for themselves in performance situations. This then contributed to choking under pressure, as they perceived they could not reach their own unrealistic demands (Gucciardi et al., 2010). Perfectionism seems to facilitate performance under pressure; however, it could also cause performance decrements.

## Reinvestment

### Definition and Background

Masters and Maxwell (2004) define reinvestment as the "manipulation of conscious, explicit, rule based knowledge, by working memory, to control the mechanics of one's movements during motor output" (p. 208). In earlier work, Masters, Polman, and Hammond (1993) viewed reinvestment as a personality trait, which suggests that trait reinvestment levels differ across individuals, which can subsequently affect performance under pressure. The majority of research within reinvestment is based around performance under psychological pressure (e.g., Laborde, Raab, et al., 2014; Mullen, Hardy, & Tattersall, 2005). Within a pressurized environment, an individual high in reinvestment will attempt to gain conscious control of their performance. This occurs as the individual reverts to the early stages of learning in an effort to control movements and decisions that are normally autonomous (Masters, 1992), which can potentially result in performance decrements. For example, Gray (2004) examined elite baseball batters performance in a movement-specific focus task and an unrelated tone task. In the first condition, batters had to identify if their bat was moving up or down when hearing a tone during the execution of the swing. The second condition prompted the batters to signal when either a high or a low tone unrelated to the batters movement was played. The study found that the movement-specific focus caused an increase in batting errors (Gray, 2004). The nature of reinvestment has been shown to cause a breakdown in skill and decision-making, particularly under pressure.

### Influence on Performance under Pressure

Within research where either pressure manipulation or self-focus instructions have been used to provoke conscious control of movement, the majority of performers have suffered a drop in performance (Masters & Maxwell, 2008). Differing levels of reinvestment can affect performance in both cognitive and motor tasks (Kinrade, Jackson, & Ashford, 2010). An example of this is the study of Mullen et al. (2005) that examined the effects of task-relevant coaching prompts and tone counting on the performance of a golf putting exercise in high-pressure conditions. They found that both the task-relevant coaching prompts and tone counting had a detrimental effect on performance in the high-pressure condition. This demonstrates the effects that conscious processing can have on performance under pressure, and findings have been consistent within similar research (i.e., Hardy, Martin, & Mullen, 2001; Mullen, Hardy, & Oldham, 2007). A study examining reinvestment in medical students under pressure found that low "reinvesters" performed better on a laparoscopic surgery task under pressure than high reinvesters (Malhotra et al., 2012). Current research is exploring decision-making and reinvestment and has shown that decision reinvestment can also cause performance decrement under pressure (e.g., Kinrade et al., 2010; Laborde, Raab, et al., 2014). A recent study by Laborde, Furley and Schempp (2015) found that in a high-pressure condition, individuals who scored higher in reinvestment showed performance decrements in a working memory task. In addition, they found that a physiological baseline (high-frequency heart rate variability, HF-HRV) could predict performance beyond the self-reported reinvestment trait, which also demonstrates the underlying importance of state measures when assessing performance under pressure.

## Resilience

### Definition and Background

Resilience can be defined as "protective factors which modify, ameliorate, or alter a person's response to some environmental hazard that predisposes to a maladaptive outcome" (Rutter, 1987, p. 316). Although some authors argue that resilience should be seen as a dynamic process rather than a stable trait (Windle, Bennett, & Noyes, 2011), alternative research has classed resilience as a trait (Block & Block, 1980; Connor & Davidson, 2003). The construct of psychological resilience has been studied to further understand why some individuals cope with or even flourish in stressful or pressurized situations (Fletcher & Sarkar, 2013). Resilience can stem from adverse life events, such as parental loss, and cause negative effects on well-being (Seery, 2011). However, the emerging concept that resilience develops through adversity (Seery, 2011) is one that has filtered through to the performance context. In this context, individuals face a variety of stressors and importantly, in some instances, that is, sport, the individuals actively put themselves in these stressful situations and are forced to develop this quality (Fletcher & Sarkar, 2012). Resilience can be developed through negative sporting experiences, such as failure (Fletcher & Sarkar, 2012; Turner & Barker, 2013), which then fosters the ability to bounce back from negative experiences (Fletcher & Sarkar, 2013), such as stress. The construct influences the stress process throughout, not only on the initial appraisal of stress, but also on the selection of coping strategies (Fletcher & Sarkar, 2013).

## Influence on Performance under Pressure

Tugade and Fredrickson (2004) explored the role of resilience while performing a stress-inducing speech task. They found that individuals that are more resilient perceived the task as a challenge rather than a threat. In a similar study by Kaczmarek (2009), resilient individuals experienced more positive affect in a stressful situation, which was mediated by a challenge appraisal of the situation. Aside from laboratory experiments, it could be suggested that resilience helps build coping resources after stressful experiences. Turner and Barker (2013) discuss this through the career experiences of Andy Murray, who had repeatedly failed to win in major grand slam finals such as the US Open. Turner and Barker (2013) suggest that these experiences helped to build resilience and develop his ability to cope under pressure. In 2012, Andy Murray lost the Wimbledon final against Roger Federer; however, two weeks later at the final of the Olympic Games, he beat him to win gold and then went on to beat the world number one Novak Djokovic and triumph at Wimbledon in 2013. If an individual experiences performing under pressure more often, the trait of resilience seemingly develops and assists in times of adversity (Seery, 2011).

## Sensation Seeking (Risk Taking)

### Definition and Background

Sensation seeking is "the need for varied, novel, and complex sensations and experiences, and the willingness to take physical and social risk for the sake of such experiences" (Zuckerman, 1979, p. 10). It is a stable personality trait, and those high in the trait actively seek out arousal and stimulation and have a higher tolerance to negative life events (Zuckerman, 1979). Furthermore, sensation seekers are attracted to competitive or opportunistic behaviors with no regard for punishment contingencies (Ball & Zuckerman, 1990). This attraction to these situations is also coupled with the propensity to take risks (i.e., reckless driving, extreme sports) as it leads to feelings that increase physiological reactions, which is experienced as the desired sensation (Zuckerman, 2007). This, in turn, could transfer to the likelihood of performing successfully under pressure.

### Influence on Performance under Pressure

Cromer and Tenenbaum (2009) conducted a laboratory study in which participants completed a motor task under pressure. They found that individuals higher in sensation seeking performed better under pressure when compared to those low on the trait. However, the performance of sensation-seeking individuals was not affected by low- or high-pressure manipulation (Cromer & Tenenbaum, 2009), thus suggesting that lower levels of sensation seeking may have greater effects on performance under pressure, due to avoidance behaviors, for instance. Within a business environment, entrepreneurs are founds to have high levels of trait risk taking as the need for profit expectations, growth, and performance outcomes are paramount (Pines, Dvir, & Sadeh, 2012). When compared with those in managerial positions, entrepreneurs had higher levels of the risk-taking trait because of the personal outcome and growth nature of the demands placed on entrepreneurs (Stewart & Roth, 2001). This research suggests that risk-taking and sensation-seeking traits are linked to the demands of the situation and may act as a motivation to engage in potentially stressful situations (Castanier, Le Scanff, & Woodman, 2010; Chirivella & Martínez, 1994; Stewart & Roth, 2001).

### PTLID Summary

Overall, each of the PTLIDs highlighted influence performance under pressure in a variety of ways. It is important to note that PTLIDs may not have a direct influence on performance or provide a definite prediction of outcome, but they possess a moderating role (Aidman & Schofield, 2004), which may affect pressurized performance. For example, a tennis player who is a set down in a final may cope with this situation more effectively if they are higher in optimism (Seligman et al., 1990). Similarly, a musician high in perfectionistic strivings may have higher intrinsic motivation to practice more to prepare for an important upcoming performance, which could lead to increased performance achievements (Stoeber & Eismann, 2007). These examples demonstrate how PTLID's have an influence over how individuals deal with the stressors faced within pressurized environments. By assessing PTLIDs in these environments, understanding can be furthered for the reasons behind successful or unsuccessful performance and help to develop traits, which may help to improve performance.

PTLIDs demonstrate the vast number of differences that may be apparent when studying those individuals within a performance context. However, as there is an array of PTLIDs, this then prompts previously highlighted issues within the trait concept; which (and how many) traits have the largest effects under pressure? Furthermore, how might a group of traits interact in differing performance situations, or even with the situation itself? Although PTLIDs provide an insight into predicting and facilitating performance under pressure, there are still many questions surrounding this area.

## FUTURE RESEARCH DIRECTIONS WITHIN PTLID RESEARCH

Building on the preceding summary, two main areas of future research have been highlighted to further the understanding of performing under pressure. The first is combining traits to understand the influence over performing under pressure and interplay between them that ultimately affects behavior. The second is to adopt an interactionist approach to personality and measure both traits and states within performance settings to understand the role of situations on personality traits.

### Integrating and Combining PTLIDs

Laborde et al. (2013) highlighted the actuality that PTLIDs overlap; for example, mental toughness shares the three Cs—commitment, control, and challenge—of hardiness (Clough & Earle, 2002). This forces the issue of clarification within the area of PTLIDs but also the role they take/have when performing under pressure. The need for integration of PTLIDs has been emphasized within Laborde's et al. (2013) work stating that studying PTLIDs together would reveal the overlapping elements and contributions to the prediction of human behavior. Building theoretical knowledge of PTLIDs and their interactions is a necessity to both understand and to identify higher order PTLIDs, that is, those that possess stronger moderating effects on performance. If this can be further understood, it may help to build on theoretical knowledge of PTLIDs and how they interact. This may help to identify if there are high-order PTLIDs that possess a stronger moderating effect over performance, which could have many applications. For example, from an applied perspective, if particular PTLIDs are associated with superior performance under pressure, they can be used as screening tools. Furthermore, these PTLIDs can be developed to help facilitate performance under pressure, as shown in Bell et al.'s (2013) longitudinal mental toughness study.

The current literature surrounding combinations of PTLIDs within research is limited. One insightful study by Gould and colleagues (2002) examined the psychological characteristics of Olympic champions. They measured a number of factors, including personality traits, that were based on previous research with elite athletes and those that are potentially linked with athletic success. Concerning personality characteristics, they found that mental toughness/resiliency, adaptive perfectionism, dispositional hope, and optimism showed links to the characterization of Olympic athletes (Gould et al., 2002). This demonstrates the range of PTLIDs that are present in the elite performers. All of these may play a role when competing on the world's biggest stage, the Olympic Games. Furthermore, this study used a mixed methods design to gain a qualitative insight to the personal beliefs and experiences behind athletic success. They found that optimism was triangulated between the psychometric scores of the Life Orientation Test Revised (one method of measuring optimism; Scheier, Carver, & Bridges, 1994) and the qualitative findings from the interviews. In other research, Hanton, Evans, and Neil (2003) found that the commitment and control counterparts of hardiness helped to increase the levels of facilitative interpretations of the competitive anxiety response. Inadvertently, the link between CTA and perfectionism is clear, and research shows that athletes with higher levels of maladaptive perfectionism tend to have higher levels of CTA (Gotwals & Dunn, 2007; Gotwals, Dunn, Causgrove Dunn, & Gamache, 2010). Similar complimentary relationships were found between hardiness and perfectionism (Sindik, Nazor, & Vukosav, 2011).

Current research suggests that combinations of traits can contribute to athletic success (Gould et al., 2002) and that particular traits complement each other to benefit/promote performance (Hanton et al., 2003; Sindik et al., 2011). This has only been achieved on a small scale, either with a small, specific sample (Gould et al., 2002) or with only two traits (Hanton et al., 2003). Therefore, there is a need for a wide range of trait screening under pressurized conditions to further understand the combinative and integrative roles of PTLIDs on performance.

## PTLIDs: An Interactionist Approach

Interactionism suggests that traits and situations interact together to affect behavior, and neither dimension alone can be considered as the cause of behavior (Carver & Scheier, 2012). Personality is not the sole contributor or predictor of pressurized performance outcomes. However, if the concept of PTLIDs is combined with other variables, such as appraisal and physiological parameters, a more rounded prediction starts to develop. Another contributor is the influence of the situation on the individual, and it is widely agreed that understanding behavior is enhanced through the interactions between the individual and the situation (Fleeson, 2001; Zuckerman, 1983). The person–environment interaction is a widely accepted conceptual paradigm within psychology (Schneider, 2001; Walsh, Craik, & Price, 1992) and this represents an interactional approach to personality (Bowers, 1973). As an example of this, Cox (2012) devised a model of estimation of importance of these factors in relation to the sporting environment.

Figure 2 demonstrates the contribution to performance behaviors from personality, situational factors, and the interaction between them; if the three areas are summed, it accounts for approximately 30% of an athlete's behavior (Cox, 2012). Although this model excludes other factors that contribute to an individual's performance, for example, motor ability, it does demonstrate how the highlighted factors may be a moderator for performance under pressure.

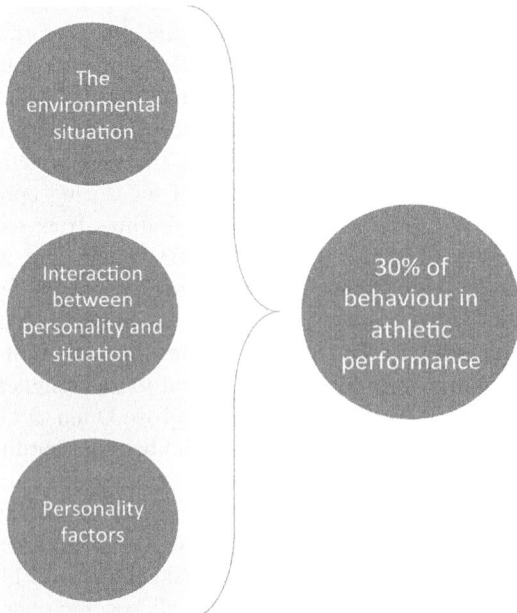

**FIGURE 2** Factors effecting athletic behavior: situation, personality, and interaction. *(Adapted from Cox, 2012).*

This area of interest is highlighted through current research trends below that are linked to situational or state factors.

## Trait Activation

Personality traits may predict behavior in particular situations, but the individual's behavior may fluctuate due to situational demands (Fleeson, 2001). Fleeson (2001) uses a density-distribution approach and believes that a person has an accumulation of traits, which are distributed among particular situations. This suggests that different traits play individual roles in particular situations, a concept recently explored within the sporting domain as trait activation (Geukes, Mesagno, Hanrahan, & Kellmann, 2012). For example, if an athlete is performing well during a training session, reinvestment may not play a role in determining behavior. However, if they were losing in the final of a competition, it may play a huge part in coping with the pressure. This is demonstrated in Geukes, Mesagno, Hanrahan, and Kellmann (2013a), who found in a private, high-pressure condition, the situational demands activated self-focus traits, that is, reinvestment, but these findings were not matched within the low-pressure condition. When assessing differing pressurized situations such as private, mixed, and public, a similar result was found in that with differing situational demands, self-focussed traits were activated (Geukes, Mesagno, Hanrahan, & Kellmann, 2013b). This shows the need to investigate together both traits and states within pressurized environments. The notion that particular traits can be activated dependent on situational demands is one of great interest. However, the situational demand may be first dependent on the individual appraisal processes.

## Appraisals

The way individuals view the situation that they are exposed to depends on a process known as cognitive appraisal (Lautenbach & Laborde, Chapter 19). This relates to the individual's perception of the stressors within the environment (Lazarus, 1984). More specifically, individuals view performance situations, where valued goals are strived toward, as either a challenge or a threat (Jones, Meijen, McCarthy, & Sheffield, 2009). Those who respond positively to potentially stressful situations are considered to have a challenge appraisal and those who respond negatively, a threat appraisal (Jones et al., 2009). This concept is present within personality not only in the composition of individual traits but also as a moderator of the resulting behavioral responses. For example, hardiness has a challenge component, which encourages individuals to see the situation as a challenge to be overcome rather than as a threat to themselves (Nelson & Simmons, 2003). This appraisal of stress promotes transformational coping, which is demonstrated in competitive anxiety research (Hanton et al., 2003). Kaczmarek (2009) found that those higher in resilience used challenge appraisals resulting in greater positive affect within stressful situations. Therefore, it may be a valuable route for future research to understand the role of appraisals and its link to PTLIDs and performance under pressure. This could be further enhanced by incorporating the objectivity of physiological measures.

## Physiological Measures

Individual differences and personality research often involve measuring constructs that we cannot directly observe, also known as latent variables (Caprara & Cervone, 2000). Therefore, by using an objective measure, such as physiological parameters, we are able to objectify how a person is reacting under a pressurized situation. Furthermore, by using an objective measure, it helps to increase the validity of measuring personality through self-report measures, given the fact that personality is considered to have links to physiological responses (Allport, 1961) and neural pathways (Davis & Panksepp, 2011). Recent PTLID research has utilized a measure known as HRV, which is a cardiac measure of activation within the autonomic nervous system (ANS). The ANS is involved with physiological activation within the stress response, and HRV represents the efficiency and adaptability of the ANS in response to environmental and situational demands (Appelhans & Luecken, 2006; Thayer & Lane, 2000). Therefore, by using HRV, researchers can understand the levels of stress an individual is experiencing; this objectifies the reaction that can be directly linked with personality's role under pressure. One trait successfully linked to HRV is EI (see also Laborde, Chapter 17; Lautenbach & Laborde, Chapter 19). Two studies (Laborde et al., 2011; Laborde, Lautenbach, et al., 2015) found that those higher in EI had a better physiological resilience to stress when exposed to laboratory stressors. Similarly, at the hormonal level when using cortisol as the physiological marker of stress (Lautenbach & Laborde, Chapter 19), a higher trait EI was associated to a lower cortisol secretion but not performance under pressure (Laborde, Lautenbach, et al., 2014). HRV has been assumed to be part of the neurophysiological basis of the reinvestment trait under pressure, as high reinvestors were found to have a higher decrease in parasympathetic activity as well as a decreased performance in comparison to low reinvestors during a pressurized decision-making task (Laborde, Raab, et al., 2014). More recently, Laborde, Furley, et al. (2015) found that HF-HRV baseline could predict performance beyond the self-reported reinvestment trait. This demonstrates that physiology helps to support findings from personality research and may even be able to predict performance beyond traits.

## THE BLANKET APPROACH: GUIDELINES FOR USE AND CONCLUDING REMARKS

So why is studying PTLIDs so pertinent within the pressurized performance domain? Not only does the stable concept demonstrate a moderating role within varied pressurized performance environments, it also stimulates the idea that PTLIDs can be developed and combined to enhance performance under pressure. The development of PTLIDs has been evidenced through interventions (i.e., mental toughness, Bell et al., 2013) and exposure to stressful environments (i.e., resilience, Turner & Barker, 2013). This demonstrates that the stable influence of PTLIDs cannot only be developed but also enhanced through psychological interventions, which ultimately can foster coping under pressure.

There is, however, a need for guidelines of use for PTLIDs within both the applied and research domains from both the administrative and interpretive fields. This is to ensure that PTLIDs are applied effectively but also to avoid overlooking the power of personality. For example, there may be some speculation surrounding personality screening through self-report measure in environments, such as talent selection, where superior ratings on desirable traits would lead to positive outcomes. However, rather than excluding the concept all together, by integrating PTLIDs with other predictors of performance, that is, appraisals and physiological reactions, it provides a more holistic approach to understanding performance under pressure.

The current chapter is not an exhaustive review of all PTLIDs that are present within today's research; however, the featured PTLIDs were selected because of their role in performance under pressure. Its aim was to act as a means of illuminating the potential moderating effects they possess over human behavior through a pressurized performance lens. A further aim was to suggest areas for future research that may be useful in further understanding the relationships between combined PTLIDs, the environment, and physiological measurements. Thus, a "blanket approach" to performing under pressure emphasizes the need to gain an informed understanding of the many elements of phenomenon. As Laborde et al. (2013) states, "people can differ greatly from the average" (p. 56), thus demonstrating the relevance of studying PTLIDs in more detail to further understand the individualized nature of performance under pressure.

## REFERENCES

Abramson, L. Y., Seligman, M. E., & Teasdale, J. D. (1978). Learned helplessness in humans: critique and reformulation. *Journal of Abnormal Psychology, 87*(1), 49–74. http://dx.doi.org/10.1037/0021-843X.87.1.49.

Aidman, E., & Schofield, G. (2004). Personality and individual differences in sport. In T. Morris & J. Summers (Eds.), *Sport psychology: Theory, applications and issues* (2nd ed.) (pp. 22–47). Milton, QLD, Australia: John Wiley & Sons Australia.

Allen, M. S., Greenlees, I., & Jones, M. (2013). Personality in sport: a comprehensive review. *International Review of Sport and Exercise Psychology, 6*(1), 184–208.

Allport, G. W. (1961). *Pattern and growth in personality*. New York: Holt, Rinehart & Winston.

Altstötter-Gleich, C., Gerstenberg, F. X. R., & Brand, M. (2012). Performing well - Feeling bad? Effects of perfectionism under experimentally induced stress on tension and performance. *Journal of Research in Personality, 46*(5), 619–622. http://dx.doi.org/10.1016/j.jrp.2012.05.010.

Appelhans, B. M., & Luecken, L. J. (2006). Heart rate variability as an index of regulated emotional responding. *Review of General Psychology, 10*(3), 229–240. http://dx.doi.org/10.1037/1089-2680.10.3.229.

Ball, S. A., & Zuckerman, M. (1990). Sensation seeking, Eysenck's personality dimensions and reinforcement sensitivity in concept formation. *Personality and Individual Differences, 11*(4), 343–353.

Barrell, G. M., & Terry, P. C. (2003). Trait anxiety and coping strategies among ballet dancers. *Medical Problems of Performing Artists, 18*(2), 59–64.

Bell, J. J., Hardy, L., & Beattie, S. (2013). Enhancing mental toughness and performance under pressure in elite young cricketers: a 2-year longitudinal intervention. *Sport, Exercise, and Performance Psychology, 2*(4), 281–297.

Block, J. H., & Block, J. (1980). *The role of ego-control and ego resiliency in the organization of behavior.* Hillsdale, NJ: Erlbaum.

Bolger, N., & Zuckerman, A. (1995). A framework for studying personality in the stress process. *Journal of Personality and Social Psychology, 69*(5), 890–902. http://dx.doi.org/10.1037/0022-3514.69.5.890.

Bowers, K. S. (1973). Situationism in psychology: an analysis and a critique. *Psychological Review, 80*(5), 307–336. http://dx.doi.org/10.1037/h0035592.

Boyle, G. J. (2008). Critique of the five-factor model of personality. In G. J. Boyle, G. Matthews, & D. H. Saklofske (Eds.), *The SAGE handbook of personality theory and assessment. Personality theories and models.* (Vol. 1) (pp. 295–312). Thousand Oaks, CA, US: Sage Publications, Inc.

Buchanan, G. M., & Seligman, M. E. P. (1995). *Explanatory style.* Hillsdale, NJ: Erlbaum.

Bull, S. J., Shambrook, C. J., James, W., & Brooks, J. (2005). Towards an understanding of mental toughness in elite English cricketers. *Journal of Applied Sport Psychology, 17*(3), 209–227.

Caprara, G. V., & Cervone, D. (2000). *Personality: Determinants, dynamics, and potentials.* Cambridge, UK; New York: Cambridge University Press.

Carver, C. S., & Scheier, M. F. (2001). Optimism, pessimism, and self-regulation. In E. C. Chang (Ed.), *Optimism & pessimism: Implications for theory, research, and practice* (pp. 31–51). American Psychological Association.

Carver, C. S., & Scheier, M. F. (2012). The trait perspective. In C. S. Carver & M. F. Scheier (Eds.), *Perspectives on personality* (7th ed.). USA: Pearson.

Castanier, C., Le Scanff, C., & Woodman, T. (2010). Beyond sensation seeking: affect regulation as a framework for predicting risk-taking behaviors in high-risk sport. *Journal of Sport & Exercise Psychology, 32*(5), 731–738.

Cerin, E., & Barnett, A. (2011). Predictors of pre- and post-competition affective states in male martial artists: a multilevel interactional approach. *Scandinavian Journal of Medicine & Science in Sports, 21*(1), 137–150.

Chemers, M. M., Hu, L.-t., & Garcia, B. F. (2001). Academic self-efficacy and first year college student performance and adjustment. *Journal of Educational Psychology, 93*(1), 55–64. http://dx.doi.org/10.1037/0022-0663.93.1.55.

Chirivella, E. C., & Martínez, L. M. (1994). The sensation of risk and motivational tendencies in sports: an empirical study. *Personality and Individual Differences, 16*(5), 777–786.

Clough, P. J., & Earle, K. (2002). When the going gets tough: a study of the impact of mental toughness on perceived demands. *Journal of Sport Sciences, 20*(1), 61.

Connor, K. M., & Davidson, J. R. T. (2003). Development of a new resilience scale: the Connor-Davidson resilience scale. *Depression and Anxiety, 18*, 76–82.

Cox, R. H. (2012). *Sport psychology: Concepts and applications* (7th ed.). New York: McGraw-Hill (International ed).

Cromer, J., & Tenenbaum, G. (2009). Meta-motivational dominance and sensation-seeking effects on motor performance and perceptions of challenge and pressure. *Psychology of Sport and Exercise, 10*(5), 552–558. http://dx.doi.org/10.1016/j.psychsport.2009.03.001.

Crust, L. (2007). Mental toughness in sport: a review. *International Journal of Sport & Exercise Psychology, 5*(3), 270–290.

Davis, K. L., & Panksepp, J. (2011). The brain's emotional foundations of human personality and the Affective Neuroscience Personality Scales. *Nuerosciences and Biobehvioural Review, 35*(9), 1946–1958.

De La Vega, R., Ruiz, R., Gomez, J., & Rivera, O. (2013). Hardiness in professional Spanish firefighters. *Perceptual & Motor Skills, 117*(2), 608–614.

Dunn, J. G. H., Gotwals, J. K., & Dunn, J. C. (2005). An examination of the domain specificity of perfectionism among intercollegiate student-athletes. *Personality and Individual Differences, 38*(6), 1439–1448. http://dx.doi.org/10.1016/j.paid.2004.09.009.

Dunn, J. G. H., Gotwals, J. K., Dunn, J. C., & Syrotuik, D. G. (2006). Examining the relationship between perfectionism and trait anger in competitive sport. *International Journal of Sport and Exercise Psychology, 4*(1), 7–24.

Fleeson, W. (2001). Toward a structure- and process-integrated view of personality: traits as density distributions of states. *Journal of Personality and Social Psychology, 80*(6), 1011–1027. http://dx.doi.org/10.1037/0022-3514.80.6.1011.

Fletcher, D., & Sarkar, M. (2012). A grounded theory of psychological resilience in Olympic champions. *Psychology of Sport and Exercise, 13*(5), 669–678.

Fletcher, D., & Sarkar, M. (2013). Psychological resilience: a review and critique of definitions, concepts, and theory. *European Psychologist, 18*(1), 12–23. http://dx.doi.org/10.1027/1016-9040/a000124.

Flett, G. L., & Hewitt, P. L. (2005). The perils of perfectionism in sport and exercise. *Current Directions in Psychological Science, 14*(1), 14–18.

Folkman, S., & Lazarus, R. S. (1985). If it changes it must be a process: study of emotion and coping during three stages of a college examination. *Journal of Personality & Social Psychology, 48*(1), 150–170.

Garbarino, S., Chiorri, C., & Magnavita, N. (2014). Personality traits of the Five-Factor Model are associated with work-related stress in special force police officers. *International Archives of Occupational & Environmental Health, 87*(3), 295–306. http://dx.doi.org/10.1007/s00420-013-0861-1.

Gentry, W. D., & Kobasa, S. C. (1984). Social and psychological resources mediating stress-illness relationships in humans. In W. D. Gentry (Ed.), *Handbook of behavioral medicine* (pp. 45–72). New York: Guilford Press.

Geukes, K., Mesagno, C., Hanrahan, S. J., & Kellmann, M. (2012). Testing an interactionist perspective on the relationship between personality traits and performance under public pressure. *Psychology of Sport & Exercise, 13*, 243–250. http://dx.doi.org/10.1016/j.psychsport.2011.12.004.

Geukes, K., Mesagno, C., Hanrahan, S. J., & Kellmann, M. (2013a). Performing under pressure in private: activation of self-focus traits. *International Journal of Sport & Exercise Psychology, 11*(1), 11–23. http://dx.doi.org/10.1080/1612197X.2012.724195.

Geukes, K., Mesagno, C., Hanrahan, S. J., & Kellmann, M. (2013b). Activation of self-focus and self-presentation traits under private, mixed, and public pressure. *Journal of Sport & Exercise Psychology, 35*(1), 50–59.

Gordon, R. A. (2008). Attributional style and athletic performance: strategic optimism and defensive pessimism. *Psychology of Sport and Exercise, 9*(3), 336–350.

Gotwals, J. K., & Dunn, J. G. H. (2007). An examination of the relationship between sport-based perfectionism and competitive trait anxiety among male intercollegiate ice hockey players. *Journal of Sport & Exercise Psychology, 29*, S164–S165.

Gotwals, J. K., Dunn, J. G. H., Causgrove Dunn, J., & Gamache, V. (2010). Establishing validity evidence for the Sport Multidimensional Perfectionism Scale-2 in intercollegiate sport. *Psychology of Sport & Exercise*, *11*(6), 423–432.

Gould, D., Dieffenbach, K., & Moffett, A. (2002). Psychological characteristics and their development in Olympic champions. *Journal of Applied Sport Psychology*, *14*(3), 172–204.

Gray, R. (2004). Attending to the execution of a complex sensorimotor skill: expertise differences, choking, and slumps. *Journal of Experimental Psychology. Applied*, *10*(1), 42–54. http://dx.doi.org/10.1037/1076-898X.10.1.42.

Gross, J. J., & Thompson, R. A. (2007). Emotion regulation: conceptual foundations. In J. J. Gross (Ed.), *Handbook of emotion regulation* (pp. 3–24). New York, NY, US: Guilford Press.

Grove, J. R., & Heard, N. P. (1997). Optimism and sport confidence as correlates of slump-related coping among athletes. *Sport Psychologist*, *11*(4), 400–410.

Gucciardi, D. F. (2011). The relationship between developmental experiences and mental toughness in adolescent cricketers. *Journal of Sport & Exercise Psychology*, *33*(3), 370–393.

Gucciardi, D. F., Gordon, S., & Dimmock, J. A. (2009). Advancing mental toughness research and theory using personal construct psychology. *International Review of Sport & Exercise Psychology*, *2*(1), 54–72.

Gucciardi, D. F., Hanton, S., Gordon, S., Mallett, C. J., & Temby, P. (2014). The concept of mental toughness: tests of dimensionality, nomological network, and traitness. *Journal of Personality*.

Gucciardi, D. F., & Jones, M. I. (2012). Beyond optimal performance: mental toughness profiles and developmental success in adolescent cricketers. *Journal of Sport & Exercise Psychology*, *34*(1), 16–36.

Gucciardi, D. F., Longbottom, J.-L., Jackson, B., & Dimmock, J. A. (2010). Experienced golfers' perspectives on choking under pressure. *Journal of Sport & Exercise Psychology*, *32*(1), 61–83.

Gucciardi, D. F., Mallett, C. J., Hanton, S., Gordon, S., & Temby, P. (2015). The Concept of Mental Toughness: Tests of Dimensionality, Nomological Network, and Traitness. *Journal of Personality*, *83*(1), 26–44. http://dx.doi.org/10.1111/jopy.12079.

Hanton, S., Evans, L., & Neil, R. (2003). Hardiness and the competitive trait anxiety response. *Anxiety, Stress & Coping*, *16*(2), 167.

Hanton, S., Neil, R., & Evans, L. (2013). Hardiness and anxiety interpretation: an investigation into coping usage and effectiveness. *European Journal of Sport Science*, *13*(1), 96–104.

Hardy, L., Martin, N., & Mullen, R. (2001). Effect of task-relevant cues and state anxiety on motor performance. *Perceptual and Motor Skills*, *92*, 943–946.

Hayes, C. T., & Weathington, B. L. (2007). Optimism, stress, life satisfaction, and job burnout in restaurant managers. *Journal of Psychology*, *141*(6), 565–579.

Jones, G., Hanton, S., & Connaughton, D. (2007). A framework of mental toughness in the world's best performers. *Sport Psychologist*, *21*(2), 243–264.

Jones, M., Meijen, C., McCarthy, P. J., & Sheffield, D. (2009). A theory of challenge and threat states in athletes. *International Review of Sport & Exercise Psychology*, *2*(2), 161–180.

Kaczmarek, Ł. (2009). Resiliency, stress appraisal, positive affect and cardiovascular activity. *Polish Psychological Bulletin*, *40*(1), 46–53. http://dx.doi.org/10.2478/s10059-009-0007-1.

Kaiseler, M., Polman, R. C. J., & Nicholls, A. R. (2012). Effects of the Big Five personality dimensions on appraisal coping, and coping effectiveness in sport. *European Journal of Sport Science*, *12*(1), 62–72.

Kaye, M. P., Conroy, D. E., & Fifer, A. M. (2008). Individual differences in incompetence avoidance. *Journal of Sport & Exercise Psychology*, *30*(1), 110–132.

Kinrade, N. P., Jackson, R. C., & Ashford, K. J. (2010). Dispositional reinvestment and skill failure in cognitive and motor tasks. *Psychology of Sport & Exercise*, *11*(4), 312–319.

Kobasa, S. C. (1979). Stressful life events, personality, and health: an inquiry into hardiness. *Journal of Personality and Social Psychology*, *37*(1), 1–11. http://dx.doi.org/10.1037//0022-3514.37.1.1.

Laborde, S., Breuer-Weissborn, J., & Dosseville, F. (2013). Personality-trait-like individual differences in athletes. In *Advances in the psychology of sports and exercise* (pp. 25–60). New York: Nova.

Laborde, S., Brull, A., Weber, J., & Anders, L. S. (2011). Trait emotional intelligence in sports: a protective role against stress through heart rate variability? *Personality and Individual Differences*, *51*(1), 23–27.

Laborde, S., Dosseville, F., & Scelles, N. (2010). Trait emotional intelligence and preference for intuition and deliberation: respective influence on academic performance. *Personality and Individual Differences*, *49*(7), 784–788. http://dx.doi.org/10.1016/j.paid.2010.06.031.

Laborde, S., Furley, P., & Schempp, C. (2015). The relationship between working memory, reinvestment, and heart rate variability. *Physiology & Behavior*, *139*, 430–436. http://dx.doi.org/10.1016/j.physbeh.2014.11.036.

Laborde, S., Lautenbach, F., & Allen, M. S. (2015). The contribution of coping-related variables and heart rate variability to visual search performance under pressure. *Physiology & Behavior*, *139*, 532–540. http://dx.doi.org/10.1016/j.physbeh.2014.12.003.

Laborde, S., Lautenbach, F., Herbert, C., Allen, M. S., & Achtzehn, S. (2014). The role of trait emotional intelligence in emotion regulation and performance under pressure. *Personality and Individual Differences*, *57*, 43–47. http://dx.doi.org/10.1016/j.paid.2013.09.013.

Laborde, S., Raab, M., & Kinrade, N. P. (2014). Is the ability to keep your mind sharp under pressure reflected in your heart? Evidence for the neurophysiological bases of decision reinvestment. *Biological Psychology*, *100*, 34–42.

Lazarus, R. S. (1984). On the primacy of cognition. *American Psychologist*, *39*(2), 124–129. http://dx.doi.org/10.1037/0003-066X.39.2.124.

Luszczynska, A., & Cieslak, R. (2005). Protective, promotive, and buffering effects of perceived social support in managerial stress: the moderating role of personality. *Anxiety, Stress & Coping*, *18*(3), 227–244.

Ma, Y. S., & Kim, H. B. (2011). The effects of psychological skills training program on anxiety, psychological skills, and performance of female Korea national shooting athletes. *Korean Journal of Sports*, *9*(4), 37–48.

Maddi, S. R. (2004). Hardiness: an operationalization of existential courage. *Journal of Humanistic Psychology*, *44*(3), 279–298.

Maddi, S. R. (2006). Hardiness: the courage to grow from stresses. *Journal of Positive Psychology*, *1*(3), 160–168. http://dx.doi.org/10.1080/17439760600619609.

Maddi, S. R., Harvey, R. H., Khoshaba, D. M., Fazel, M., & Resurreccion, N. (2009). Hardiness facilitates performance in college. *Journal of Positive Psychology*, *4*(6), 566–577.

Maddi, S. R., Harvey, R. H., Resurreccion, R., Giatras, C. D., & Raganold, S. (2007). Hardiness as a performance enhancer in firefighters. *International Journal of Fire Services Leadership and Management*, *1*(2), 3–9.

Maddi, S. R., Matthews, M. D., Kelly, D. R., Villarreal, B., & White, M. (2012). The role of hardiness and grit in predicting performance and retention of USMA cadets. *Military Psychology*, *24*(1), 19–28. http://dx.doi.org/10.1080/08995605.2012.639672.

Malhotra, N., Poolton, J., Wilson, M., Ngo, K., & Masters, R. (2012). Conscious monitoring and control (reinvestment) in surgical performance under pressure. *Surgical Endoscopy*, *26*(9), 2423–2429. http://dx.doi.org/10.1007/s00464-012-2193-8.

Martens, R., Vealey, R. S., & Burton, D. (1990). *Competitive trait anxiety*. Champaign, IL: Human Kinetics.

Masters, R. S. W. (1992). Knowledge, knerves and know-how – the role of explicit versus implicit knowledge in the breakdown of a complex motor skill under pressure. *British Journal of Psychology, 83*, 343–358.

Masters, R. S. W., & Maxwell, J. P. (2004). Implicit motor learning, reinvestment and movement disruption: what you don't know won't hurt you? In A. M. Williams & N. J. Hodges (Eds.), *Skill acquisition in sport: Research, theory and practice* (pp. 207–288). London: Routledge.

Masters, R., & Maxwell, J. (2008). The theory of reinvestment. *International Review of Sport & Exercise Psychology, 1*(2), 160–183.

Masters, R. S. W., Polman, R. C. J., & Hammond, N. V. (1993). 'Reinvestment': a dimension of personality implicated in skill breakdown under pressure. *Personality and Individual Differences, 14*(5), 655–666.

Miller, S. R., & Chesky, K. (2004). The multidimensional anxiety theory: an assessment of and relationships between intensity and direction of cognitive anxiety, somatic anxiety, and self-confidence over multiple performance requirements among college music majors. *Medical Problems of Performing Artists, 19*(1), 12–20.

Mullen, R., Hardy, L., & Oldham, A. (2007). Implicit and explicit control of motor actions: revisiting some early evidence. *British Journal of Psychology, 98*(1), 141–156. http://dx.doi.org/10.1348/000712606X114336.

Mullen, R., Hardy, L., & Tattersall, A. (2005). The effects of anxiety on motor performance: a test of the conscious processing hypothesis. *Journal of Sport & Exercise Psychology, 27*(2), 212–225.

Murray, N. P., & Janelle, C. M. (2007). Event-related potential evidence for the processing efficiency theory. *Journal of Sports Sciences, 25*(2), 161–171.

Nelis, D., Quoidbach, J., Hansenne, M., & Mikolajczak, M. (2009). Increasing emotional intelligence: (How) is it possible? *Personality and Individual Differences, 47*(1), 36–41. http://dx.doi.org/10.1016/j.paid.2009.01.046.

Nelson, D. L., & Simmons, B. L. (2003). Health psychology and work stress: a more positive approach. In J. C. Quick & L. E. Tetrick (Eds.), *Handbook of occupational health psychology* (pp. 97–119). Washington, DC, US: American Psychological Association.

Nicholls, A. R., Levy, A. R., Polman, R. C. J., & Crust, L. (2011). Mental toughness, coping self-efficacy, and coping effectiveness among athletes. *International Journal of Sport and Exercise Psychology, 42*(6), 513–524.

Parkes, J. F., & Mallett, C. J. (2011). Developing mental toughness: attributional style retraining in rugby. *Sport Psychologist, 25*(3), 269–287.

Pervin, L. A. (2003). *The science of personality* (2nd ed.). New York, NY, US: Oxford University Press.

Pervin, L. A., & Cervone, D. (2010). *Personality: Theory and research* (11th ed.). Hoboken, NJ: Wiley (International student version/Lawrence A. Pervin, Daniel Cervone).

Peterson, C. (2000). The future of optimism. *American Psychologist, 55*(1), 44–55. http://dx.doi.org/10.1037/0003-066X.55.1.44.

Petrides, K. V., Pita, R., & Kokkinaki, F. (2007). The location of trait emotional intelligence in personality factor space. *British Journal of Psychology, 98*(2), 273–289. http://dx.doi.org/10.1348/000712606X120618.

Pines, A. M., Dvir, D., & Sadeh, A. (2012). Dispositional antecedent, job correlates and performance outcomes of entrepreneurs' risk taking. *International Journal of Entrepreneurship, 16*, 95–112.

Qualter, P., Gardner, K. J., Pope, D. J., Hutchinson, J. M., & Whiteley, H. E. (2012). Ability emotional intelligence, trait emotional intelligence, and academic success in British secondary schools: a 5 year longitudinal study. *Learning and Individual Differences, 22*(1), 83–91.

Quick, J. C., Wright, T. A., Adkins, J. A., Nelson, D. L., & Quick, J. D. (2013). Individual differences in the stress response. In *Preventive stress management in organizations* (2nd ed.) (pp. 43–57). American Psychological Association.

Rutter, M. (1987). Psychosocial resilience and protective mechanisms. *American Journal of Orthopsychiatry, 57*(3), 316–331.

Sagar, S. S., & Stoeber, J. (2009). Perfectionism, fear of failure, and affective responses to success and failure: the central role of fear of experiencing shame and embarrassment. *Journal of Sport & Exercise Psychology, 31*(5), 602–627.

Scheier, M. F., & Carver, C. S. (1985). Optimism, coping, and health: assessment and implications of generalized outcome expectancies. *Health Psychology, 4*(3), 219–247. http://dx.doi.org/10.1037/0278-6133.4.3.219.

Scheier, M. F., & Carver, C. S. (1992). Effects of optimism on psychological and physical well-being: theoretical overview and empirical update. *Cognitive Therapy and Research, 16*(2), 201–228. http://dx.doi.org/10.1007/BF01173489.

Scheier, M. F., Carver, C. S., & Bridges, M. W. (1994). Distinguishing optimism from neuroticism (and trait anxiety, self-mastery, and self-esteem): a reevaluation of the Life Orientation Test. *Journal of Personality and Social Psychology, 67*(6), 1063–1078. http://dx.doi.org/10.1037/0022-3514.67.6.1063.

Scheier, M. F., Carver, C. S., & Bridges, M. W. (2000). Optimism, pessimism, and psychological well-being. In E. C. Chang (Ed.), *Optimism and pessimism: Implications for theory, research, and practice* (pp. 189–216). Washington, DC: American Psychological Association.

Schneider, B. (2001). Fits about fit. *Applied Psychology: An International Review, 50*(1), 141.

Seery, M. D. (2011). Resilience: a silver lining to experiencing adverse life events? *Current Directions in Psychological Science, 20*(6), 390–394.

Seligman, M. E. P. (1990). *Learned optimism*. New York: A.A. Knopf.

Seligman, M. E. P., Nolen-Hoeksema, S., Thorton, N., & Thornton, K. M. (1990). Explanatory style as a mechanism of disappointing athletic performance. *Psychological Science (Wiley-Blackwell), 1*(2), 143–146.

Sindik, J., Nazor, D., & Vukosav, J. (2011). Correlation between the conative characteristics at top senior basketball players. *Sport Science, 4*(1), 78–83.

Smoll, F. L., & Smith, R. E. (1990). Psychology of the young athlete. Stress-related maladies and remedial approches. *Pediatric Clinics of North America, 37*(5), 1021–1046.

Stewart, W. H., Jr., & Roth, P. L. (2001). Risk propensity differences between entrepreneurs and managers: a meta-analytic review. *Journal of Applied Psychology, 86*(1), 145–153. http://dx.doi.org/10.1037//0021-9010.86.1.145.

Stoeber, J. (2011). The dual nature of perfectionism in sports: relationships with emotion, motivation, and performance. *International Review of Sport and Exercise Psychology, 4*(2), 128–145.

Stoeber, J., & Becker, C. (2008). Perfectionism, achievement motives, and attribution of success and failure in female soccer players. *International Journal of Psychology, 43*(6), 980–987.

Stoeber, J., & Eismann, U. (2007). Perfectionism in young musicians: relations with motivation, effort, achievement, and distress. *Personality and Individual Differences, 43*(8), 2182–2192. http://dx.doi.org/10.1016/j.paid.2007.06.036.

Stoeber, J., & Kersting, M. (2007). Perfectionism and aptitude test performance: Testees who strive for perfection achieve better test results. *Personality and Individual Differences, 42*(6), 1093–1103.

Stoeber, J., Uphill, M. A., & Hotham, S. (2009). Predicting race performance in triathlon: the role of perfectionism, achievement goals, and personal goal setting. *Journal of Sport & Exercise Psychology, 31*(2), 211–245.

Strutton, D., & Lumpkin, J. T. (1993). The relationship between optimism and coping styles of salespeople. *Journal of Personal Selling & Sales Management, 13*(2), 71–82.

Thayer, J. F., & Lane, R. D. (2000). A model of neurovisceral integration in emotion regulation and dysregulation. *Journal of Affective Disorders, 61*(3), 201–216. http://dx.doi.org/10.1016/S0165-0327(00)00338-4.

Tugade, M. M., & Fredrickson, B. L. (2004). Resilient individuals use positive emotions to bounce back from negative emotional experiences. *Journal of Personality and Social Psychology, 86*(2), 320–333. http://dx.doi.org/10.1037/0022-3514.86.2.320.

Turner, M. J., & Barker, J. B. (2013). Resilience: lessons from the 2012 Olympic Games. *Reflective Practice, 14*(5), 622–631. http://dx.doi.org/10.1080/14623943.2013.835724.

Walsh, W. B., Craik, K. H., & Price, R. H. (1992). In W. B. Walsh, K. H. Craik, & R. H. Price (Eds.), *Person-environment psychology: Models and perspectives*. Hillsdale, NJ: L. Erlbaum Associates.

Weinberg, R. S., & Genuchi, M. (1980). Relationship between competitive trait anxiety, state anxiety, and golf performance: a field study. *Journal of Sport Psychology, 2*(2), 148.

Weinberg, R. S., & Gould, D. (2011). *Foundations of sport and exercise psychology* (5th ed.). Champaign, IL; Leeds: Human Kinetics.

Westman, M. (1990). The Relationship between stress and performance: The moderating effect of hardiness. *Human Performance, 3*(3), 141.

Windle, G., Bennett, K. M., & Noyes, J. (2011). A methodological review of resilience measurement scales. *Health and Quality of Life Outcomes, 9*, 8. http://dx.doi.org/10.1186/1477-7525-9-8.

Zeng, H. Z., Leung, R. W., & Wenhao, L. I. U. (2008). An examination of competitive anxiety and self-confidence among college varsity athletes. *Journal of Physical Education & Recreation (10287418), 14*(2), 6–12.

Zuckerman, M. (1979). *Sensation seeking: Beyond the optimal level of arousal*. Hillsdale, NJ: L. Erlbaum Associates. New York: distributed by the Halsted Press Division of Wiley, 1979.

Zuckerman, M. (1983). Sensation seeking and sports. *Personality and Individual Differences, 4*(3), 285–292.

Zuckerman, M. (2007). *Sensation seeking and risky behavior*. Washington, DC: American Psychological Association.

Chapter 19

# The Influence of Hormonal Stress on Performance

Franziska Lautenbach[1], Sylvain Laborde[1,2]
[1]*Department of Performance Psychology, Institute of Psychology, German Sport University, Cologne, Germany;* [2]*UFR STAPS, EA 4260, University of Caen, Caen, France*

In today's competitive societies, performing well is one of the most essential elements of success. Winning or losing is often a reflection of one's performance, and performing well can be particularly challenging in high-pressure situations. Such situations are stressful because the stakes are high and there may be a lot to lose, say, a client, a job, a lover, or a sports competition. Performing well under pressure leads to winning: Delivering an excellent presentation wins the client over; staying calm when answering a recruiter's questions wins the job; showing one's best possible side wins the heart of another; and playing the best tennis wins the final.

## INSTRUCTIONS FOR GETTING YOUR DREAM JOB BASED ON CORTISOL RESEARCH

Consider one example of how cortisol can influence performance. Imagine you have been looking for a job for about 8 weeks, and you see an offer for a great position. If you are planning on doing your Ph.D., it is probably a three-year government-subsidized position without teaching; or if you are a post doc, it may be a tenure-track position in a city in which you and your family have wanted to live. The following example applies for every other job interview, as well (we are just better at describing dream jobs for scientists).

In short, you apply, and they ask you to come in for an oral presentation and interview. According to cortisol research—and wisdom from your granny—the day before your interview, you should go to bed early and get enough sleep. Moreover, cortisol research would also advise you to have a sip of something "strong" (Stalder, Huckleridge, Evans, & Clow, 2009).

As you wake up on the morning of your job interview, your body is already reacting differently at the physiological level: The cortisol awakening response (CAR; see Suay & Salvador, 2012), also known as the stress marker in the morning, is showing a higher amplitude than usual—but if you went to bed early and had a sip of the strong stuff (Stalder, Hucklebridge, Evans, & Clow, 2009), it is not as high as it might have been. It makes sense to get up at least 2–3 h before your interview, so there is time for the CAR to decrease, which may reduce your level of stress (in accordance to Suay & Salvador, 2012).

On your way to the interview you will be thinking, "Am I good enough? Do I have what it takes? What will they ask?" You have a good chance of performing well in the interview if, first of all, you possess high emotional intelligence, as it serves as a protector again acute stressors (in this case, your interview). Second, you have an even better chance if you perceive this interview as a challenge rather than a threat. Third, you will improve your chances if you plan enough time to get to the place of the interview. Even though you will probably not reach the physical exhaustion limit that is necessary to increase cortisol, it is just not nice to arrive late and/or be sweaty.

In general, during your oral presentation and the following interview, you should keep in mind that your potential new boss is not threatening your ego with the questions he or she is asking. So, there is no need for you to feel threatened. Let us first talk about your oral presentation. This is when your motor performance becomes important: Either do not use a pointer or follow the preceding instructions to keep your cortisol levels low and your hand steady. Regarding your interview, your attention should be on the questions asked by the person in front of you and not on the sexy eyes of the student assistant also sitting in the room (Putman & Berling, 2011). Also, blame it on your cortisol level when you forget things or cannot recall them too well (Schoofs, Preuβ, & Wolf, 2008). Depending on the position you are applying for, try telling an appropriate funny joke, a good trick for decreasing your cortisol level very quickly (see Buchanan, al'Absi, & Lovallo, 1999).

In the end, remember it is hard to infer directionality. In other words, if we mentioned here mainly an influence of cortisol on your job interview performance, it is also possible that your cognitive and motor performance influence your cortisol level as well. Also, it is useful –for your self-esteem–to differentiate between the process and the outcome. Just because you did an excellent job (read: realized a very good performance) does not mean you will get the job (read: obtain a positive outcome). There might still be someone that the selection committee found more suited for the job than you. Nevertheless, best of luck!

As you can imagine, the situation we just described involves many complex interactions among the phenomena we mentioned. To help shed light on the underlying mechanisms and also to provide structural guidelines for future research and interventions, in this chapter, we propose an integrative framework of the cortisol–performance relationship. We now invite you to reflect on the scientific evidence on which the preceding instructions for getting your dream job were based.

## CORTISOL—WHAT IS IT AND WHAT DOES IT DO?

Cortisol is the end product of a stimulation of the hypothalamus-pituitary-adrenal (HPA) axis that is released as a result of psychosocial stress and/or physical effort. Receiving a stimulus, the hypothalamus releases corticotrophin-releasing hormones from the paraventricular nucleus. Via the hepatic portal system, these hormones pass to the anterior pituitary, where they stimulate the production and release of adrenocorticotropic hormones, which enter the bloodstream and finally stimulate the adrenal cortex to release glucocorticoids, cortisol being one of them.

Psychosocial stressors include ambiguous, novel, uncontrollable, unpredictable stimuli and situations with high-ego involvement (Hellhammer, 2008). The physical effort necessary to increase cortisol depends on the intensity/load and duration (for details, see the review by Crewther, Keogh, Cronin, & Cook, 2006). Hill and colleges (2008) showed that an intensity of 60–80% of the maximal oxygen uptake is needed to increase cortisol levels. Others found evidence that the intensity has to be higher than 70% (Davies & Few, 1973). An overview by Gatti and De Palo (2011) explored whether cortisol is already increased due to physical effort in different performance areas (e.g., soccer, basketball, tennis, video games) and, thus, can be used as a starting point for the present investigation.[1]

Cortisol docks on glucocorticoid receptors, which are found everywhere in the body. Thus, it is widely accepted that cortisol has a physiological effect (e.g., on metabolism), thereby potentially influencing motor performance (see Salvador & Costa, 2009; for a review). Second, as these glucocorticoid receptors are present particularly in the prefrontal cortical structures (Putman & Berling, 2011), it does not come as a shock that cortisol has an influence on cognitive performance (e.g., attention control; see Putman & Roelofs, 2011; working memory; see Schoofs, Wolf, & Smeets, 2009).

## STATE AND TRAIT INFLUENCES ON CORTISOL

Individuals can show different psychophysiological reactions to the same stressful situation (Selye, 1975) depending on how they appraise the stressor (Lazarus, 2000). Lazarus (2000) noted two important components of such appraisals in particular: the evaluation of the importance of the situation (*primary appraisal*) and the ability to cope with or control the stressor (*secondary appraisal*). An individual appraises a competitive situation as either a threat or a challenge and will reappraise the situation over and over again, because a competition is an ongoing process. For example, when the situation is particularly important for an individual—that is, when there is high-ego involvement—but that person's self-perceived abilities to cope or control the stressor are low, a threat appraisal is most likely to happen (Salvador & Costa, 2009). Because the HPA system reacts to situations with high-ego involvement and perceived uncontrollability, a cortisol increase follows, as meta-analytic findings have confirmed (Dickerson & Kemeny, 2004). In this meta-analysis, the authors took into account 208 acute psychological laboratory

---

1. When interested in only cortisol increase due to physical effort, data collected before and after matches (i.e., competition days) should be ignored because of psychosocial stress.

stressors[2] and found that appraising a situation as threatening led to a higher increase in cortisol. The influence of cognitive appraisal on cortisol changes was corroborated by Gaab, Rohleder, Nater, and Ehlert (2005). They found that cognitive appraisal explained 35% of the cortisol response variance.

It is likely that the psychophysiological reaction to stressful situations depends not only on state characteristics but also on trait and stable characteristics. For example, *trait emotional intelligence* (Petrides, 2009) has a positive effect on emotion regulation (e.g., Laborde, Brüll, Weber, & Anders, 2011; Mikolajczak, Nelis, Hansenne, & Quoidbach, 2008). More specifically, it was associated with a lower increase in cortisol after an acute stressor, such as an arithmetical task (Laborde, Lautenbach, Allen, Herbert, & Achtzehn, 2014) or the TSST (Mikolajczak, Roy, Luminet, Fillee, & de Timary, 2007). On the other hand, perfectionism, particularly concerns over mistakes and doubts (CMD scale) was found to be positively associated with an increase in cortisol after the TSST and accounted significantly for 18% of the cortisol response variance (Wirtz et al., 2007). In this study, the authors defined perfectionism as setting excessively high standards for performance accompanied by overly critical self-evaluation. It seems that certain traits (e.g., trait emotional intelligence) are related to more adaptive emotional regulation responses, which can lower the cortisol response to an acute stressor, whereas other traits facilitate the level of ego involvement (e.g., perfectionism, CMD scale), thereby fueling the cortisol response.

In summary, trait characteristics, through their influence on cognitive appraisal, may have an impact on cortisol level before, during, and after competitive situations. Yet the question of how cortisol influences performance during a competition remains. In the following section, we discuss the role of cortisol in performance.

## CORTISOL AND PERFORMANCE

To better understand the influence of emotions on performance, assessing subjective reports (e.g., questionnaires) and physiological measures (e.g., cortisol, heart rate variability, galvanic skin resistance) is necessary (Laborde, Raab, & Dosseville, 2013), as they reflect different components of emotions (Scherer, 2005). So far, studies have examined cortisol primarily as a passive marker, one that indicates, for example, how objectively stressed athletes are during their first day of a competition (Filaire, Alix, Ferrand, & Verger, 2009). However, increasing empirical evidence shows that cortisol may be not only a stress marker but also a component of mechanisms linked to performance. Researchers assume that cortisol's influence on performance occurs not only in the field of sports but also, for example, in examination settings (Huwe, Hennig, & Netter, 1998), police work (Akinola & Mendes, 2012), and medical settings (McGraw et al., 2013).

---

2. The Trier Social Stress Test (TSST; Kirschbaum, Pirke, & Hellhammer, 1993) is the most efficient method of increasing cortisol levels in the laboratory.

## Cortisol and Sports Performance

Several studies have focused on the cortisol–performance relationship in sports. In a basketball study, Robazza et al. (2012) assessed cortisol immediately before several basketball games during one season and explored its relation to athletes' and coaches' technical performance evaluations. They found no significant correlation. The first experimentally controlled study showing a link between cortisol and sports performance focused on a particular performance parameter (i.e., the second tennis serve) before and after anxiety induction (i.e., the second part of the TSST). The correlation between cortisol and service performance was negative (Lautenbach, Laborde, Achtzehn, & Raab, 2014). Doan, Newton, Kraemer, Kwon, and Scheet (2007) looked at whether the same link between cortisol and performance could be found during an actual competition. The authors investigated the cortisol response during a 10-hr, 36-hole golf competition. They found a negative correlation, implying that the poorer golfing performance was associated to higher cortisol levels. A similar negative relationship between cortisol and most performance parameters was found in a single case analysis of two opponents in a tennis match (Lautenbach, Laborde, Klämpfl, Achtzehn, & Herbert, submitted). This study showed in detail that cortisol increased and decreased rapidly over the course of a 70-min match, which can be understood as a reaction to the ongoing processes linked to different but continuing stressors during a competition. Overall, a picture emerges that there is a negative relationship between cortisol and performance in sport.

### Cortisol and Outcome in Sports

Because performance is a main indicator of outcome (you are more likely to get the job if you perform well in the interview), we review the literature on the cortisol–outcome relationship as well. Once again, most of the research in this area has been done in the sports context (for a review, see Salvador & Costa, 2009) and has focused particularly on the different cortisol responses of winners and losers before and after a competition. The commonly acknowledged cortisol response pattern shows an anticipatory rise before competition (e.g., Salvador, Suay, González-Bono, & Serrano, 2003) and a second increase shortly after competition due to intense mental and athletic effort (e.g., Kraemer, Fry, & Rubin, 2001). In brief, the results of comparisons between winners and losers are inconsistent. Findings suggest no differences (Suay et al., 1999), higher levels in losers before and after competition (Filaire et al., 2009), and no difference before a competition but a higher increase in cortisol in losers after competition (Jiménez, Aguilar, & Alvero-Cruz, 2012).

### Model of Neuroendocrine and Mood Responses to a Competitive Situation

Salvador and Costa (2009) proposed a theoretical approach to explain neuroendocrine and mood responses before, during, and after a competitive situation. They argued that previous experience of victory and defeat, social status, personality, and generic abilities as well as the cognitive appraisal of a competition (pre-competition phase) play a part in activating the coping response during competition, because they cause changes in testosterone and cortisol, which additionally influence changes in mood. When an individual appraises a competition as a challenge, an active coping response is more likely to develop, resulting in better mood, an increase in testosterone, and a decrease in cortisol (competition phase). Consequently, the probability of winning (transition from competition to post-competition phase) increases.

### Critique of the Model of Neuroendocrine and Mood Responses to a Competitive Situation

Salvador and Costa (2009) model is an initial attempt to explain the cortisol response to competition, and certain aspects need critical consideration. Previous experience and personality influence not only the pre-competition phase but also the cortisol response during and after a competition, a process that Salvador and Costa may have oversimplified in their model. During a competitive situation, which is an ongoing process, individuals continually appraise and reappraise potential new stressors, partial results, or their own performance. For instance, if your oral presentation went well, you might appraise the interview more as a challenge because you now realize how well you can do. In contrast, if you did poorly in the presentation, the prospect of still having to do the interview might simply terrify you. You might change your behavior (answer questions with more or less confidence) depending on your evaluation of the situation as a challenge or as a threat. You will keep reappraising the competition (Lazarus, 2000)—or in this case the job interview—doing this not only once beforehand and once after the outcome but throughout the encounter.

This model further assumes that the hormonal changes are rather linear, which has been empirically falsified (Doan et al., 2007; Lautenbach et al., submitted). Moreover, it is concerned primarily with outcome and not performance. Even though performance is an indicator for outcome, other factors such as the opponent in sports, the other candidates in a job interview, or the questions posed in an exam can change the outcome even though the athlete, the applicant, or the student performed well. Most importantly, the underlying mechanisms for the cortisol–outcome relationship presented in this model are changes in mood due to increases and decreases in cortisol and testosterone. Mood, defined as a rather long-lasting phenomenon (at least hours to days; Scherer, 2005), cannot change as quickly as proposed in the model during the course of a competition. In summary, to understand the influence of hormones (i.e., cortisol) on outcome, it is necessary to look closely, first, at performance and, second, at other potential mechanisms, such as cognition and motor performance.

## Cortisol and Cognitive Performance

Looking at cognition can reveal possible underlying factors that explain the cortisol–performance relationship. It is generally accepted that cortisol has an impact on cognitive functions (Suay & Salvador, 2012) because cortisol can pass the blood–brain barrier, and glucocorticoid receptors are found in almost every organ in the body, particularly in the prefrontal cortical structures (Putman & Berling, 2011) and the hippocampus (Kirschbaum, 2001). Cortisol, both at baseline and when increased after an acute stressor or single administration of cortisol pills, affects decision-making, attention, self-infiltration (Quirin, Koole, Baumann, Kazén, & Kuhl, 2009), and memory (Kirschbaum, Wolf, Wippich, & Hellhammer, 1996) by inhibiting and altering information processing. In this chapter, we focus on attention; attention is believed to reflect the *common executive function* because it is needed for all cognitive processes (according to Miyake & Friedman, 2012).

When assessing attentional changes, studies have focused mainly on selective attention to threatening stimuli (for a review, see Putman & Roelofs, 2011). Van Honk et al. (2003) found that the baseline level of a high-cortisol subgroup was related to an allocation of attention away from threat stimuli. Similar results were found when 40 mg (Putman, Hermans, & van Honk, 2010) and 10 mg (Breitberger et al., 2013) of cortisol were administered. In a double-blind study, 10 mg showed the same increase in inhibition for angry faces as the previous studies, but the effect was not found for 40 mg (Taylor, Ellenbogen, Washburn, & Joober, 2011), in contrast to the results of Putman et al. study (2010). Taylor et al. (2011) concluded that "moderate glucocorticoid elevations may have adaptive effects on emotional information processing, whereas high glucocorticoid elevations appear to attenuate this effect" (p. 17). Overall, the effects are complex, but cortisol is believed to increase the focus on task-relevant information (Taylor et al., 2011).

Thus, while you are being interviewed, very high levels of cortisol will drive your attention away from the sexy eyes of the student assistant. This is supported by Putman and Berling (2011) study using 40 mg of cortisol. They found a reduction in selective attention for erotic words using a Stroop task.

## Cognitive-Processing Hypothesis

Putman and Roelofs (2011) cognitive-processing hypothesis offers an understanding of the connection between cortisol and cognition. It states that within the cognitive system, elaborated and goal-directed behavioral strategies are superior to the processing of emotional information. In stressful situations, this superiority gives way to a driven and reflex-like behavior, which is followed by a change in attentional selection toward threat-related emotional stimuli (*threat-biased attention*; thoughts such as "Oh no, I will never get this job!") that are irrelevant or can even interfere with immediate performance (Vuilleumier, 2002). According to this hypothesis, stress-released cortisol elevation influences cognitive functions and, thereby, promotes a more effective coping process. More specifically, cortisol is helpful for re-establishing proper goal-directed and approach-related behavior in healthy people. Thus, cortisol elevation is beneficial because it helps people regain focus on task-relevant information. Cortisol is not only an indicator of stress but also simultaneously part of active coping processes.

## Critique of the Cognitive-Processing Hypothesis

The cognitive-processing hypothesis could help explain the influence of cortisol on cognition, but so far, no connections between this hypothesis and performance have been made. Implementing the cognitive-processing hypothesis in a cortisol–performance framework requires keeping some of its limitations in mind. Indeed, evidence supporting this hypothesis has been obtained only in clinical trials. This support should be viewed with caution given the differences in results obtained in clinical trials and those obtained in real competitive settings. First, the level of cortisol induced by competition is usually around 20–40 nmol/l, whereas cortisol administration of 20–40 mg can increase cortisol up to 120 nmol/l. A nonlinear dose–response relation between cognitive effects of cortisol has been assumed (e.g., Putman & Roelofs, 2011). Thus, it seems feasible that the lower level of cortisol concentration observed in sports in comparison with clinical trials is not enough to foster adaptive and goal-oriented coping processes in this domain. Second, performing a computer task is certainly not as complex as performing a fine-tuned motor skill in sports. Thus, even though an athlete has a tendency toward task-oriented focus and approach-related behavior as a coping process due to higher cortisol, this does not necessary mean that the athlete is able to perform accordingly with respect to potential skills and the requirements of the situation. Third, clinical trials neglect the previously mentioned impact of cognitive appraisal, because cortisol increase is initially due to physiological changes. Last, the hypothesis does not take into account the impact of physical exercise on cortisol level (for a review, see Gatti & De Palo, 2011), because it is not essential for clinical trials. Despite those acknowledged limitations, the cognitive-processing hypothesis still provides a perspective that can help explain cortisol–performance relations by integrating research findings that propose an influence of cortisol on cognitive functions.

At first glance, this is where it gets tricky for your job interview. Results from research on the cortisol–outcome relation are inconsistent. Research investigating the cortisol–performance relationship suggests that low cortisol is good for performance, and now, research on cortisol and cognition suggests that a high level of cortisol could be helpful for focusing on task-relevant information.

## Cortisol and Motor Performance[3]

In the beginning of this chapter, we focused on physical effort and its impact on cortisol. We now turn to how cortisol influences motor performance.

---

3. Given that outside the sports context motor performance is, to a certain extent, of little concern, and given that this book seeks to give a broader perspective on performance psychology, this section on motor performance and cortisol is shortened. For a more detailed description of the relationship between cortisol and motor performance, see Lautenbach, Laborde, Crewther, Putman, and Raab (in prep) and the review by Crewther et al. for the resistance-training model (2011).

Most research on cortisol and motor performance, also referred to as physical performance, has focused on speed, strength, and endurance (see review by Crewther, Cook, Cardinale, Weatherby, & Lowe, 2011), which have been tested using different test batteries. An overall picture has emerged of a higher baseline cortisol level being related to higher speed and strength (e.g., Crewther, Lowe, Weatherby, Gill, & Keogh, 2009; Passelergue & Lac, 2012). A different result was found for endurance performance, showing no interaction in a 20-km time-trial cycling task of performance and a higher cortisol level (Baume, Steel, Edwards, Thorstensen, & Miller, 2008). Given that motor performance is related to (sports) performance, cortisol could also have an indirect impact on performance through *motor performance*.

## CORTISOL–PERFORMANCE FRAMEWORK

We designed the cortisol–performance framework, presented in Figure 1, to address the limitations of the models we reviewed earlier and to provide a way to understand how cortisol might influence performance. In the next paragraphs, we describe the model in more detail, followed by a section on its limitations.

In terms of the elements that underlie the cortisol–performance relation, the framework suggests that in a competitive situation, an individual's *trait characteristics* (e.g., trait emotional intelligence) influence the way the individual appraises a situation and will influence cortisol changes through this *cognitive appraisal*. *Demographic parameters* can have an impact on cognitive appraisal and reappraisal, as well, and they are directly connected to the cortisol response itself. Studies have shown that factors such as sex (e.g., women's use of oral contraceptives or menstrual cycle phase; Kirschbaum, Kudielka, Gaab, Schommer, & Hellhammer, 1999), age (Netherton, Goodyer, Tamplin, & Herbert, 2004), smoking (Rohleder & Kirschbaum, 2006), and physical fitness

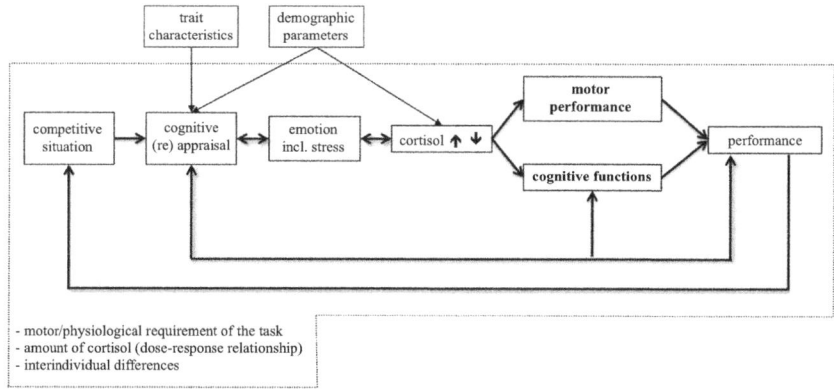

**FIGURE 1** The cortisol–performance framework.

(Rimmele et al., 2007) influence the cortisol response to an acute stressor. So far, it cannot be clearly untangled whether the differences in cortisol responses are mediated by cognitive appraisal (due to a stress induction) or are purely physiological. Given this ambiguity, in our framework, demographic parameters are connected directly to both cortisol changes and cognitive appraisal.

In a competitive situation, an individual's cognitive appraisal applies to situation-specific stressors that arise during the competition and to his or her perceived ability to cope. This appraisal leads to the experience of stress and, consequently, emotions, because stress is part of every emotion (Lazarus, 2000). Many physiological reactions (e.g., change in heart rate, pulse, or blood pressure) to an emotion will take place (for a review, see Kreibig, 2010). Although this framework focuses on cortisol, the general concept can be used for a variety of physiological changes in stressful situations.

During a competition, the two main influences on cortisol level are psychosocial stress (e.g., Filaire et al., 2009) and physical effort (Cadore et al., 2012). Studies have shown that physical exercise can provoke an increase in cortisol (Gatti & De Palo, 2011). Because our framework describes the effects of cortisol increase due to psychosocial stress and cognitive appraisal, the presented arrow is unidirectional, that is, we show a one-way influence of cortisol on physiological conditions. If the influence of cortisol on *cognitive performance* (e.g., attention) has been shown when the cortisol rise originates from psychosocial stress (e.g., Putman & Roelofs, 2011), there is, to the best of our knowledge, no study providing empirical evidence for an influence on cognitive functions when cortisol is increased by physical effort (McMorris, 2009).

Due to the impact of cortisol on cognitive performance, a change in *coping processes* is possible (e.g., attention shift away from threatening stimuli). Therefore, cortisol may drive coping processes and, hence, have an indirect impact on performance as well. A change in cognitive performance can also have a direct influence on *performance*; for example, an athlete showing a faster reaction time will potentially experience an increase in performance.

After considering the differential influence of cortisol on cognitive performance according to the causes provoking its rise, we need to consider, as well, the various consequences of a cortisol rise for motor performance, especially in the sports context. The performance displayed during a competition is understood as a dynamic process, as the performer faces a series of different stressors until the competition ends (in victory or defeat). Therefore, cognitive reappraisal is expected to happen continuously during a competition. In this process, cortisol is expected to fluctuate and influence cognitive and motor performance continuously.

Note that we do not claim that the framework is linear. Rather, the processes described should be understood as occurring in parallel. Finally, the model assumes no hierarchical order of systems (e.g., cognitive or physiological).

## SUMMARY AND OUTLOOK

The cortisol–performance framework is the first step in showing that cortisol is not merely a passive marker of stress or anxiety but has an active role in cognitive and motor performance. The majority of cortisol studies suggest that losers have a higher level of cortisol, which leads to poorer performance. It is unclear, however, whether this poorer performance by a losing athlete or individual is due to a lack of performance ability or to the potentially negative impact of cortisol on cognitive performance. Even now, most studies focus on the differences between winners and losers. To our knowledge, no study has yet investigated individuals competing several times and examined their cortisol levels related to victories and defeats. Moreover, it would be interesting to study the cortisol levels of an individual during (1) a competition in which she or he is ahead and later wins and (2) a competition in which she or he is behind and later wins.

To test and improve the framework, future research should aim to answer several challenging questions: Is there an optimal cortisol level that fosters cognitive and motor performance? Does this relationship have an inverted U-shape? Does it depend on the task? Does it differ between individuals? How can people reach an optimal cortisol level? From a theoretical perspective, finding answers to these questions will further our understanding of how cortisol influences performance and of the cognitive mechanisms underlying this relationship. From an applied perspective, a greater understanding of the link between cortisol and performance will help people face life's challenges, whether they be exams, important meetings, job interviews, or tournaments, which all take place in the competitive setting of society.

## REFERENCES

Akinola, M., & Mendes, W. B. (2012). Stress-induced cortisol facilitates threat-related decision making among police officers. *Behavioral Neuroscience*, *126*, 167–174.

Baume, N., Steel, G., Edwards, T., Thorstensen, E., & Miller, B. F. (2008). No variation of physical performance and perceived exertion after adrenal gland stimulation by synthetic ACTH (Synacthen) in cyclists. *European Journal of Applied Physiology*, *104*, 589–600.

Breitberg, A., Drevets, W. C., Wood, S. E., Mah, L., Schulkin, J., Sahakian, B. J., et al. (2013). Hydrocortisone infusion exerts dose- and sex-dependent effects on attention to emotional stimuli. *Brain and Cognition*, *81*, 247–255.

Buchanan, T. W., al'Absi, M., & Lovallo, W. R. (1999). Cortisol fluctuates with increases and decreases in negative affect. *Psychoneuroendocrinology*, *24*, 227–241.

Cadore, E. L., Izquierdo, M., dos Santos, M. G., Martins, J. B., Rodrigues Lhullier, F. L., Pinto, R. S., & Kruel, L. F. M. (2012). Hormonal responses to concurrent strength and endurance training with different exercise orders. *Journal of Strength and Conditioning Research*, *26*, 3281–3283.

Crewther, B. T., Cook, C., Cardinale, M., Weatherby, R. P., & Lowe, T. (2011). Two emerging concepts for elite athletes. The short-term effects of testosterone and cortisol on the neuromuscular system and athe dose-response training role of these endogenous hormones. *Sports Medicine*, *41*, 103–123.

Crewther, B. T., Keogh, J., Cronin, J., & Cook, C. J. (2006). Possible stimuli for strength and power adaptation. Acute hormonal responses. *Sports Medicine*, *36*, 215–238.

Crewther, B. T., Lowe, T., Weatherby, R. P., Gill, N., & Keogh, J. (2009). Neuromuscular performance of elite rugby union players and relationships with salivary hormones. *Journal of Strength and Conditioning Research, 23*, 2046–2053.

Davies, C. T., & Few, J. D. (1973). Effects of exercise on adrenocortical function. *Journal of Applied Physiology, 35*, 887–891.

Dickerson, S. S., & Kemeny, M. E. (2004). Acute stressors and cortisol responses: a theoretical integration and synthesis of laboratory research. *Psychological Bulletin, 130*, 355–391. http://dx.doi.org/10.1037/0033-2909.130.3.355.

Doan, B. K., Newton, R. U., Kraemer, W. J., Kwon, Y.-H., & Scheet, T. P. (2007). Salivary cortisol, testosterone, and T/C ratio responses during a 36-hole golf competition. *International Journal of Sports Medicine, 28*, 470–479. http://dx.doi.org/10.1055/s-2006-924557.

Filaire, E., Alix, D., Ferrand, C., & Verger, M. (2009). Psychophysiological stress in tennis players during the first single match of a tournament. *Psychoneuroendocrinology, 34*, 150–157. http://dx.doi.org/10.1016/j.psyneuen.2008.08.022.

Gaab, J., Rohleder, N., Nater, U. M., & Ehlert, U. (2005). Psychological determinants of the cortisol stress response: the role of anticipatory cognitive appraisal. *Psychoneuroendocrinology, 20*, 599–610. http://dx.doi.org/10.1016/j.psyneuen.2005.02.001.

Gatti, R., & De Palo, E. F. (2011). An update: salivary hormones and physical exercise. *Scandinavian Journal of Medicine & Science in Sports, 21*, 157–169. http://dx.doi.org/10.1111/j.1600-0838.2010.01252.

Hellhammer, D. (2008). Principles of the crosstalk between brain and body—Glandotropy, ergotropy, and trophotropy. Key Issues in Mental Health. In D. H. Hellhammer, & J. Hellhammer (Eds.), *Stress. The brain-body connection* (Vol. 147) (pp. 21–38). Basel, Switzerland: Karger.

Huwe, S., Hennig, J., & Netter, P. (1998). Biological, emotional, behavioral, and coping reactions to examination stress in high and low state anxious subjects. *Anxiety, Stress & Coping: An International Journal, 11*, 47–55.

Jiménez, M., Aguilar, R., & Alvero-Cruz, J. R. (2012). Effects of victory and defeat on testosterone and cortisol response to competition: evidence for same response patterns in men and women. *Psychoneuroendocrinology, 37*, 1577–1581.

Kirschbaum, C. (2001). Das Stresshormon Cortisol – Ein Bindeglied zwischen Psyche und Soma? [The stresshormone cortisol – a link between psych and soma?]. In H. Süssmuth (Ed.), *Jahrbuch der Heinrich-Heine-Universität Düsseldorf*. Düsseldorf: Heinrich-Heine-Universität Düsseldorf e.V.

Kirschbaum, C., Kudielka, B. M., Gaab, J., Schommer, N., & Hellhammer, D. H. (1999). Impact of gender, menstrual cycle phase, and oral contraceptives on the activity of the hypothalamus-pituitary-adrenal axis. *Psychosomatic Medicine, 61*, 154–162.

Kirschbaum, C., Pirke, K.-M., & Hellhammer, D. H. (1993). The 'Trier Social Stress Test'—a tool for investigating psychobiological stress responses in a laboratory setting. *Neuropsychobiology, 28*, 76–81.

Kirschbaum, C., Wolf, O. T., Wippich, W., & Hellhammer, D. H. (1996). Stress- and treatment-induced elevations of cortisol levels associated with impaired declarative memory in healthy adults. *Life Sciences, 58*, 1475–1483.

Kraemer, W. J., Fry, A. C., & Rubin, M. R. (2001). Physiological and performance responses to tournament wrestling. *Medicine and Science in Sports and Exercise, 33*, 1367–1378.

Kreibig, S. D. (2010). Autonomic nervous system activity in emotion: a review. *Biological Psychology, 84*, 394–421. http://dx.doi.org/10.1016/j.biopsycho.2010.03.010.

Laborde, S., Brüll, A., Weber, J., & Anders, L. S. (2011). Trait emotional intelligence in sports: a protective role against stress trough heart rate variability? *Personality and Individual Differences, 51*, 23–27.

Laborde, S., Lautenbach, F., Allen, M. S., Herbert, C., & Achtzehn, S. (2014). The role of trait emotional intelligence in emotion regulation and performance under pressure. *Personality and Individual Differences, 57,* 43–47. http://dx.doi.org/10.1016/J.Paid.2013.09.013.

Laborde, S., Raab, M., & Dosseville, F. (2013). Emotions and performance: valuable insights from the sports domain. In C. Mohiyeddini, M. Eysenck, & S. Bauer (Eds.), *Handbook of psychology of emotions: Recent theoretical perspectives and novel empirical findings* (Vol. 1) (pp. 325–358). New York, NY: Nova.

Lautenbach, F., Laborde, S., Achtzehn, S., & Raab, M. (2014). Preliminary evidence of salivary cortisol predicting performance in a controlled setting. *Psychoneuroendocrinology, 42,* 218–224.

Lautenbach, F., Laborde, S., Crewther, B. T., Putman, P., & Raab, M. (in prep). Cortisol: Optimizing or hampering human performance? A position paper integrating cognitive and motor components

Lautenbach, F., Laborde, S., Klämpfl, M., Achtzehn, S., & Herbert, C. (submitted). The role of cortisol on performance during a tennis match—An exploratory case study. Manuscript submitted for publication.

Lazarus, R. S. (2000). How emotions influence performance in competitive sports. *The Sport Psychologist, 14,* 229–252.

McGraw, L. K., Out, D., Hammermeister, J. J., Ohlson, C. J., Pickerin, M. A., & Granger, D. A. (2013). Nature, correlates, and consequences of stress-related biological reactivity and regulation in army nurses during combat casualty simulation. *Psychoneuroendocrinology, 38,* 135–144.

McMorris, T. (2009). Exercise and cognitive function: a neuroendocrinological explanation. In T. McMorris, P. D. Tomporowski, & M. Audiffren (Eds.), *Exercise and cognitive function* (pp. 41–67). Chichester, England: Wiley.

Mikolajczak, M., Nelis, D., Hansenne, M., & Quoidbach, J. (2008). If you can regulate sadness, you can probably regulate shame: associations between trait emotional intelligence, emotion regulation and coping efficiency across discrete emotions. *Personality and Individual Differences, 44,* 1356–1368. http://dx.doi.org/10.1016/j.paid.2007.12.004.

Mikolajczak, M., Roy, E., Luminet, O., Fillee, C., & de Timary, P. (2007). The moderating impact of emotional intelligence on free cortisol responses to stress. *Psychoneuroendocrinology, 32,* 1000–1012. http://dx.doi.org/10.1016/j.psyneuen.2007.07.009.

Miyake, A., & Friedman, N. P. (2012). The nature and organization of individual differences in executive functions: four general conclusions. *Current Directions in Psychological Science, 21,* 8–14.

Netherton, C., Goodyer, I., Tamplin, A., & Herbert, J. (2004). Salivary cortisol and dehydroepiandrosterone in relation to puberty and gender. *Psychoneuroendocrinology, 29,* 125–140. http://dx.doi.org/10.1016/S0306-4530(02)00150-6.

Passelergue, P. A., & Lac, G. (2012). Salivary hormonal responses and performance changes during 15 weeks of mixed aerobic and weight training in elite junior wrestlers. *Journal of Strength and Conditioning Research, 26,* 3049–3058.

Petrides, K. V. (2009). *Technical manual for the trait emotional intelligence questionnaires (TEIQue).* London: London Psychometric Laboratory.

Putman, P., & Berling, S. (2011). Cortisol acutely reduces selective attention for erotic words in healthy young men. *Psychoneuroendocrinology, 36,* 1407–1417. http://dx.doi.org/10.1016/j.psyneuen.2011.03.015.

Putman, P., Hermans, E. J., & van Honk, J. (2010). Cortisol administration acutely reduces threat-selective spatial attention in healthy young men. *Physiology & Behavior, 99,* 294–300.

Putman, P., & Roelofs, K. (2011). Effects of single cortisol administrations on human affect reviewed: coping with stress through adaptive regulation of automatic cognitive processing. *Psychoneuroendocrinology, 36,* 439–448. http://dx.doi.org/10.1016/j.psyneuen.2010.12.001.

Quirin, M., Koole, S. L., Baumann, N., Kazén, M., & Kuhl, J. (2009). You can't always remember what you want: the role of cortisol in self-ascription of assigned goals. *Journal of Research in Personality, 43*, 1026–1032. http://dx.doi.org/10.1016/j.jrp.2009.06.001.

Rimmele, U., Zellweger, B. C., Marti, B., Seiler, R., Mohiyeddini, C., Ehlert, U., et al. (2007). Trained men show lower cortisol, heart rate and psychological responses to psychosocial stress compared with untrained men. *Psychoneuroendocrinology, 32*, 627–635. http://dx.doi.org/10.1016/j.psyneuen.2007.04.005.

Robazza, C., Gallina, S., D'Amico, M. A., Izzicupo, P., Bascelli, A., Di Fonso, A., & Di Baldassarre, A. (2012). Relationship between biological markers and psychological states in elite basketball players across a competitive season. *Psychology of Sport and Exercise, 13*, 509–517. http://dx.doi.org/10.1016/j.psychsport.2012.02.011.

Rohleder, N., & Kirschbaum, C. (2006). The hypothalamic-pituitary-adrenal (HPA) axis in habitual smokers. *International Journal of Psychophysiology, 59*, 236–243. http://dx.doi.org/10.1016/j.ijpsycho.2005.10.012.

Salvador, A., & Costa, R. (2009). Coping with competition: neuroendocrine responses and cognitive variables. *Neuroscience & Biobehavioral Reviews, 33*, 160–170. http://dx.doi.org/10.1016/j.neubiorev.2008.09.005.

Salvador, A., Suay, F., González-Bono, E., & Serrano, M. A. (2003). Anticipatory cortisol, testosterone and psychological responses to judo competition in young men. *Psychoneuroendocrinology, 28*, 364–375. http://dx.doi.org/10.1016/s0306-4530(02)00028-8.

Scherer, K. R. (2005). What are emotions? and how can they be measured? *Social Science Information, 44*, 695–729. http://dx.doi.org/10.1177/0539018405058216.

Schoofs, D., Preuß, D., & Wolf, O. T. (2008). Psychosocial stress induces working memory impairments in an *n*-back paradigm. *Psychoneuroendocrinology, 33*, 643–653.

Schoofs, D., Wolf, O. T., & Smeets, T. (2009). Cold pressor stress impairs performance on working memory tasks requiring executive functions in healthy young man. *Behavioral Neuroscience, 213*, 1066–1075.

Selye, H. (1975). Confusion and controversy in the stress field. *Journal of Human Stress, 1*, 37–44.

Stalder, T., Hucklebridge, F., Evans, P., & Clow, A. (2009). Use of a single case study design to examine state variation in the cortisol awakening response: relationship with time of awakening. *Psychoneuroendocrinology, 34*, 607–614.

Suay, F., & Salvador, A. (2012). Cortisol. In F. Ehrlenspiel, & K. Strahler (Eds.), *Psychoneuroendocrinology of sport and exercise: Foundations, markers, trends* (pp. 43–60). Oxford, England: Routledge.

Suay, F., Salvador, A., Gonzalez-Bono, E., Sanchis, C., Martinez, M., Martinez-Sanchis, S., & Montoro, J. B. (1999). Effects of competition and its outcome on serum testosterone, cortisol and prolactin. *Psychoneuroendocrinology, 24*, 551–566.

Taylor, V. A., Ellenbogen, M. A., Washburn, D., & Joober, R. (2011). The effects of glucocorticoids on the inhibition of emotional infromation: a dose–response study. *Biological Psychology, 86*, 17–25.

Van Honk, J., Kessels, R. P. C., Putman, P., Jager, G., Koppeschaar, H. P. F., & Postma, A. (2003). Attentionally modulated effects of cortisol and mood on memory for emotional faces in healthy young males. *Psychoneuroendocrinology, 28*, 941–948. http://dx.doi.org/10.1016/s0306-4530(02)00116-6.

Vuilleumier, P. (2002). Facial expression and selective attention. *Current Opinion in Psychiatry, 15*, 291–300.

Wirtz, P. H., Elsenbruch, S., Emini, L., Rudisuli, K., Groessbauer, S., & Ehlert, U. (2007). Perfectionism and the cortisol response to psychosocial stress in men. *Psychosomatic Medicine, 69*, 249–255. http://dx.doi.org/10.1097/PSY.0b013e318042589.

# Chapter 20

# Performing under Pressure: High-Level Cognition in High-Pressure Environments

**K. Werner**
*Department of Performance Psychology, Institute of Psychology, German Sport University Cologne, Cologne, Germany*

The successful practice of high-level cognition has ensured human survival for over thousands of years. Imagine you are hungry and the only weapon you have is a wooden spear. Your prey is 3-m high and weighs 6 tons—enough to feed your entire clan for weeks. The problem you have to solve is how to hunt down the mammoth without being injured or killed. Obviously, most of our ancestors successfully solved this problem. Our challenges are different nowadays, but the use of high-level cognition is still an important tool for dealing with life's requirements. For example, imagine you are the chief engineer of a global car company. Your marketing department wants creative innovation in the new car you are engineering right now, to make it competitive in the future and to increase the company's sales volume. Yet few of us work in engineering, design, or similar jobs, so I will introduce decision-making with a more everyday example: crossing a street. At first glance, this is a simple go or no-go decision, hardly a complex task needing high-level cognitive functions. However, if I add some cars moving along the street in both directions, this task becomes more complex. For a correct decision, that is, to cross without being run over by a car, we have to consider the speed of the cars, the distance to the cars, our walking speed, and the distance to the other side, and a few other environmental factors (e.g., whether the ground is slippery). Despite all these factors, most of us are able to make a correct decision and cross the street.

In this chapter, I review the current literature on performing so-called high-level cognitive tasks in stressful situations. My focus is on problem solving, creativity, and decision-making in complex situations, as these are the most sophisticated cognitive tasks humans can perform. Following the general purpose of this book, I explore these high-level cognitions in different domains. I end with suggestions for future directions in research on high-level cognition and affective states.

**FIGURE 1** Katonas' five-square problem (you will find the solution at the end of this chapter, Figure 3).

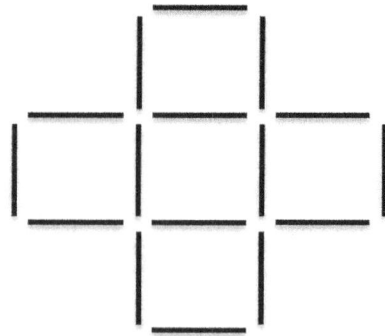

## PROBLEM SOLVING

Problem solving is a stepwise process that consists of two basic phases: creating the problem space and generating solutions to the problem.[1] The problem space takes into account the task environment, which includes the givens (e.g., a wooden spear, the landscape, and the number of clan members), the actual state (e.g., being hungry), and the goal state (e.g., being alive and full of mammoth meat for the next few weeks). Once a person creates a problem space, the next step is to generate solutions to the problem (Newell & Simon, 1972). For instance, in the opening example, the problem space can be built up and you can start creating solutions to the mammoth problem given that you will not achieve the goal state with a simple frontal attack.

Essential for problem solving is a gap between the actual and the goal state. A predetermined program cannot easily overcome this gap. Thus, an open math equation, such as "1 + 1 =," is not an actual problem-solving task. After finishing first grade, you can easily run your addition program to calculate the correct result. In contrast, Katona's (1940) five-square problem cannot be solved out of hand. In this problem, you have to reduce the number of squares from five to four, moving only three matchsticks (Figure 1).

## CREATIVITY

Creativity is based on a stepwise process that consists of four phases (see Lubart, 2001; for a review). It starts with the preparation phase, which is similar to the first phase in the problem-solving process and consists of defining the actual and the goal state. Note, that the goal state is more ambiguous than in most problem tasks (e.g., create something new versus hunt down the mammoth).

---

1. In the problem-solving research, different models explain the process with different numbers of phases. The different phases of these models can easily relate to one of the two phases used in the Newell and Simon (1972) model that I refer to in this chapter. For a more complex model, see Öllinger, Jones, and Knoblich (2014).

The incubation phase follows, in which the creator steps back from the task and does not consciously think about the task until the third phase is reached. This illumination phase is characterized by an "aha" experience. In the final phase of this creativity process, the ideas generated in the illumination phase are verified regarding their fit to the predetermined goal state.

In addition to describing the processes of creativity, it is necessary to define creativity. The most accepted definition states that creativity requires originality and effectiveness (Runco & Jaeger, 2012; also for a critical review). Regarding originality, something has to be novel, unique, or unusual to be creative. However, unusual things by themselves are not necessarily creative. Perhaps, there is a good reason for being unusual, like using paper to build a container ship. Therefore, creative things have to be effective (in the sense of appropriate) for specific requirements. Consider the car design example. You have designed a car that is state of the art, but to be competitive in the future, something new is needed, something that is novel and unique. After defining the actual and goal states, you collect further information (reading scientific articles, asking consumers about their needs, etc.). Now, you let this information work in your mind, hoping that after a while, some ideas will occur to you. Maybe you are thinking about a wireless steering wheel that is not attached to the dashboard, so everyone in the car is able to drive it. Although such a thing would be novel, unusual, and unique (at least for real cars), it is not creative by definition, because it is not appropriate to drive a car from the backseat without having control over the brakes.

## DECISION-MAKING

In contrast to problem solving and creativity, for decision-making, I will focus on the last step of a stepwise process: the selection of one of at least two options (Laborde, Dosseville, & Raab, 2013). Further, decisions can be differentiated between short-term and long-term; in this chapter, I focus on short-term decisions, such as playing a pass in football. And, last but not least, as indicated in the title of this chapter, I focus on decisions that are made under pressure.

Summing up, the decision-making I discuss in this chapter refers to a quick selection of a previously generated option in a pressure situation (e.g., time pressure). Thus, I use the simple heuristics approach as the theoretical base for this chapter[2] (Gigerenzer, Todd, & the ABC Research Group, 1999). To put this in the context of one of the opening examples, you have to make a quick (given you want to cross the street as soon as possible) selection between options (go or no go) in a pressure situation (you have to catch the bus or a wrong decision could lead to heavy injuries).

---

2. For a nice overview about decision-making especially in sports I recommend *Judgment, Decision-making and Success in Sport* by Bar-Eli, Plessner, and Raab (2001).

## COMPARING PROBLEM SOLVING AND CREATIVITY

A look at the underlying mechanisms of problem solving and creativity makes their differences obvious. For problem solving, the underlying mechanism is based on convergent thinking, whereas for creativity, it is based on divergent thinking (Akbari Chermahini & Hommel, 2012). Convergent thinking is characterized by looking for one (possible) solution to reach the goal state (e.g., hunting down the mammoth). It requires focused thinking about the solution. In contrast, divergent thinking is characterized by looking for as many options as possible (i.e., to find something new) before evaluating the options regarding their appropriateness. Thus divergent thinking—finding novel and unique options—requires an open mind. Let us take a closer look at this by using an example from the sports domain.

Imagine you are the coach of a national soccer team and your team will play in the World Cup final. On the basis of the scouting reports, you know that you have to stop your opponent's best player to prevent the opposing team from scoring. As he is probably the best player in the world, you cannot defend him one-on-one; rather, you have to find a solution that is based on the givens (i.e., the players of your team). Finding the best solution to this problem requires focused thinking. Now, imagine you are the playmaker of this soccer team. Your major task is to coordinate the offense to overcome the defense. You can do this with a pass to a teammate or dribbling, or you can shoot. Given this is the final of the world championship, there will be no simple option to score. You might think about playing a pass that nobody expects (except your teammate) to set up a one-on-one situation with the goalkeeper for one of your teammates. To find this creative option of playing this pass requires open-minded thinking.[3] Looking at these two examples, one might argue that creativity could also help to solve tactical problems or that playing the pass is just the solution to a specific problem. These arguments can be integrated by mapping problem solving and creativity and the underlying mechanisms (convergent and divergent thinking) as the two ends of a continuum regarding the level of definition (Figure 2).

**FIGURE 2** Continuum of thinking processes to accomplish goal states (GS) from convergent (left) represented by problem solving to divergent (right) represented by creativity.

---

3. For an overview about divergent thinking in sports see the work of Raab and Laborde (e.g., Johnson & Raab, 2003; Laborde & Raab, 2013) as well as Memmert (e.g., Memmert, 2010, 2012).

## COMPARING DECISION-MAKING AND PROBLEM SOLVING/CREATIVITY

Just as there are differences in the underlying mechanisms of problem solving and creativity, there are also differences in the underlying mechanisms between those processes and decision-making. From the simple heuristics perspective, the main differences between decision-making and problem solving/creativity are the amount of information and the amount of time available. In particular, decisions are often made with a lack of information and/or in a short time span (a couple of seconds). This becomes clearer in the next example.

Suppose you live on the African savanna. Your scouts have told you that you will be confronted with a lion later that day at a specific place, so you can prepare yourself. You can think about a strategy, which is a type of problem solving or an exercise in creative thinking. But, suppose as you are running through the savanna, you suddenly come across a lion in your path. You have mere seconds—or less—to make a quick and good decision (i.e., a short time span). In this case, you might rely on a simple heuristic, a rule of thumb that is based on experience or knowledge (Bar-Eli et al., 2001, p. 36). I now briefly introduce two heuristics, one used when lacking information and one for time pressure.

Imagine you have to predict the outcome of a tennis game (Roger Federer versus Daniel Brands) and you are not familiar with tennis (neither the game nor the players). Thus, you need a heuristic that takes account the lack of information. Luckily, you recently heard about Federer, who is the most successful tennis player of all time (17 Grand Slam victories), whereas you have never heard of Brands (number 163 in the tennis ranking).[4] In such a situation, you would probably decide in favor of Federer, based on the *recognition heuristic*, which states that when one of two alternatives is recognized and the other is not, infer that the recognized object has the higher value with respect to the criterion. Despite the heuristic's simplicity, you have a high probability of making a correct prediction (Federer defeats Brands) if you use it.

An example of a heuristic used in time pressure situations is the *take-the-first heuristic*. Imagine, you are on the field in the World Cup final. You are in a center position and receive a pass from a teammate at a distance of 17 m from the goal. Now, you have to decide whether to pass or to shoot within less than a second or you will lose the ball. The take-the-first heuristic states that as an expert in your domain you should go for the first option that comes to mind. This decision has a high probability of being of good quality because of your expertise (Johnson & Raab, 2003).

## EMOTIONAL INFLUENCE ON HIGHER COGNITION

We have seen the commonalities and differences of high-level cognitions and how they are related to each other and now come to the second focus of this chapter.

---

4. Standing as of July 28, 2014.

What do all of the examples that I introduced so far have in common? Given the title of this section, this question seems superfluous. In all of the examples, the high-level cognitive task has to be performed in a high-pressure environment in different pressure situations (time, precision, success) resulting in stress or in different affective states (e.g., emotional or mood states; for clarifications on the concepts see Laborde, Chapter 17). How do the aforementioned factors affect these high-level cognitions?

## Problem Solving

First, we have a look at problem solving. Where is the pressure in our mammoth problem? There is mainly a pressure of success, as killing the animal is essential for the survival of your clan, and perhaps, there is time pressure as well, as your clan has had no food for a week. Additionally, there are emotions in this problem, namely anxiety about trying not to be killed by the mammoth. Unfortunately, there is, to my knowledge, very little literature about problem solving in relation to mood and emotions and no literature about problem solving and stress. Problem solving, as I defined it here in this chapter, and hence convergent thinking, benefits from negative moods (e.g., sadness; see Pham, 2007). In other words, when people are sad, the ability to engage in focused and analytic thinking is increased. Two possible reasons for this effect are the focus of attention and the vigilance in a situation. Regarding attention, negative moods are known to narrow one's focus (Rowe, Hirsh, Anderson, & Smith, 2007). This narrowing triggers the depth of information processing (Pham, 2007) that is needed to find a solution to a specific problem task. Regarding vigilance, sadness can be seen as a source of information (affect-as-information hypothesis; Schwarz, 1990) about the importance of the current situation (here, the problem task; Schwarz, 2002). From this, it follows that people are more attentive to situations when they are sad, and this again triggers a deeper information processing. In contrast to negative moods, negative emotions (see Chapter 17 in this book for a definition) such as anxiety are known to decrease the depth of information processing and, subsequently, the performance for convergent thinking (Pham, 2007).

As there is little to no literature on pressure and problem solving, I can only speculate about their interactions. Given the curvilinear relation between performance and pressure for most cognitive tasks, I would assume the same inverted-U-shaped relationship (Yerkes & Dodson, 1908) also for problem-solving performance. What does this mean for our example? If on the day you are out hunting you are sad, maybe because of bad weather conditions, your problem-solving performance will increase in contrast to days when your mood state is neutral or positive. On the other hand, negative emotions (e.g., anxiety about being killed) will impair your problem-solving performance. Given the assumed inverted-U-shaped relationship, no pressure (the clan already has food for the next month) or high pressure (the clan has had no food for two weeks) will decrease your problem-solving performance, whereas a moderate amount of pressure would result in an increase in problem-solving performance.

## Creativity

Regarding creativity and pressure, imagine once again that you are an engineer who has to come up with a creative innovation for a new car. We can easily assume two kinds of pressure. First, there is time pressure, as the marketing department does not want to wait a month to present the new car. Second, you are under pressure to succeed, as your company wants to sell a lot of cars, which is much easier with new features. In addition, we can find anxiety in this example, as well, although the consequences of failure here are not as dramatic as in the mammoth example.

In contrast to problem solving, there is literature about the relationship between pressure, emotions, mood, and creativity. Starting with pressure, there are three relationships reported in the current literature: (1) pressure decreases creative performance; (2) pressure increases creative performance; and (3) pressure and creativity have a curvilinear relationship. Byron, Khazanchi, and Nazarian (2010) performed a meta-analytic review. The first result of their meta-analysis (with 76 studies) was the detection of a curvilinear U-shaped relationship between pressure and creative performance, meaning that a moderate level of activation induced by pressure is needed for the best creative performance, whereas when pressure creates no or too much activation, this decreases creative performance.

In the next step of their meta-analysis, the authors investigated whether trait anxiety moderates this pressure–creativity relationship. They assumed that highly anxious people perceive stressors as a threat (Pearson & Thackray, 1970). They found people with low trait anxiety scores to have an increased creative performance in comparison to people with high trait anxiety scores.

Whereas the results on pressure and emotions seem to be quite similar to those for problem solving, the opposite has been found for mood. Several studies revealed empirical evidence for increased creative performance when people are in a positive mood (see Baas, De Dreu, & Nijstad, 2008; for a review). It has been argued that a positive mood broadens the focus of attention (i.e., divergent thinking) and decreases the depth of information processing (i.e., convergent thinking; Pham, 2007), which enhances the cognitive flexibility by building more associations (Bar, 2009). Moreover, recent studies provided empirical findings regarding a bidirectional interaction between mood and creativity (Akbari Chermahini & Hommel, 2012). More specifically, the authors found that performing a creative task (alternate-uses task) led to an increase of positive mood.

In summary, what does this mean for the car company? To have optimal creative performance, the right dose of pressure is needed: Given none or too much pressure, creative performance will decrease. So, imagine now that you are the creative director of this company. If you want optimal creative performance, do not ask your chief engineer one week before the release of the new car for new features. Furthermore, do not pressure your chief engineer with job loss, as this will increase anxiety, and heightened anxiety decreases creative performance. Finally, a positive mood can increase creative performance. Thus, as a senior executive of this car company, you should care about the mood of your employees, especially those whose creative contributions are difficult to replace.

## Decision-Making

Finally, I focus on decision-making and the links to different emotional and mood states. As my focus is on short-term decision-making, I will leave out time pressure from this overview. If we look at our sports example, we can assume that negative emotions (e.g., anxiety) would change the decision-making behavior. The same is true for different mood states.

For a detailed treatment of the influence of emotions on decision-making, I direct the reader to the special issue "Emotions and decision making in sports" of the *International Journal of Sport and Exercise Psychology* (2013). Here, I will briefly summarize the most important findings (see also the introduction to the special issue, Laborde et al., 2013). Damasio (1996) is well known for his work on the link between emotions and decision-making, showing that emotions are needed in order to make the best decisions. Given his findings were mainly obtained from a card game (i.e., the Iowa Gambling Task), future research has to consider the investigation of the relationship between emotions and decision-making in more ecologically valid environments (see Laborde et al., 2013).

In terms of mood, so far negative (for problem solving) and positive (for creativity) moods have been shown to have beneficial effects. Another recent study with sports experts showed that a neutral (deactivating) mood is best for decision-making performance, in comparison to a positive (activating) or a negative (activating) mood (Laborde & Raab, 2013). Due to a lack of further studies on mood and decision-making (most studies were on mood and option generation), it is not clear whether neutral mood is the optimal mood state for decision-making performance. What are the implications for daily life? If you are afraid of cars, big streets, or being killed by a car, you will perform poorly in the crossing-the-street decision task. In more detail, you will make more wrong decisions with different (negative) consequences. If you decide for no-go in a go situation, you will only lose time and miss the bus, but if you decide to cross with the wrong timing, the consequences might be more dramatic. In terms of mood states, neutral moods seem to be optimal for decision-making because attention is neither too focused (negative mood) nor too broad (positive mood).

## CONCLUSION AND DIRECTIONS FOR FUTURE RESEARCH

This short review of high-level cognition in high-pressure environments showed similarities and differences between problem solving, creativity, and decision-making in how they are influenced by emotion, pressure, and mood. Summing up, all of these complex cognitive tasks are negatively affected by anxiety and follow an inverted-U-shaped relation with pressure. In contrast, mood affects each of the cognitive tasks in a different way. These differences are explained by the different mechanisms that underlie the three tasks (i.e., problem solving, creativity, and decision-making). Support for this is provided by the affect-as-information hypothesis (Schwarz, 1990). In conclusion, a lot of work has been done to build a bridge between emotion and cognition. Nevertheless, a lot of work is still needed to understand the emotion–cognition interaction. In the next paragraph, I discuss the most crucial aspects that, in my opinion, need to be considered for future research.

## Future Research on Pressure and High-Level Cognition

As mentioned earlier, pressure and performance in high-level cognitive tasks display an inverted-U-shaped relationship. Research so far has investigated mainly two types of pressure, that is, time pressure and social pressure. In addition, the research itself has been lacking a systematic investigation; no other types of pressure (e.g., pressure to succeed or social pressure) have been studied, nor combinations with other types of pressure. Social pressure, for example, is of main interest in creativity research given its practical relevance to economics. My engineering example shows the common practice of evaluating creative results in a committee and to decide which would be the best idea to go for. The review by Byron et al. (2010) showed that announcing evaluation led to increased creative performance, whereas overemphasizing evaluation, as well as threatening with the expectation of negative consequences, led to decreased performance.

Time pressure has been investigated primarily as part of decision-making research, and indeed, there is an entire branch of research focusing on decisions under time pressure (heuristics). Given that social pressure can also occur in problem-solving and decision-making tasks, and time pressure can occur in creative and problem-solving tasks, a systematic investigation of their interaction is needed. It is easy, for instance, to put social pressure into our problem-solving sports example (i.e., national coach in the World Cup final). As the coach, you will find a solution to the problem that leads to a specific formation or tactic. Especially if your team loses, you know your solution will be relentlessly discussed afterward in the media. There is not much empathy needed to imagine the social pressure in this situation.

A systematic investigation of the effects of pressure on high-level cognition would have an impact on both the theoretical and the practical level. On a theoretical level, combining different tasks and different kinds of pressure would reveal whether all kinds of pressure have an inverted-U-shaped relationship with the performance in cognitive tasks, and whether all kinds of pressure affect cognition in the same way. Depending on the results, it could be interesting to consider arousal as a key factor affecting cognition. In addition, differentiating the pressure conditions will be a strong argument to prove that different mechanisms underlie the different kinds of pressure conditions.

## Future Research on Affective States and High-Level Cognition

In this chapter, I have reviewed the commonalities and differences between different affective states (e.g., moods and emotions) and three high-level cognitive functions, which represent the higher level of executive functioning described by Laborde (Chapter 17). First, regarding emotions, the commonalities have been shown for the emotional state of anxiety. For all three functions, anxiety appears to decrease performance, but little is known about other negative emotions and how they affect cognitive performance, or about the connections between positive emotions and the three cognitive functions I reviewed in this chapter (see the discrete approach in Chapter 17). Just as I argued for systematic research regarding the link between pressure and high-level cognition, I argue the same is needed for emotions, especially on the dimensions of valence and arousal (see the dimensional approach in Chapter 17). Second, regarding mood, the effects of mood on high-level cognition are inconsistent. Whereas problem-solving performance seems to increase during negative mood, creative performance is facilitated by positive mood. Based on the affect-as-information hypothesis (Schwarz, 1990), Pham (2007) argued that mood can affect the depth of information processing, and this has a close link to performance in problem solving, which relies more on deep information processing, and in creative tasks, which rely more on broad information processing. For decision-making, a neutral mood leads to higher decision quality, especially in time-pressure environments. Researchers have found that a neutral mood does not interfere with the decision-making process, whereas positive and negative moods do. On the other hand, decision-making is based on the options that are generated early in the process, and option generation is associated with creativity, which is positively affected by positive mood. In addition, a study on the bidirectional link between affective states and cognitive functions showed not only an effect of positive mood on creative performance but also higher scores for positive mood after performing a creative task. This finding raises several questions: Is this bidirectional link also true for problem solving? Will a problem solver have a negative mood state after solving a problem? If yes, what is the mechanism behind this? If not, why does this link only work for positive mood? Answers to these questions would provide a better understanding of how to increase performance on high-level cognitive tasks given their interaction with specific affective states (Figure 3).

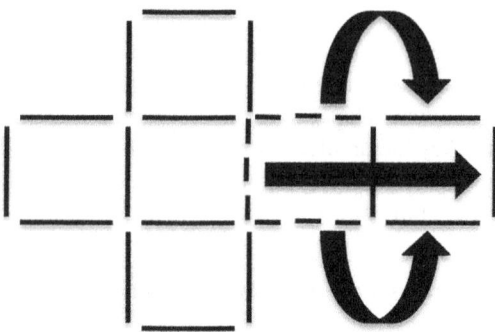

**FIGURE 3**  Solution to Katonas' five-square problem.

# REFERENCES

Akbari Chermahini, S., & Hommel, B. (2012). Creative mood swings: divergent and convergent thinking affect mood in opposite ways. *Psychological Research, 76*, 634–640.

Baas, M., De Dreu, C. K. W., & Nijstad, B. A. (2008). A meta-analysis of 25 years of mood-creativity research: hedonic tone, activation, or regulatory focus? *Psychological Bulletin, 134*, 779–806.

Bar, M. (2009). A cognitive neuroscience hypothesis of mood and depression. *Trends in Cognitive Sciences, 13*, 456–463.

Bar-Eli, M., Plessner, H., & Raab, M. (2001). *Judgment, decision-making and success in sport*. Malden, MA: Wiley.

Byron, K., Khazanchi, S., & Nazarian, D. (2010). The relationship between stressors and creativity: a meta-analysis examining competing theoretical models. *Journal of Applied Psychology, 95*, 201–212.

Damasio, A. (1996). The somatic marker hypothesis and the possible functions of the prefrontal cortex. *Philosophical Transactions of the Royal Society of London. Series B, Biological Sciences, 351*, 1413–1420.

Gigerenzer, G., Todd, P. M., & the ABC Research Group. (1999). *Simple heuristics that make us smart*. New York, NY: Oxford University Press.

Johnson, J. G., & Raab, M. (2003). Take the first: option generation and resulting choices. *Organizational Behavior and Human Decision Processes, 91*, 215–229.

Katona, G. (1940). *Organizing and memorizing: Studies in the psychology of learning and teaching*. New York, NY: Columbia University Press.

Laborde, S., Dosseville, F., & Raab, M. (2013). Introduction, comprehensive approach, and vision for the future. *International Journal of Sport & Exercise Psychology, 11*, 143–150.

Laborde, S., & Raab, M. (2013). The tale of hearts and reason: the influence of mood on decision making. *Journal of Sport & Exercise Psychology, 35*, 339–357.

Lubart, T. I. (2001). Models of the creative process: past, present, and future. *Creativity Research Journal, 13*, 295–308.

Memmert, D. (2010). Creativity, expertise, and attention: exploring their development and their relationships. *Journal of Sport Science, 29*, 93–104.

Memmert, D. (2012). Kreativität im Sportspiel. *Sportwissenschaft, 42*, 38–49.

Newell, A., & Simon, H. A. (1972). *Human problem solving*. Engelwood Cliffs, NJ: Prentice-Hall.

Öllinger, M., Jones, G., & Knoblich, G. (2014). The dynamics of search, impasse, and representational change provide a coherent explanation of difficulty in the nine-dot problem. *Psychological Research, 78*, 266–275.

Pearson, D. W., & Thackray, R. I. (1970). Consistency of performance change and autonomic response as a function of expressed attitude toward a specific stress situation. *Psychophysiology, 6*, 561–568.

Pham, M. T. (2007). Emotion and rationality: a critical review and interpretation of empirical evidence. *Review of General Psychology, 11*, 155–178.

Rowe, G., Hirsh, J. B., Anderson, A. K., & Smith, E. E. (2007). Positive affect increases the breadth of attentional selection. *Proceedings of the National Academy of Sciences of the United States of America, 104*, 383–388.

Runco, M., & Jaeger, G. J. (2012). The standard definition of creativity. *Creativity Research Journal, 24*, 92–96.

Schwarz, N. (1990). Feelings as information: Informational and motivational functions of affective states. In E. T. Higgins & R. Sorrentino (Eds.), *Handbook of motivation and cognition: Foundations of social behavior* (Vol. 2) (pp. 527–561). New York, NY: Guilford Press.

Schwarz, N. (2002). Situated cognition and the wisdom of feelings: cognitive tuning. In L. F. Barrett & P. Salovey (Eds.), *The wisdom in feelings: Psychological processes in emotional intelligence* (pp. 144–166). New York: Guilford Press.

Yerkes, R. M., & Dodson, J. D. (1908). The relation of strength of stimulus to rapidity of habit-formation. *Journal of Comparative Neurology and Psychology, 18*, 459–482.

# Index

*Note*: Page numbers followed by "f", and "t" indicate figures, and tables, respectively.

## A

Acculturation phenomenon, 40
Action, 3–4, 5t–6t, 7, 16
   and cognition, link between, 68–70
   cognitive mechanisms guiding, neurophysiological implementation of, 136–142
   competence, 19
   execution, 283–285
   functional architecture of, 20–23. *See also* Functional architecture, of action
   mirror neuron theory of, 210–212
   paradigm, 15–17
   perception affected by, 213–215
   perception affecting, 208–213
   selection, 283–285
   simulation, 224
   space, 18–19
   valence, 19
Action–perception coupling, bidirectional, 236–237, 239f
Action phases, 20–21
   anticipation phase, 21
   interpretation phase, 21
   realization phase, 21
Action system
   functional network of, 17f
   functional systems of, 22–23
Action theory perspectives, 11–12, 15–16
   action paradigm, 15–17
   action space, 18–19
   emotional processes. *See* Emotional processing
   functional architecture of actions, 20–23. *See also* Functional architecture, of action
   functional disturbances, 24–25
   intention, 17–18
   multifacetedness, 23–24
   peak performance, 14–15
   performance and psychology, 12–13
   performance orientation, 13–14
Activities-Specific Balance Confidence Scale, 125

Activity analysis, 54
Adaptation process, 37–40
Adaptive-Reactive Theory, 90–91
Advanced Cognitive Training for Independent and Vital Elderly (ACTIVE), 191–192
Affective states, high-level cognition and, 337–338
Affects, 276
Agent causality, 14
Alternative models, 36
Anterior cingulate cortex (ACC)
   gaining attention, 155
   as motor control filter, 154
   performance monitoring and, 152–153
   stimulus processing following an error, 155–156
Anticipation, 21–23
Applied performance psychology, 45–64
   functions of, 46–47
   measurements. *See* Performance psychology, measurement considerations in
   in music, 50–53
   performance excellence, 47–50
Applied science, 261–265
Appraisal, 305–306
   cognitive, 305–306, 323–324
   primary, 317–318
   secondary, 317–318
Approach behavior, 37–38
Artificial movement sounds, auditory action perception with, 243–246
   motor control and learning, 245–246
   motor perception, 244
Athlete perceptions, of successful coaching, 94–96
Athlete performance, 92
   coach expectations and, 96–98
Attention
   selective. *See* Selective attention
   threat-biased, 321
Attentional control theory, 282–285, 284t
Attentional selection, 137–140

**341**

Auditory action perception, 236
  with artificial movement sounds, 243–246
  empirical evidence of, 239–247
  future research of, 247–248
  internal model and reafferences, 238–239
  with natural movement sounds, 241–242
  outlook in applied fields, 246–247
Auditory perception, 235–236
Automatic action control system, 22
Autonomic nervous system (ANS), 106, 111
Avoidance behavior, 37–38
Aware perception, 179–182

# B

Basal ganglia
  action-selection process, 140–141
  dopamine system of, 154
  response-selection process, 141
Basketball shooting, visual control of, 261–265
  methodological considerations, 264–265
  previous studies, reinterpretation of, 263–264
Behavioral adaptation, predicting, 160–162
Behavioral theories, 90
Bidirectional action–perception coupling, 236–237, 239f
Biopsychosocial model of challenge and threat, 279–280, 284t
Blanket approach, 306–307
Blindness
  change, 173–176, 174f, 179–181
  inattentional, 176–181, 181f
Body–environment relations, 225. *See also* Motion capture
Botulinum toxin, for musician's dystonia, 111–112
Boxes and arrows framework, 280–281
Brain activity, measurement tools for capturing, 41t
Brain computer interfaces (BCI), 161–162
Burnout, 39–40

# C

Case selection, importance of, 35
Central nervous system, 40–41, 104, 238–239
Change blindness, selective attention and, 173–176, 174f
Chill responses, 105
Choking under pressure (CuP), 72, 106–108, 111–114
Coach behavior
  expectations and, 97
  member characteristics and, 92–93

perceptions of, athletes affected by, 97–98
  and performance, 94, 96–97
  and satisfaction, 93
Coach education, 75–77, 77t
Coaching
  defined, 94–95
  expertise, development of, 89–94
  profession of, 88–89
  purpose of, 89
  successful, athlete perceptions of, 94–96
Cognition, 3–4, 5t–6t, 7–8, 277
  and action, link between, 68–70
  embodied, 7, 223–234. *See also* Motion capture
  high-level cognition, in high-pressure environments. *See* High-level cognition, in high-pressure environments
Cognitive action control system, 22
Cognitive–affective–behavioral linkage
  in performance psychology, two-parameter model for capturing, 37–40, 38f
Cognitive appraisal, 305–306, 323–324
Cognitive behavioral therapy (CBT), 110–111
Cognitive control, 151
Cognitive leisure activity, 191–198
  cognitive training, 191–195
  physical training, 195–198
Cognitive map, 26
Cognitive-motivational-relational theory, 278, 284t
Cognitive performance
  cortisol and, 320–322
  in older adults, lifestyle and interventions for improving, 189–204
Cognitive-processing hypothesis, 321
  critique of, 322
Cognitive restructuring, 70
Cognitive training (CT), 191–195
Competitive trait anxiety (CTA), 293–294
  definition and background of, 293
  influence on performance under pressure, 293–294
Computerized cognitive training (CCT), 191–195, 197–198
Conflict monitoring theory, 153–154
Construct-related validity, 33
Construction function, 16
Content-related validity, 33
Contingency Theory of Leader Effectiveness, 90–91
Controllability of Motor Imagery (CMI) test, 123–124, 126

Correct-response negativity (CRN), 157–161
Cortisol, 315–316
   and cognitive performance, 320–322
   defined, 316–317
   functionality of, 316–317
   and motor performance, 322–323
   and performance, 318–323
   and sports performance, 319–320. *See also* Sports performance
   state and trait influences on, 317–318
Cortisol awakening response (CAR), 315–316
Cortisol–performance framework, 323–324, 323f
Creativity, 330–331
   comparison with decision-making, 333
   comparison with problem solving, 332, 332f
   emotional influence on, 335
Criterion-related validity, 33
Criterion sampling, 35
Cybercycling, 198

## D

Dance, motion capture in, 223, 226, 230–231
Dartism, 109–110
Decision-dependent measure, 33
Decision-directed measure, 33
Decision making, 7–8, 280–281, 331
   comparison with creativity, 333
   comparison with problem solving, 333
   emotional influence on, 336
Deductive reasoning, 35
Deliberate practice, 48, 103–104
Developing the Potential of Young People in Sport, 49
Dimension approach, 276–277
Discrepancy Theory, 90–91
Discrete approach, 276–277
Distraction theories, 107
Dopamine, role of, 142
Drift diffusion model (DDM), 137, 142–146
Dynamic stereotype (DS), 72–73, 107–108, 110–114

## E

Embodied cognition, 7, 223–234
Embodiment, 225
Embouchure dystonia (ED), 108
Emotion, 3–4, 5t–6t, 8–9, 276–277
   functional disturbances, 24–25
   functional relevance in action regulation, 24f
   functional role of, 11–12
   multifacetedness, 23–24
Emotional action control system, 22
Emotional influence, on higher cognition, 333–336
   creativity, 335
   decision-making, 336
   problem solving, 334
Emotional processing
   functionality of, 23
   options in, 25–26
Equivalent models, 36
Error(s)
   commission, 138–139, 141–142, 144–146
   monitoring, 162–163
   potential (ErrP), 162
   processing, theories of, 152–156
   serious, 169–170
Error prediction, 160–163
   behavioral adaptation, predicting, 160–162
   individual differences and error monitoring, 162–163
Error-related negativity (ERN), 152–153, 212
Escitalopram, for musician's dystonia, 111
Event-related potential (ERP), 163, 179–181, 181f, 192–194, 212
Executive function, 277
Exergaming, 198
Expectancy
   effects, in competitive sport, 96–98
   theory, 96
Expert action, visual perception in, 253–272
   basketball shooting, visual control of, 261–265
   future research of, 266–267
   spatial aspects of, 258–261
   temporal aspects of, 254–258
Expert performance, 68, 70, 81
Exploration function, 16

## F

Feedback-related negativity (FRN), 154–155, 162, 164, 212
Flicker paradigm, 175
Focal dystonia, task-specific, 71–74
Focal hand dystonia (FHD), 108
Forced choice detection paradigm, 175
Forward genetics, 42
Forward models, 238
Functional architecture, of action, 20
   functional systems of, 22–23
   phase structure of action, 20–22. *See also* Action phases
   system levels of action organization, 20

Functional autonomy, 21
Functional complexity, 23–24
Functional magnetic resonance imaging (fMRI), 152–153, 197

## G

Gaining attention, 155
Gait Efficacy Scale, 125
Gaze behavior, 262–263, 265
Generalized anxiety disorder, 106
Generalized Self-Efficacy Scale, 125
Genetics, 31, 40, 42, 103–104, 106, 110, 114
Goal-directed performance, 180
Goal-terminating intention, 18

## H

Hardiness, 295
　definition and background of, 295
　influence on performance under pressure, 295
Head-Up Display (HUD), 176–178
Heart rate variability (HRV), 299–300, 305–306
Heuristic model, for musician's performance failure, 112–114, 113f
Heuristics, 8
High-level cognition, in high-pressure environments, 329–340
　creativity, 330–331
　decision-making, 331
　decision-making and problem solving/creativity, comparison of, 333
　emotional influence, 333–336
　future research of, 336–338
　problem solving, 330
　problem solving and creativity, comparison of, 332, 332f
Hormonal stress on performance, influence of, 315–328
Human action, 15
Human Genome Resource, 42

## I

Ideomotor theory, 224
Imitation learning, 210–211
Inattentional blindness, 176–181, 181f
Individual differences, 162–163
　personality trait–like, 291–314
Individual/team assessment, 54–55
Individual zone of optimal functioning theory (IZOF), 35–36, 111, 278–279, 284t

Inductive reasoning, 35
Information-processing system, 4–6, 258
Intention, 15–17
　goal, 18
　implementation, 18
　outcome, 17
　value, 17
Internal forward models, 122
Internal model, 238–239
Interpretative pluralism, 35
Interruption intention, 18
Inverse models, 238
Item calibration, 32
Item response theory (IRT), 37–38

## J

Judgment, 7–8

## K

Katonas' five-square problem, 330, 330f, 338f

## L

Leadership
　multidimensional model of, 91–92
　in sport, history of, 89–90
Leadership Scale for Sport (LSS), 92
Learning
　ideomotor, 224
　imitation, 210–211
　motor, 245–246
　observational, 208–210
Learning-based sensorimotor training (LBST), 111
Lifestyle interventions, for improving cognitive performance in older adults, 189–204
Long-Term Athlete Development (LTAD) model, 45–46, 55–56

## M

Manifest variables, 32
Maximin-principle, 14
Maximizing intention, 14–15
Measurement considerations, in performance psychology, 31
Medial prefrontal cortex (mPFC). *See also* Prefrontal cortex (PFC)
　gaining attention, 155
　stimulus processing following an error, 155–156
Memory, 3–4

Mental balance training, for postural control, 124–130
Mental functioning, 11
Mental skills training, 49–52, 56–57
Mental toughness, 295–296
  definition and background of, 295–296
  influence on performance under pressure, 296
Mental training
  for balance tasks, 127
  motor imagery as prerequisite for, 122–123
  in older adults, 123–124
Methodological triangulation, 35
MIDI (Musical Instrument Digital Interface) standard, 243
Mirror neuron theory of action, 210–212
Mobility, 121, 124–125
Model of Great Coaching, 95–96, 95f
Mood, 276
Mood responses to competitive situation, model of, 319–320
  critique of, 320
Motion capture, 223–234
  empirical research of, 225–230
  theoretical approaches of, 224–225
Motor control, 245–246
Motor execution (ME), 122–123
Motor imagery (MI)
  ability, 126
  future research of, 130
  in older adults, 123–124
  as prerequisite for mental training, 122–123
Motor learning, 245–246
Motor perception, 244
Motor performance, cortisol and, 322–323
Motor system, 214–215
Movement, 224, 226–232. *See also* Motion capture
Movement sound, auditory action perception with
  artificial, 243–246
  natural, 241–242
Multidimensional Model of Leadership (MML), 91–94, 91f, 95f
Multifacetedness, 23–24
Musicians, 48–53
Musicians' dystonia (MD), 108–112, 109f, 114
Music performance, 103–120
  failures, 105–110, 112–114
  improving, 110–112
  motion capture, 223, 230–231
  prevention of failures, implications for, 114–115
  and sports performance, communalities and differences between, 104–105
Music performance anxiety (MPA), 105–107, 110–112

# N

Natural movement sounds, auditory action perception with, 241–242
Negative affect, 163
Negative emotionality, 163
Neuro-cognition, 40–41
Neuroendocrine to competitive situation, model of, 319–320
  critique of, 320
Neurovisceral integration model, 282–283, 284t
Nintendo Wii™ exergame, 198
Non-equivalent models, 36
Non-executive function, 277
Nutrition, 189–190

# O

Observational learning, 208–210
Occupation, 190
Older adults
  cognitive performance in, lifestyle and interventions for improving, 189–204
  mental imagery and training in, 123–124
One-to-many relation, 40–41
One-to-one relation, 40–41
Optimism, 296–298
  definition and background of, 296–297
  influence on performance under pressure, 297–298
Orientation, 53–54

# P

Paradoxical performance, 68, 70, 75
Parsimonious principle, 36–37
Path-Goal Theory, 90–91
Peak performance characteristics, 14–15
Perception, 3–7, 5t–6t
  action affected by, 208–213
  action affecting, 213–215
  auditory, 235–236
  auditory action. *See* Auditory action perception
  aware, 179–182
  cognitive mechanisms guiding, neurophysiological implementation of, 136–142
  motor, 244

Perception–action link
  bidirectional link, 208, 213, 215, 217
  online and offline effects of, 215–217
Perceptual errors, under inattentional blindness, 176–181, 181f
Perceptual judgments, 216–217
Perceptual resonance, 214–215
Perfectionism, 298–299
  definition and background of, 298
  influence on performance under pressure, 298–299
Performance, 11
  building blocks, 1–10, 5t–6t
  cortisol and, 318–323
  excellence, 47–50
  functional aspects of, 13
  functionality of emotional processes, 23
  goal-directed, 180
  hormonal stress on, influence of, 315–328
  intention, 14
  measures, administering and interpreting, 34
  mechanism, 3–4
  orientation, 13–14, 13f. *See also* Performance-oriented action; Performance-oriented emotion–cognition theories
  phenomena of, 5t–6t
  variables, 11
Performance-oriented action
  situated action for, 19f
  triadic phase structure of, 21f
Performance-oriented emotion–cognition theories, 280–283, 284t
  attentional control theory, 282–285, 284t
  boxes and arrows framework, 280–281
  neurovisceral integration model, 282–283, 284t
  theory of reinvestment, 281, 283–285, 284t
Performance psychology, 11–13
  activity analysis, 54
  agents, 12–13
  applied. *See* Applied performance psychology
  conceptualization, 55–56
  considerations for implementing, 53–58
  evaluation, 57–58
  implementation, 56–57
  individual/team assessment, 54–55
  (meta-) theoretical foundation, 11–12
  orientation, 53–54
  psychological skills training, 56
Performance psychology, measurement considerations in, 31

  cognitive–affective–behavioral linkage, two-parameter model for capturing, 37–40, 38f
  new trends, 40–42
  qualitative measures, 34–36
  quantitative measures, 32–34
  theory-measurement development-advanced theory process, 36–37
Performance under pressure, 291–314
  competitive trait anxiety, 293–294
  hardiness, 295
  mental toughness, 296
  optimism, 297–298
  perfectionism, 298–299
  pessimism, 297–298
  reinvestment, 299–300
  resilience, 301
  sensation seeking, 301–302
  trait emotional intelligence, 294
Performing arts, motion capture in, 226–229
Periodization, 47
Peripheral nervous system, 40–41
Personality trait–like individual differences (PTLIDs), 291–314, 293f
  competitive trait anxiety, 293–294
  hardiness, 295
  as interactionist approach, 304–306, 304f
  mental toughness, 295–296
  optimism, 296–298
  perfectionism, 298–299
  pessimism, 296–298
  reinvestment, 299–300
  research, future research directions within, 302–306
  resilience, 300–301
  sensation seeking, 301–302
  trait emotional intelligence, 294
Pessimism, 296–298
  definition and background of, 296–297
  influence on performance under pressure, 297–298
PETTLEP model, 123
Phenomena, 4, 5t–6t
Physical activity, 190
Physical training, 195–198
Postural control, mental balance training for, 124–130
  intervention procedures, 126–128, 128t
  method, 125–126
  results and discussion, 128–130, 129f
Predicted-response outcome (PRO) model, 153, 155
Prediction system, 214–215

Prefrontal cortex (PFC)
    action-selection process, 140–141
    attentional selection and, 137–140
    medial. *See* Medial prefrontal cortex (mPFC)
    task-goal representations in, 142, 143f
Presentation function, 16
Pressure, 275, 280–281, 283–285
    high-level cognition and, 337
    performance under, predicting, 291–314
Problem solving, 330
    comparison with creativity, 332, 332f
    comparison with decision-making, 333
    emotional influence on, 334
Processes, 4–9
Progressive muscle relaxation, 74
Protection function, 16
Psychological Characteristics for Developing Excellence (PCDEs), 49, 55–56
Psychological interventions, 110–111
Psychological skills training, 56
Psychology, 31
    performance. *See* Performance psychology
Psychometric properties, 36

## Q

Qualitative measures, 34–36
Quantitative measures, 32–34

## R

Reafferences, 238–239
Recognition heuristic, 333
Reentrant processing, 183–184
Reference standards, 13–14
Reinforcement learning theory (RFL), 153–155, 162
Reinvestment, 299–300
    definition and background of, 299
    influence on performance under pressure, 299–300
Reliability, 33
Resilience, 300–301
    definition and background of, 300
    influence on performance under pressure, 301
Response conflict, 136
Response monitoring, 151–152, 154–156, 160–161, 164
Response-related EEG activity, 152, 152f
    methodological issues in measuring, 156–159, 157f–158f
Reversal Theory (RT), 39
Reverse genetics, 42

Risk taking, 163, 301–302
    definition and background of, 301
    influence on performance under pressure, 301–302
Rules of thumb, 8

## S

Sample size, 33–34
Sampling, 35
Selective attention
    as biased competition, 170–173, 172f
    and change blindness, 173–176, 174f, 179–181
    and visual awareness, 181–184
Self-efficacy, 125, 128–129
Self-focus theories, 107
Self perception theory, 12–13
Self-talk, 74
Sensation seeking, 163, 301–302
    definition and background of, 301
    influence on performance under pressure, 301–302
Sensitivity to reward, 163
Sensory channels, 6–7
Serious errors, 169–170
Simulation theory, 237
Situated action, for performance-oriented action, 19f
Situational theories, 90–91
Snow-ball sampling, 35
Social anxiety, 106
Social interaction, 11
Solomon Expectancy Sources Scale (SESS), 96–97
Sonification, auditory action perception with, 243–246
    motor perception, 244
    motor control and learning, 245–246
Spatial aspects of visual perception and action, 258–261
Sports, 8
    motion capture in, 229–230
    psychology, 11
SportScotland, Developing the Potential of Young People in Sport, 49
Sports performance, 8
    cortisol and, 319–320
    and music performance, communalities and differences between, 104–105
S-states, 214
Statistical power, 33–34
Stress, 190, 276

Striatum, action-selection process, 140–141
Stroop task, 140
Supplementary motor area (SMA), 211–212
Sustained posterior contralateral negativity (SPCN), 180–181
Systems approach, 24–25

## T

Take-the-first heuristic, 333
Talent
 development, 79–81, 79f
 identification, 78–81
 selection, 81
Task-specific focal dystonia (TSFD), 71–74, 108
Task-switching task (TST), 194
Temporal aspects of visual perception and action, 254–258
 offline use, 254–255
 online use, 255–257
 optical information pick-up, timing of, 257–258
Theory integration, 35–36
Theory of challenge and threat states in athletes, 280, 284t
Theory of event coding (TEC), 213–216
Theory of reinvestment, 281, 283–285, 284t
Threat-biased attention, 321
Trait
 activation, 305
 anxiety, 106
 emotional intelligence, 318
 influences on cortisol, 317–318

Trait emotional intelligence, 294
 definition and background of, 294
 influence on performance under pressure, 294
Transcranial direct current stimulation (tDCS), 285
Transcranial magnetic stimulation (TMS), 183, 215, 285
Transferability, of measures and research outcomes, 35–36
Transtheoretical Model, 53–54
Trier Social Stress Test (TSST), 318–319

## V

Validity, 33
Value orientation, 14
Visual awareness, selective attention and, 181–184
Visual perception, in expert action, 253–272
 basketball shooting, visual control of, 261–265
 future research of, 266–267
 spatial aspects of, 258–261
 temporal aspects of, 254–258
Visuo-motor priming, 215–216

## W

"The winner take all" (WTA) mechanism, 140–141
Wisconsin Card Sorting Test (WCST), 195–196

## Y

Yips, the, 70–75, 109–110, 113–114
 diagnosing and treating, 73–75

Lightning Source UK Ltd.
Milton Keynes UK
UKOW06n0015221215

265165UK00003B/17/P